Changing Climate over India

Neloy Khare
Editor

Changing Climate over India

Past to Present

Springer

Editor
Neloy Khare
Ministry of Earth Sciences (MoES)
Government of India
New Delhi, India

ISBN 978-981-92-1097-8 ISBN 978-981-92-1098-5 (eBook)
https://doi.org/10.1007/978-981-92-1098-5

This Springer imprint is published by the registered company Springer Nature Singapore Pte Ltd.
The registered company address is: 152 Beach Road, #21-01/04 Gateway East, Singapore 189721, Singapore

This book is dedicated to my loving brother
(Late) Rakesh Khare and Nephew
(Late) Prateek Khare

—Neloy Khare

Foreword

Since the middle of the twentieth century, India has witnessed notable changes in its weather conditions a decrease in monsoon precipitation; an increase in extreme temperature and rainfall events, droughts and sea levels; and an increase in the intensity of severe cyclones, alongside other changes in the monsoon system.

Undoubtedly, a significant increase has been reported in weather and climate extremes globally (e.g. heat waves, droughts, heavy precipitation and severe cyclones), changes in precipitation and wind patterns (including shifts in the global monsoon systems), warming and acidification of the global oceans, melting of sea ice and glaciers, rising sea levels, and changes in marine and terrestrial ecosystems owing to the warming since the 1950s. It is non-debatable that the present climate change is imposing huge challenges. Extreme events are now more frequent. The adverse consequences of extreme events may alter the spatio-temporal characteristics (e.g. frequency, intensity and duration) and the distribution of water availability. Persistent extreme weather events, namely droughts, floods and heatwaves, can significantly alter natural ecosystems and impose enormous societal challenges, including food security and water resources for a population of 1.2 billion in India.

Further, several studies pointed out that the evident extremes are mostly due to climate change rather than the influence of internal variability. There is compelling scientific evidence that human activities have influenced these changes in regional climate. Human-induced climate change is expected to continue at pace during the twenty-first century. To improve the accuracy of future climate projections, particularly in the context of regional forecasts, it is essential to develop strategic approaches for improving the knowledge of Earth system processes and to continue enhancing observation systems and climate models. To further understand these climate models, it is significant to get insights into how the climate has behaved in the past. Climate is a dynamic system, and therefore its changing pattern with time is inevitable. The preparedness through climate scenarios in India is of paramount significance, owing to the fact that it is vital to understand the risks of climate change to development in India, which is already experiencing a warming climate. Unusual and unprecedented spells of hot weather are expected to occur far more frequently and cover much larger areas. Under 4°C warming, the west coast and southern India are projected to shift to new, high-temperature climatic regimes with significant impacts on agriculture. Climate change is expected to have major health impacts in India.

Given above, it is a commendable effort to bring out a book *Changing Climate over India: Past to Present* which aptly reviews the climate dynamics with changing time. The book begins with an assessment of climate impacts in modern times, with the first chapter on Lightning in India and its Trend with Climate Change by Srivastava, followed by a chapter on Arsenic in groundwater, causing water security issues for Bhagalpur: A comparative study on the extent of contamination on either bank of River Ganga by Akhouri et al. The climate changes in the recent times have been evaluated by Bandyopadhyay in her chapter. The geodetic mass balance of the High Mountain Asia glaciers: a critical analysis of the estimates coupled with another chapter of Suresh and Suresh Gandhi on Ecological risk assessment from the distribution of Recent Benthic Foraminifera from Palar Estuarine Sediments, Tamil Nadu, India.

On the other hand, Chadong et al. have studied climates on a shorter scale through Morphometric and Morphotectonic characterization of the Kopili River Basin, North-east India: Integrating Geospatial and Remote Sensing Techniques, followed by chapters on Climate changes in the Himalayan region based on tree ring proxy- a review by Bhattacharya et al. and on Multiproxy sediment core record of precipitation/ drought Cycles and hydrodynamic changes in Gangapur Reservoir, Maharashtra, India by Humane et al.

Geoscience plays an important role in solving many archaeological and historical riddles through scientific evidence and explanations. In this direction Nigam et al. have put forth their chapter highlighting Geological tools deciphering tsunami as the guiding factor for shaping the ancient Harappan city, Dholavira, Gujarat, India, duly supported by Sabale who performed Site catchment analysis of Math Pimpri Archaeological site, Deccan: Past and present land use patterns to understand the palaeoenvironment coupled with another interesting chapter on Geotourism in South Bihar, India: Sustainability of a unique geological and historical heritage amidst apparent imprints of Climate Change by Akhouri et al.

The book covers various facets of climate change on longer scales through a total of six important chapters on a geochemical investigation from tropical India, which reveals the way the monsoon climate affects sedimentary processes at different longitudes by Singh et al., and on Phytolith-based paleoclimate studies in the Indian subcontinent by Qader et al. Similarly, Banerji unravelled the climate during the Meghalaya Stage from the mudflats of the southern Saurashtra coast, western India. At a further larger time scale, Roy and Vyasulu studied climate change in India through geothermal records, Niranjan et al. studied the Climate regime during glauconite formation in the Adhaura block, Rohtas district, Bihar, its search as a potash fertilizer and Shrey and Akhouri evaluated the Paleoclimate and mass extinction from the Gondwana Basin in Jharkhand.

I appreciate the commendable efforts of Dr. Neloy Khare for being thoughtful in addressing the issue of how climate has changed at different time scales over India in the present book. This book's collated data and insights are vital for preparing futuristic models and scenarios of climate conditions over the Indian subcontinent.

Lucknow, India
March 2026

Prof. A. R. Bhattacharya

Preface

Climate change significantly influences the global hydrological cycle and consequently affects climatic extremes. Climate over the Indian subcontinent has varied significantly in the past century in response to natural variations and anthropogenic forcing. Considerable progress in understanding the influence of anthropogenic climate change over the Indian subcontinent, particularly the regional monsoon, has been made. Notably, India's average temperature has risen by around 0.7 °C during 1901–2018 owing to GHG-induced warming, partially offset by forcing due to anthropogenic aerosols and changes in Land Use Land Cover (LULC). According to an estimate, by the end of the twenty-first century, the average temperature over India is expected to rise by approximately 4.4 °C relative to the recent past, as compared to the global average temperature, which is reported to have risen by around 1 °C since pre-industrial times.

Indubitably, these parameters do have substantially altered the atmospheric composition, and consequently the planetary energy balance, and are thus primarily responsible for the present-day climate change. Nevertheless, the natural variations in climate also equally contribute to the present climate scenario. The climate dynamics cannot be explained alone, based on a single cause, and they must necessarily take into account changes due to both natural and anthropogenic factors, making climate change a complex puzzle and offering a challenge to climatologists and modellers, including paleoclimatologists. In the recent years, increasing catastrophic losses due to the adverse consequences of extreme climatic events have aroused public awareness

It is therefore of paramount importance to investigate the future variability of extreme weather events related to precipitation, as this will provide insight into probable changes in the occurrence and magnitude of floods and droughts. This will then guide suitable adaptation strategies and disaster mitigation policies. To prepare any climate scenario, we need to know both the past and present patterns of the climate over any region. It is the need of the hour to collate all available data on the climatic conditions at different intervals in the past, vis-à-vis what is currently undergoing.

The present book, therefore, focuses on the varying patterns of climate in the past as well as its impact and pattern in modern times across the Indian subcontinent to help assess the future trend of climate scenarios. The past climate conditions could be reconstructed through various proxies duly preserved in various geological archives.

Having understood well the significance of climate data both in space and time, collating important insights into the climate dynamics over India in the present book, *Changing Climate over India: Past to Present*, is a significant attempt.

The book is composed of a total of sixteen chapters covering varied facets of climate change through the ages. The book begins with a chapter on Lightning phenomena over India with changing climates in modern times. Srivastava put forth that the captivating yet perilous natural phenomenon of lightning has held humanity in awe for centuries. This chapter embarks on a comprehensive journey, unveiling the multifaceted dimensions of lightning in India. On the other hand, in the following chapter on Arsenic in groundwater, of Bhagalpur, in Bihar, India, is discussed by Akhouri et al.

Bandyopadhyay, in the third chapter, provides a comparison between different studies, highlighting both the utility of geodetic mass balance estimates as well as remaining issues and proposes combining different methods to circumvent the challenges. On the contrary, Suresh and Suresh Gandhi, in their chapter, inventorized the foraminiferal fauna and evaluated the ecological relationship of benthic foraminiferal assemblages with physicochemical characteristics that govern the living population of the Palar estuary of Tamil Nadu, India.

Similarly, the Kopili River Basin in Northeast India represents a vital region for conducting morphometric and morphotectonic studies, which is studied by Chadong et al. Their chapter integrates geospatial and remote sensing techniques to analyse the basin's geomorphic and tectonic characteristics.

Undoubtedly, the Himalaya has a great role in controlling the climate dynamics of the Indian subcontinent. Major shifts in climatic changes are linked closely to the rise and fall of cultures and civilizations of the Indian subcontinent. The few instruments of climate observations and the absence of lengthy records from this area make it difficult to explain historical climatic anomalies that were highlighted in disasters and to anticipate future climate change. There are ongoing efforts to use various climatic proxies to reconstruct the past climate. Tree ring proxy in understanding climate has rapidly progressed since the end of the twentieth century, with the result that there are several studies on climate reconstruction covering the diversified geographical sites of the Himalaya. In the sixth Chapter, Bhattacharyya et al. gave a general description of a climatic scenario that was obtained from Himalayan tree ring investigations. Humane et al. studied a multiproxy study of a 70cm long sediment core and provided the signatures of watershed changes and envisaged further the transformations caused by anthropogenic activities in the Gangapur Reservoir between ~1931CE and ~2012CE.

Dholavira, located in the Rann of Kutch region of Gujarat, India, is an ancient archaeological site that provides a fascinating glimpse into the Indus Valley Civilization (IVC). It is one of the five largest Harappan sites and gained further attention after it was declared by UNESCO as the 40th site of world cultural heritages in India. Nigam et al. reported the oldest known evidence of a possible impact and protective measures employed to thwart tsunami, from the Harappan port town of Dholavira, in the state of Gujarat, India.

While Sabale conducted a site catchment analysis of Math Pimpri. The interrelation of environment and hunter gatherers societies is the integral part of research in geoarchaeological investigation. The behaviour of such societies one can be reconstruction through the detailed catchment analysis of any sites.

Interestingly, the Nalanda-Gaya-Jehanabad districts in Bihar exhibit exciting face of Geotourism interwoven intricately with the archeological heritage through the sands of time. Geotourism encompasses the roles of geosciences in decision and approaches of wide public interest including geological heritage, geotourism and archaeological treasure with its foundation on a geological platform. A glimpse on the geoheritage wealth inscribed in the sector includes, the recently reported volcano-sedimentary sequences juxtaposed with Rajgir meta-sediments extended over known ~40 km aerial distance from Bathani to Churi-Jagannathpur in Gaya District of Bihar. This finding puts the Munger-Rajgir belt in an entirely new litho-stratigraphic perspective. Akhouri et al. detailed Geotourism in South Bihar in their chapter.

Singh et al. examined the variability of chemical weathering and sediment maturity under tropical savannah climate in the Hassan region and tropical monsoon climate in the Dakshina Kannada region of southern India. The Plagioclase Index of Alteration (PIA) and Rb/Sr ratio are used to assess the chemical weathering of the studied area.

The Indian subcontinent has experienced significant climatic and environmental changes mostly influenced by tectonic movements and the evolution of the monsoon system. These changes have had profound impacts on the fauna and the distribution and patterns of vegetation. At the moment, the region is home to a diverse range of vegetation zones, such as grasslands, thorn forests and scrubs, tropical deciduous forests, and tropical evergreen forests. Qader et al., in their chapter, present an overview of the application of phytoliths, microscopic silica structures formed within plants, as a proxy for understanding the past environmental conditions and the monsoon dynamics in the Indian subcontinent.

Generally, the growth and decay of ice sheets have been the prime indicative of high-latitude climate variability, while monsoon intensities predominantly govern the low-latitude climate system. Despite such close association of high latitude with the low-latitude climates, the signals of abrupt climatic events of high latitudes are seldom addressed and investigated in the low-latitude climatic regime. Therefore, Banerji attempted to overview the signatures of abrupt climatic events of the Meghalayan stage (~4.2 ka to present) from the Indian Summer Monsoon (ISM) dominated region of western India (Saurashtra Peninsula, Gujarat). Here, the climatic signals emanating from the geochemical study of the mudflat sediments from the Southern Saurashtra coast have been collated and compared to depict the abrupt climatic events of the Meghalayan Stage and the possible global teleconnection for such events are decoded. While Roy and Vyasulu reviewed in their chapter the Geothermal record of climate change in India. Niranjan et al. detailed the Climate regime during glauconite formation. Search for glauconite as a fertilizer mineral was undertaken in the Adhaura area (to its south), Rohtas district, Bihar.

The book ends with an interesting chapter on paleoclimate and mass extinction by Shrey and Akhouri. According to them, mass extinctions have obliterated the fossil record across geological periods along the span of stratigraphic time scale, marking the transition junctures in geological history. Fossils from one period to another script the terminal phase of one species and the initiation of another.

The present book is expected to be a ready reference for anyone interested in understanding climate, its changing pattern in the geological past and new insights into the changing climate over the Indian subcontinent. The policymakers, economic planners and scientific institutions are among the main beneficiaries of the information this book contains. Hope the present efforts meet the expectations of the readers.

New Delhi, India Neloy Khare

Acknowledgements

Kind support and help received from all contributing authors are gratefully acknowledged. Without their valuable inputs on various facets of the past climate changes reconstructed based on multiple proxies over the Indian subcontinent, the book would not have been possible. Various learned experts who have reviewed different Chapters are sincerely acknowledged for their timely, essential and critical reviews.

I sincerely thank Dr. M. Ravichandran, Secretary, the Ministry of Earth Sciences, Government of India, New Delhi for various inputs, support and encouragement. My wife, Dr. Rajni Khare, and my sons, Flight Lieutenant (Dr.) Akshat Khare and Ashmit Khare, have unconditionally supported during various stages of this book. Dr. Upasana. S. Banerjee, Shri Avinash Gola and Shri Rohan Singh of the Ministry of Earth Sciences, New Delhi, have helped immensely in formatting this book's text and figures and bringing them to their present form. Publishers have done a commendable job and are sincerely acknowledged.

September, 2025 Neloy Khare
New Delhi, India

Contents

Editor and Contributors

About the Editor

Dr. Neloy Khare is currently the Adviser to the Government of India at Ministry of Earth Sciences (MoES). He has extensive research experience in Paleoclimatology, Oceanography, and Polar Science, with contributions to the field of palaeoclimatology of Southern high-latitude regions (the Antarctic and the Southern Ocean) using Micropaleontology as a tool. Dr. Khare received a Ph.D. in tropical marine regions and holds a Doctor of Science degree in Southern High-latitude marine regions with environmental/climatic implications. He has been conferred Honorary Professorship and Adjunct Professorship by many Indian Universities, and has an extensive list of publications in both National and International journals. Dr. Khare is the recipient of the Rajiv Gandhi National Award 2013 and Rajbhasha Gaurav Puraskar 2022, conferred by the Honourable President of India and Honourable Union Home Minister. His other accolades include ISCA Young Scientist Award, BOYSCAST Fellowship, CIES French Fellowship, Krishnan Gold Medal, Eminent Scientist Award, ISCA Platinum Jubilee Lecture, IGU Fellowship, Prof. S.N. Bhalla Gold Medal–2022 and Anil Chandra Award among others.

Contributors

Rajan Agarwal Geological Survey of India, Kolkata, West Bengal, India

Malik Zubair Ahmad Department of Earth Sciences, University of Kashmir, Srinagar, India

Sajad Ahmad Department of Earth Sciences, University of Kashmir, Srinagar, India

Ajit Deccan College, Pune, India

Vyasulu V. Akkiraju Ministry of Earth Sciences, Borehole Geophysics Research Laboratory, Karad, India

Ringphami Awungshi Geological Survey of India, Itanagar, Arunachal Pradesh, India

Debmita Bandyopadhyay Ministry of Earth Sciences, New Delhi, India

Upasana Swaroop Banerji National Centre for Earth Science Studies (NCESS), Thiruvananthapuram, Kerala, India

Amalava Bhattacharyya Birbal Sahni Institute of Palaeosciences, Lucknow, India

Akhouri Bishwapriya Geological Survey of India, Ranchi, Jharkhand, India

Ibitmeki Chadong Department of Geology, North-Eastern Hill University, Shillong, India

Abhishek Kumar Chaurasia Geological Survey of India, Patna, Bihar, India

Reyaz Ahmad Dar Department of Earth Sciences, University of Kashmir, Srinagar, India

Pankaj K. Das Geological Survey of India, Kolkata, West Bengal, India

Piu Dhibar Geological Survey of India, Kolkata, West Bengal, India

Rupesh Dhyani Department of Geography, Justus-Liebig-University, Gießen, Germany

Divya Geological Survey of India, Patna, Bihar, India

Rupal Dubey National Institute of Oceanography, Goa, India

Sanjay Kumar Dutta Geological Survey of India, Patna, Bihar, India

Pranit B. Gajbhiye Department of Geology, Rashtrasant Tukadoji Maharaj Nagpur University, Nagpur, India

M. Suresh Gandhi Department of Geology, University of Madras Guindy Campus, Chennai, India

Ashok Kumar Ghosh Mahavir Cancer Sansthan & Research Centre, Patna, Bihar, India

Samaya S. Humane Department of Geology, Rashtrasant Tukadoji Maharaj Nagpur University, Nagpur, India

Sumedh K. Humane Department of Geology, Rashtrasant Tukadoji Maharaj Nagpur University, Nagpur, India

Arun Kumar Mahavir Cancer Sansthan & Research Centre, Patna, Bihar, India

Shikhar Kumar Department of Geology, North-Eastern Hill University, Shillong, India

Jagdish Chandra Kuniyal Department of Geography, Justus-Liebig-University, Gießen, Germany;
College of Forestry, Veer Chandra Singh Garhwali Uttarakhand University of Horticulture and Forestry, Ranichauri, Tehri Garhwal, Uttarakhand, India

Victor J. Loveson National Institute of Oceanography, Goa, India

Ishfaq Ahmad Mir Geological Survey of India, Bengaluru, Karnataka, India

Jehangeer Ahmad Mir Department of Earth Sciences, University of Kashmir, Srinagar, India

Ranbir Nandan Department (Geology), B. N. College, Patna University, Patna, India

Rajiv Nigam National Institute of Oceanography, Goa, India

Avinash Niranjan Geological Survey of India, SU: Arunachal Pradesh, Itanagar, Arunachal Pradesh, India

Kalyan Pal Geological Survey of India, Bengaluru, Karnataka, India

Janardan Prasad Geological Survey of India, Kolkata, India

Waseem Qader Department of Earth Sciences, University of Kashmir, Srinagar, India;
Centre for Disaster Risk Reduction, Islamic University of Science and Technology, Kashmir, Awantipora, India

Irfan Rashid Department of Earth Sciences, University of Kashmir, Srinagar, India;
Department of Botany, University of Kashmir, Srinagar, India

Asaad Ahsan Raza Geological Survey of India, Ranchi, Jharkhand, India

Sukanta Roy Ministry of Earth Sciences, Borehole Geophysics Research Laboratory, Karad, India

P. D. Sabale Department of Archaeology, Deccan College, Pune, India

Rajeev Saraswat National Institute of Oceanography, Goa, India

Suhail Hussain Sheikh Department of Earth Sciences, University of Kashmir, Srinagar, India

Mayank Shekhar Birbal Sahni Institute of Palaeosciences, Lucknow, India

Vasant S. Shinde Deccan College, Pune, India

Shreya Shrey Geological Survey of India, SU: Jharkhand, Ranchi, India

Prateek Singh Geological Survey of India, Kolkata, West Bengal, India

Shreya Singh Geological Survey of India, Patna, Bihar, India

Shubhra Singh Geological Survey of India, Bengaluru, Karnataka, India

Sanjay Kumar Srivastava Chairman, Climate Resilient Observing Systems Promotion Council, New Delhi, India

N. Suresh Department of Geology, University of Madras Guindy Campus, Chennai, India

M. Thara Geological Survey of India, Bengaluru, Karnataka, India

Rajeshree R. Ukey Geological Division, Central Design Organization, Maharashtra Engineering Research Institute (MERI), Nashik, India

Anamika Upadhaya Department of Environmental Studies, North-Eastern Hill University, Shillong, India

Sandhya Verma Geological Survey of India, State Unit Bihar, Patna, India

Devesh Walia Department of Geology, North-Eastern Hill University, Shillong, India

Part I
Impacting Climate in Modern Time

Chapter 1
Lightning in India and its Trend with Climate Change

Sanjay Kumar Srivastava

1 Introduction

The captivating yet perilous natural phenomenon of lightning has held humanity in awe for centuries. Its erratic and powerful displays of atmospheric electricity transcend geographical boundaries, impacting regions across the globe. In the recent times, India has witnessed a significant increase in both the frequency and intensity of lightning, prompting a meticulous examination of its patterns and a probing inquiry into potential links with climate change. This chapter embarks on a comprehensive journey, unveiling the multifaceted dimensions of lightning in India. It encompasses far-reaching impacts, recent advancements, the evolution of early warning systems, ongoing research and development initiatives, and the intricate interplay between lightning trends and the changing climate.

2 Lightning in India: An Overview

2.1 Impact of Lightning

India's diverse climatic regions, intricate geographical features, and seasonal variations converge to create a landscape highly susceptible to the vagaries of lightning. Beyond the immediate and tragic loss of human life, lightning inflicts far-reaching consequences. Agricultural sectors bear the brunt of crop damage, power and communication infrastructures suffer debilitating blows and disruptions to daily life become pervasive. Vulnerable populations, particularly those engaged in outdoor pursuits

S. K. Srivastava (✉)
Chairman, Climate Resilient Observing Systems Promotion Council, New Delhi, India
e-mail: sanjaysonisa@gmail.com

N. Khare (ed.), *Changing Climate over India*,
https://doi.org/10.1007/978-981-92-1098-5_1

like farming, find themselves at an elevated risk, necessitating a holistic approach to lightning risk management that goes beyond immediate casualties.

2.2 Recent Developments in Lightning Management

In response to the escalating threat posed by lightning, India has taken significant strides in lightning management, orchestrating efforts to fortify forecasting, detection, early warning systems and public awareness. Collaborative endeavours between the India Meteorological Department (IMD) and scientific institutions such as the Indian Institute of Tropical Meteorology (IITM) Pune have culminated in initiatives aimed at accurately predicting and monitoring lightning activities. Establishing a lightning detection network and research projects by institutions like the National Remote Sensing Centre (NRSC), ISRO has augmented the country's capabilities in forecasting and detecting lightning.

2.2.1 Lightning Forecasting and Detection

The IMD's ground-breaking lightning forecasting initiative, launched in 2019, relies on a strategically placed network of lightning detection sensors spanning the nation. The synergistic collaboration with IITM Pune has significantly elevated the accuracy of lightning predictions. Complementary efforts from the NRSC's Lightning Detection Network, as showcased in the Annual Lightning Report 2022–2023, underscore the efficacy of these joint initiatives in comprehensively addressing lightning-related risks.

2.2.2 Early Warning Systems

Recognizing the pivotal role of timely warnings in mitigating the impact of lightning, the National Disaster Management Authority (NDMA) introduced a protocol in 2021 for the streamlined dissemination of early warnings. This protocol has substantially improved the reach of warnings to vulnerable communities, empowering them to take prompt safety measures and minimize potential damages.

2.2.3 Mobile Applications

The advent of mobile applications, such as 'DAMINI' and 'SACHET,' has emerged as a pivotal component in the dissemination of lightning early warnings and safety instructions to the general public. These applications have not only heightened public awareness but have also played a pivotal role in enhancing preparedness for lightning hazards, fostering a sense of collective responsibility towards safety.

2.2.4 State-Level Initiatives

Several states across India, including Jharkhand, Odisha, Andhra Pradesh, West Bengal, Karnataka, Bihar, Uttar Pradesh, Assam and Kerala, have taken independent initiatives to address the challenges posed by lightning. These initiatives involve the formulation of state action plans, investments in early warning systems and concerted efforts in capacity building to equip local communities with the knowledge and tools necessary to navigate lightning hazards effectively.

2.2.5 Expert Group on Lightning

The formation of the NDMA's Expert Group on Lightning stands as a testament to India's commitment to effective lightning hazard management. This expert group has been instrumental in devising comprehensive strategies and guidelines, encompassing the preparation of informational materials and the sensitization of the public about lightning risks. Their tireless efforts contribute significantly to creating a culture of safety and preparedness.

2.2.6 Reduction of Fatalities

Government initiatives and capacity building efforts have yielded positive outcomes, contributing to a decline in lightning-related fatalities in specific states. The proactive measures taken by Jharkhand, including the installation of lightning arresters in government facilities, serve as a noteworthy example of how strategic investments can translate into tangible reductions in human casualties, exemplifying the potential for broader nationwide success.

In the intricate dance between nature's awe-inspiring forces and human ingenuity, the strides made in India's lightning management landscape underscore the collective determination to not only understand and predict this natural phenomenon but also to protect lives and livelihoods from its unpredictable and potentially devastating consequences (Fig. 1). As we delve deeper into the layers of lightning intricacies, the following sections will unravel the evolving landscape of lightning research, the impact on vulnerable populations and the symbiotic relationship between lightning trends and the changing climate in the Indian context.

2.3 Lightning Trends in India

The study of lightning trends in India holds paramount importance in the context of climate change and its potential ramifications. By delving into the data provided by the Lightning Resilience Programme and other meteorological sources, we gain valuable insights into the frequency, intensity and geographical distribution of lightning

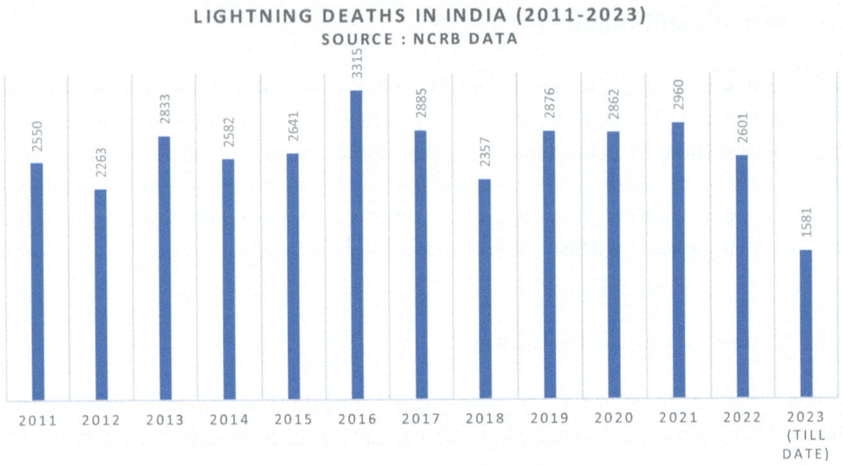

Fig. 1 Lighting-induced deaths in India

events in the country. This extended discussion aims to explore regional variations, the impact of climate change on lightning incidence and the measures undertaken for lightning risk management education and awareness in India.

2.3.1 Regional Variations

India's vast and diverse climate zones contribute to notable regional variations in lightning patterns. Some areas experience a higher frequency of lightning strikes, while others face more severe consequences due to factors such as topography, land use and population density. These regional nuances highlight the complexity of lightning occurrences and emphasize the need for targeted strategies to address specific challenges faced by different regions.

2.3.2 Climate Change and Lightning Incidence

The observed increase in lightning activity in India aligns with global trends associated with climate change. The rise in temperatures, changes in atmospheric moisture content and alterations in weather patterns contribute to the intricate interplay that influences lightning occurrences. Understanding the relationship between climate change and lightning is crucial for developing effective strategies to mitigate the impacts and enhance resilience.

3 Lightning Risk Management Education and Awareness

Public awareness emerges as a pivotal factor in reducing the impact of lightning hazards. The Lightning Resilience Programme recognizes the significance of education, sensitization and community engagement in fostering a culture of preparedness. The State Disaster Management Authority (SDMA) plays a lead role in organizing awareness programs, mock exercises and lightning risk audits to enhance community resilience.

3.1 Lightning Capacity Building Programme

A focused initiative, the Lightning Capacity Building Programme, aims to impart risk knowledge, educate stakeholders about early warning systems, disseminate lightning safety dos and don'ts and build capacity on prevention and mitigation. Creating awareness about weather hazards and utilizing corporate social responsibility (CSR) funds for lightning-safe initiatives are integral components of this programme. By empowering communities with knowledge and resources, the programme contributes to a proactive approach in managing lightning risks.

3.2 Research and Development

Despite notable progress in the recent years, the research and development landscape for lightning in India faces certain limitations. Existing infrastructure, while valuable, needs expansion to accommodate more comprehensive studies. Initiatives by institutions such as IITM Pune, IMD, CPRI Bangalore and DRDO represent significant steps, but there is a pressing need for increased investments and collaborative efforts to bolster capabilities.

3.2.1 Existing Status of Research and Development

Institutions under the Ministry of Earth Science, including IITM Pune and IMD, have established labs and testing facilities for lightning research. CPRI Bangalore contributes with a high voltage lab, and DRDO focuses on high voltage engineering for aircraft safety. However, the overall research and development capabilities are limited, highlighting the urgency for expanding infrastructure and allocating additional resources.

3.2.2 Research on Lightning and Climate Change

The impact of climate change on lightning trends is an emerging area of research. The observed increase in lightning activity in India aligns with global studies attributing the surge to rising temperatures, altered atmospheric conditions and increased moisture content. Understanding these relationships is pivotal for developing strategies to mitigate the impacts of climate change on lightning occurrences. Collaborative efforts between research institutions, government bodies, and private entities are essential to drive forward this crucial research agenda.

4 Lightning and Climate Change: A Complex Nexus

The intricate relationship between lightning trends and climate change requires a nuanced understanding of atmospheric processes, temperature variations and environmental factors. The following sections delve into the multifaceted connections between lightning activity and different aspects of climate change.

The dance of lightning across the skies has long captivated human imagination, but its intricate relationship with climate change presents a challenging puzzle that demands a comprehensive exploration. This extended discussion aims to unravel the multifaceted connections between lightning activity and various aspects of climate change, delving into atmospheric processes, temperature variations and environmental factors.

4.1 Atmospheric Moisture and Lightning

One crucial aspect of this intricate relationship lies in the interplay between atmospheric moisture and lightning activity. The Intergovernmental Panel on Climate Change's (IPCC) Fifth Assessment Report has brought to light the profound impact of anthropogenic forcing on the warming of Earth's upper oceans. This warming, in turn, results in an increase in atmospheric moisture content, a factor that plays a pivotal role in influencing lightning occurrences.

Understanding the dynamics of atmospheric moisture, cloud formations and the subsequent development of lightning events underscores the complex nature of the climate change-lightning nexus. As global temperatures continue to rise, it becomes imperative to dissect these interconnections to formulate effective strategies for mitigating the potential risks associated with increased lightning activity.

4.2 *Heat and Lightning*

Global warming, driven by various anthropogenic factors, significantly contributes to the surge in lightning activity, particularly in regions like India. The mechanism involves the intensification of evaporation processes, leading to elevated atmospheric moisture levels. As warm air rises, it interacts with cooler air within cumulonimbus clouds, creating conditions conducive to the generation of lightning discharges.

Recognizing the correlation between rising temperatures and increased lightning occurrences is paramount for developing climate-resilient strategies. The implications extend beyond meteorological phenomena, influencing various sectors such as agriculture, infrastructure and public safety. In the pursuit of sustainable solutions, understanding the intricacies of this relationship becomes essential.

4.3 *Drought and Lightning*

Drought conditions introduce a unique challenge to the complex web of climate change and lightning dynamics. While drought periods are often associated with a reduction in the number of lightning days, there is a paradoxical intensification of lightning activity during such times. Analysing lightning trends during droughts reveals a complex interplay of environmental factors.

Data from the Lightning Resilience Programme highlights specific years and regions experiencing a surge in lightning occurrences during drought conditions. This emphasizes the need for targeted mitigation strategies during these challenging periods. The seemingly contradictory nature of reduced lightning days coupled with intensified activity demands a nuanced approach to understanding and managing lightning risks in the context of climate change (Fig. 2).

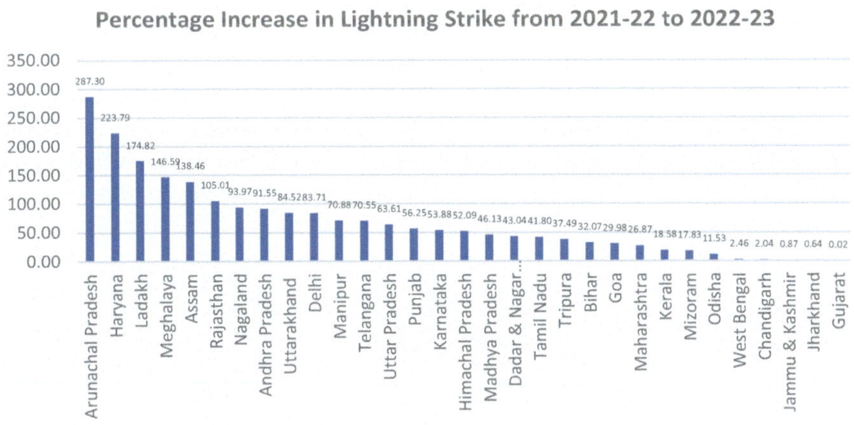

Fig. 2 Percentage Increase in lightning strike

4.4 Lightning and Cosmic Rays

An area of active scientific exploration lies in the relationship between cosmic rays and the initiation of lightning. Although the exact mechanisms remain elusive, emerging evidence suggests that cosmic rays may influence the initiation of lightning discharges. The complexities involved in this relationship necessitate ongoing research to unravel the potential connections between cosmic rays and lightning activity.

This section underscores the need for a multidisciplinary approach, combining atmospheric science with astrophysics, to shed light on the elusive mechanisms governing the interplay between cosmic rays and lightning. Understanding these intricate relationships is vital for developing a holistic understanding of the factors influencing lightning occurrences in the context of a changing climate.

5 Conclusions

In conclusion, the increasing trend of lightning occurrences in India poses a multifaceted challenge that intersects with the broader canvas of climate change. The impact of lightning extends beyond immediate hazards, affecting crucial sectors such as agriculture, infrastructure and public safety. While advancements in forecasting, detection, and early warning systems show promise, there is an urgent need for further research and development.

Understanding regional variations, the influence of climate change on lightning trends, and the intricate relationships between atmospheric processes is essential for formulating effective mitigation and adaptation strategies. Public awareness, education and community engagement remain integral components of lightning risk management.

As India grapples with the complexities of a changing climate and the challenges posed by increased lightning activity, concerted efforts at the national, state and community levels become imperative. Fostering a culture of resilience, investing in research and implementing proactive measures can enhance India's preparedness and reduce the impact of lightning disasters in the face of evolving climatic conditions. This extended exploration serves as a call to action, urging stakeholders to collaborate in deciphering the complex web of climate change and lightning dynamics for a safer and more resilient future.

Chapter 2
Arsenic in Groundwater, Causing Water Security Issues for Bhagalpur: A Comparative Study on the Extent of Contamination on Either Bank of River Ganga

Akhouri Bishwapriya, Shreya Singh, Arun Kumar, Divya, Pankaj K. Das, Piu Dhibar, Prateek Singh, Rajan Agarwal, and Ashok Kumar Ghosh

With world's increasing population, scarcity of clean water has assumed a precarious situation. Consequently, in many parts of the world, the easily accessible groundwater, when consumed, sometimes causes serious health issues under the influence of contaminants in groundwater (Mukherjee et al., 2019; Lone et al., 2019). One such groundwater pollutant, responsible for a high mortality rate worldwide, is Arsenic. Because of its carcinogenic properties, globally, over 300 million people are affected by consumption of water contaminated with Arsenic, out of which India and Bangladesh alone contribute 180 million people (Kumar et.al, 2018).

Eighteen states and three Union Territories of India have reported arsenic contamination in groundwater with a fast increasing rate (Mishra et.al. 2016). Out of reported eighteen states, Bihar, Uttar Pradesh and West Bengal have seen severe impact on the livelihoods of the stakeholders due to arsenic menace. Bihar stands third in the 'Top five Arsenic Affected States' with 1.023% affected habitations to total habitations (Data from DDWS, April 2011, UNICEF Report on "Water in India: Situation and Prospects", 2013).

A. Bishwapriya (✉)
Geological Survey of India, Ranchi, Jharkhand, India
e-mail: akhouri.bishwapriya@gsi.gov.in

S. Singh · Divya
Geological Survey of India, Patna, Bihar, India

A. Kumar · A. K. Ghosh
Mahavir Cancer Sansthan & Research Centre, Patna, Bihar, India

P. K. Das · P. Dhibar · P. Singh · R. Agarwal
Geological Survey of India, Kolkata, West Bengal, India

N. Khare (ed.), *Changing Climate over India*,
https://doi.org/10.1007/978-981-92-1098-5_2

Bhagalpur district lies within the Middle Ganga plain, reportedly having high arsenic incidence in groundwater. Arsenic in groundwater is thus a major health concern for the populace dependent on groundwater, mainly to meet their drinking water and domestic needs. The geological setting of the Middle Ganga plain is responsible for high As contamination. (Acharyya and Shah 2004). The city has been identified as a smart city contender, wherein the availability of fresh water will be an integral component for planners and those implementing the same.

The River Ganga forms the principal median drainage dividing the district across its banks. The tributaries of Ganga debouches its sediments from two contrasting sources, i.e. Himalayan geogeny (left bank) and peninsular provenance (right bank). In addition, the area constitutes a zone of oscillation of River Ganga where it not only flows from west to east but also oscillates in north-south direction thus constituting the zone of miscibility of inflow of sediments from contrasting sources. This sinuosity defines the active flood plain and older flood plain.

The overall quantity of groundwater is determined by geochemical mobility of elements in groundwater which is affected not only by recharging water composition and the mineralogy of geological formation forming the aquifer but impact of human activity and environment as well. (Kumar et al. 2006, 2007; Bhattacharya et al. 2009; Hasan et al. 2009). Several studies show reductive dissolution of Fe (III) oxyhydroxide of younger alluvial sediments in strongly reducing conditions. (Ahmed et al. 2004; Bhattacharya et al. 1997, 2002; Harvey et al. 2002; McArthur et al. 2001; Nickson et al. 1998; Nickson et al. 2000). The reducing environment is favourable for the dissolution of arsenic in groundwater from the sediments. The lowland flood-delta basin of the Ganga and Brahmaputra Rivers is entrenched over the Pleistocene terraces, where sedimentation was influenced by sea-level fall and rise during the Holocene (~12,000–6000 year B.P.) as observed in the Bengal Delta (Umitsu 1993; Goodbred and Kuehl 2000).

As per the guidelines of WHO (1996), the permissible limit of As in drinking water is 10 ppb, while many affected nations (including India) still have adopted 50 ppb as permissible limit depending on availability of groundwater. However, the acceptable limit of As in drinking water as per Bureau of Indian Standard has also been revised to10ppb (IS 10500 (2012): Drinking water). According to BIS, acceptable limit for arsenic in drinking water is adopted everywhere in India, except for the regions where there is no alternate source. In a water surplus zone like Bhagalpur with established endemicity, the lower limit (<10 ppb) has been taken as the benchmark to define anomalous concentration.

In the recent years, arsenic occurrences have assumed an alarming proportion along the north and south bank of the Ganga River, covering districts like Saran, Vaishali (Chaturvedi, R et.al; F.S. 2007-08 and 2008-09), Buxar (Mitra, I. et. al; F.S. 2004-05), Bhojpur (Haque M.W. et. al; F.S. 2003-04) and Patna districts of Bihar. Its distribution in different parts of Gangetic plain is a matter of serious concern and calls for concerted, multidisciplinary integrated approach towards correlation and prioritization of critically affected areas to understand the process responsible for arsenic concentration and suggest sustainable solution to the challenge.

Prominent results of the work carried out by Geological Survey of India, Patna, suggest that in the southern bank of Ganga River in Buxar, Bhojpur and Patna district, water is extracted from tubewell depth ranging 6.096 m to 213.36 m. These comprise (i) down to 12.19 m from the surface (arsenic content mostly <50 ppb and visually low incidence of Fe determined by yellow coloration) (ii) 12.19 m to 57.91 m (arsenic content 10 ppb to >500 ppb with visually high incidences of Fe determined by yellow coloration), (iii) 57.91 m to 89.92 m (arsenic content <10 ppb to 50 ppb and very clear water without any yellow coloration indicating absence of Fe) and (iv) below 89.92 m (arsenic content <10 ppb).

Thus, population residing in the levee portion within the oscillation zone is exposed to arsenic poisoning which has led to health-related problems. The most common health effects of arsenicosis include retarded development in children, cases of skin disorders (melanosis, leucomelanosis, keratosis, hyperkeratosis and skin cancer), non-pitting oedema, gangrene, hypertension, cancer (bladder, lung, kidney and liver), peripheral vascular disorders, ischaemic heart disease, cardiovascular diseases, restrictive pulmonary disease, non-cirrhotic portal fibrosis, respiratory problems, hypertension and diabetes mellitus. The problem stems from the fact that the signs of arsenicosis develop into symptoms very slowly (Kumar et al. 2023; 2022[a,b,c]; 2021[a,b,c,d]; 2020[a,b]; 2016: Edmunds et al. 2015).

Though prevalent in more than 20 districts in Bihar, mostly in districts lying to the north Ganga River, the predominance, secondary enrichment and health manifestations has been observed more along the oscillation zone of Ganga River. Bhagalpur offers one such example of a district administrative layout interspersed on either bank of the Ganga River, thus defining two contrasting sources of sediments and hence underlining the possible source of arsenic in sediments, helped by a comparison study.

This study focusses on As contamination in surface/ groundwater on either bank of River Ganga in Bhagalpur district. The paper deals with aspects of comparative river bank study for contamination, helped by an engineering geological investigation approach, accentuated by drilling androle of geology to identify pockets with severe health manifestations.

1 Study Area and Geological Set Up

The study area lies between latitude 25°11′00″ to 25°19′00″ N and longitude 86°54′ 00″ to 87°15′ 00″ E, falling in parts of Survey of India toposheet Nos. 72 K/15, 72 K/ 16, 72O/3 and 72O/4 (Fig. 1). The territorial limits of the district extend on either bank of the Ganga River. To its south, the study area is represented by a flattish topography, with a gentle slope towards north, and having a few isolated hillocks as observed near Shahkund village. The southern part of the area falls in the Chandan-Badua sub-basin and the area to north of Ganga falls in the Baghmati—Kosi sub-basin.

These two basins are parts of Mid-Ganga basin. Predominantly, undulating alluvial fill deposits (older flood plain) of the Ganga River encompasses the area to the

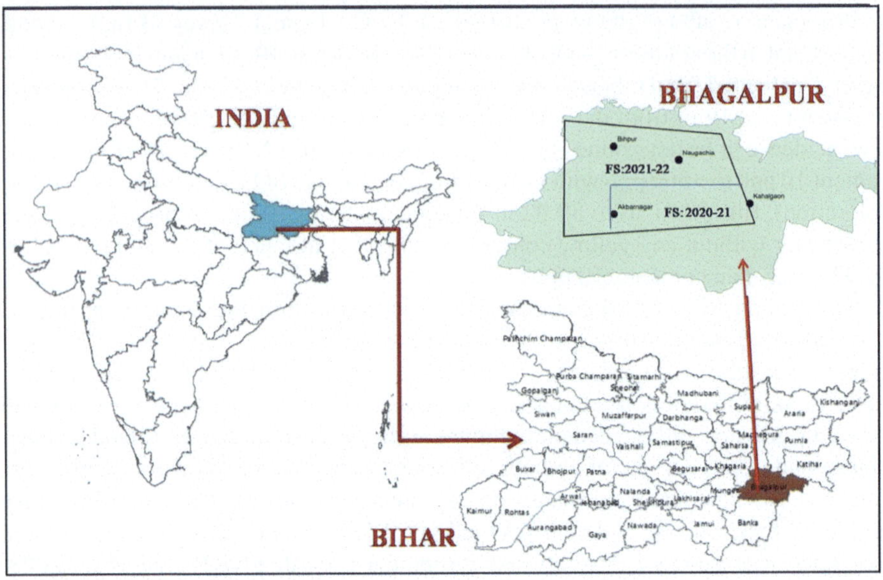

Fig. 1 Location map of the study area

north of Ganga. To the south of the Ganga River, alluvial upland with older alluvium is extensively present (Fig. 2).

Quaternary Alluvium predominates the study area. It mostly comprises sediments of quaternary succession ranging in age from Pleistocene to Holocene. Quaternary sediments include the older and younger alluvium sediments comprising a polycyclic sequence of sand, silt, clay and kankar. Geological connotation ascribes to alluvial sediments of the Present-day Formation, Diara Formation, Purnea Formation and Kosi-Ganga Formation of Younger Alluvium and Jamui and Nawada Formation of Older Alluvium, ranging in age from Pleistocene to Holocene. The stratigraphic succession of the area is mentioned in Table 1.

2 Methodology

Broadly, the methodology of study includes collection of water samples and delineating their spatial distribution on various geomorphic elements exposed in the area. A village-wise determination of groundwater arsenic, by arsenic kit (E-MERCK test kit), with at least 3–4 measurements of tubewells per village or as needed, was found adequate to establish the distribution pattern with population distribution depicting a random stratified orientation.

The location of each sample has been determined with a hand-held Global Positioning System (GPS) receiver; the well depth has been recorded from the estimation

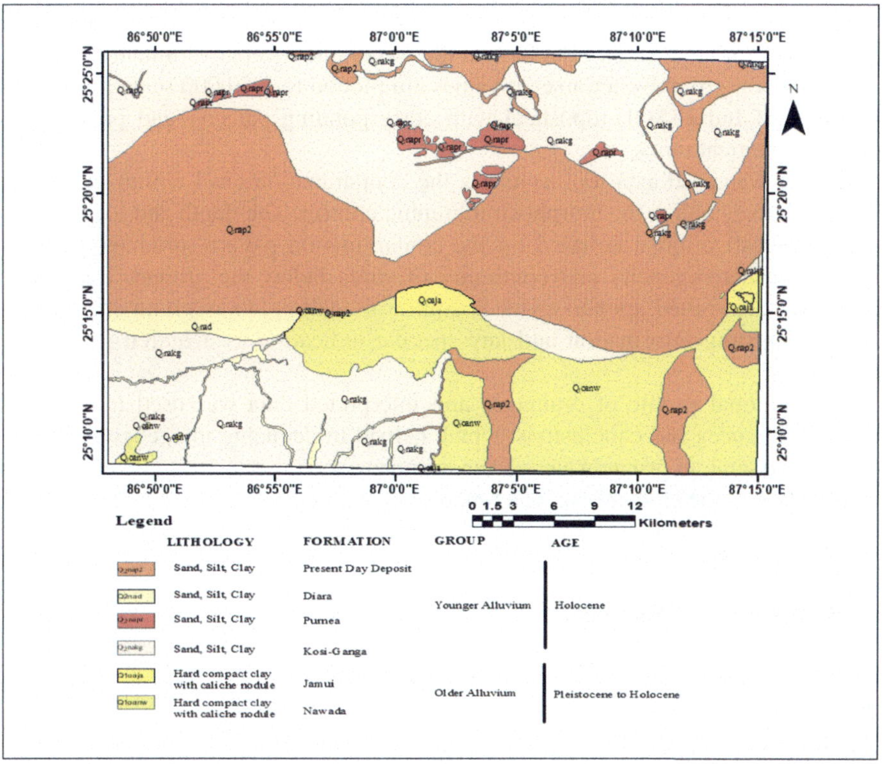

Fig. 2 Lithological map of study area (1:50000 geological map series, GSI)

Table 1 Generalized stratigraphic succession of the area. (All India unified legend, 2017)

Age	Super group	Group	Formation	Lithology
Holocene—recent		Younger alluvium	Present day formation	Sand, silt, clay
			Diara formation	Sand, silt, clay
Holocene			Purnea formation	Sand, silt, clay
			Kosi-ganga formation	Sand, silt, clay
Pleistocene		Older alluvium	Jamui formation	Hard compact clay with caliche nodule
			Nawada formation	Hard compact clay with caliche nodule

based on the number of pipes that were used in the construction of the well, and reconfirmed with the data available with the owner and local government agencies.

The data on groundwater arsenic values are plotted to 1:50,000 scale base maps of Survey of India (SOI) toposheet with corresponding palaeo—and present-day geomorphological units.

Drilling was used as a tool to assess the contamination level within sediments with depth to establish the morphostratigraphic control. The depth and location for invasive investigation was based on the contamination pattern, drainage control, topographic heterogeneity and continuity of strata below the ground. Resistivity survey was also relied upon to arrive at aquifer modelling, to ascertain change vis-a-vis the existing information and any specific indication as acquired from such data.

Longitudional profile of drainages and interpreted data was used to arrive at various inferences and conclusions, which helped in deducing the geogeny for the source of Arsenic in groundwater. A few samples were also tested to determine the valence state of the element but was found to be inconclusive.

3 Results and Discussion

Groundwater Samples:

Out of the 858 samples, 368 samples were collected from northern bank of Ganga River and remaining 490 samples were collected from the southern bank. Out of the total samples collected, 29.15% of samples indicate Arsenic values exceeding the acceptable limit (>10 ppb).

3.1 Sampling on the Right Bank of the River Ganga

In the area of interest, River Ganga forms the median drainage and receives water through rivers from two contrasting source, i.e. extra peninsular (left bank) and peninsular (right bank) source. The part of the district on the right bank is drained by rivers originating from the peninsular region. Several ephemeral streams such as Ghoga, Chandan, Kadwa, Gerua and Bhena from Chhotnagpur plateau, Bihar Mica Belt and Rajmahal Traps join the higher order drainage River Ganga forming an amphitheatre.

Altogether, 490 water samples from surface as well as groundwater have been analysed from the right bank of River Ganga. Nearly, 85% of the samples were found to be safe with less than 10 ppb of arsenic complying to acceptable standards of BIS. The contamination range of 10–49 ppb is uniformly distributed along the southern bank within the levee portion of River Ganga. Moving southward from the river, the contamination level sharply and conspicuously decreases with falling contamination level.

Mirzapur and Champanagar villages located towards west of Bhagalpur on the right bank of River Ganga were found affected with arsenic contamination. Maximum contamination is found to be 500 ppb in Mirzapur and Dougachi villages. Altogether, 28 groundwater samples were collected in and around these two localities, out of which 7 samples (25%) indicated As values >200 ppb.

Masaaru, Ghoga and Ghospur villages also showed high As levels in groundwater. These villages lie on the active flood plain of River Ganga. Value as high as 100 ppb has been determined near Masaaru village.

Out of the 15% of groundwater samples with high As values, nearly 85% of the samples with high contamination lie within the active and younger alluvium sediments. The area of study has a heterogeneous mix of alluvium with dual source of extra peninsular and peninsular region owing to its occurrence within the oscillation zone of Ganga River. The rocky uplands to the south, east and west of area also contribute to the influx of sediments by the drainages emanating from the peninsular part. Moreover, the area comprises formation that ranges from Late Pleistocene to Recent. It is observed that the samples falling on the Sautadih surface (Late Pleistocene) do not show high value of As concentration with a predominantly peninsular source.

3.2 Sampling on the Left Bank of the River Ganga

The part of the district on the left bank is drained by rivers originating from the extra peninsular region. Here, 368 nos. water samples were collected and145 nos of samples were analysed to be contaminated with >10 ppb As, constituting ~39% samples. The contamination range of 10–35 ppb (56 nos.) is uniformly distributed throughout the area. Out of 39%, ~9% are falling in the contamination range between 50 and 200 ppb and 5% samples have values >200 ppb. Higher As concentration (>100 ppb) is mostly observed in pockets along the left-over channels or where the lower order drainages are coalescing to form higher order drainages.

There is a higher concentration of arsenic in the groundwater (> = 100 ppb) in and around Naugachiya, Singhia Makandpur, Sahuparbatta, Marwa, Bhawanipur, Jagatpur and Gopalpur villages. These are identified as vulnerable pockets for the incidence of arsenic. Maximum contamination ~500 ppb is analysed from Singhia Makandpur and Gopalpur localities.

High values of arsenic is confined to pockets characterized by younger alluvium or proximal regimes of present-day channels. This provides a favourable environment for the release of arsenic in water from the aquifer sediment by reductive dissolution or desorption. The affected area falls in the interfluves of Ganga and Kosi River. A comparative study of Arsenic contamination in groundwater on the left and right bank of Ganga River has been depicted in Figs. 3, 4, 5 and 6.

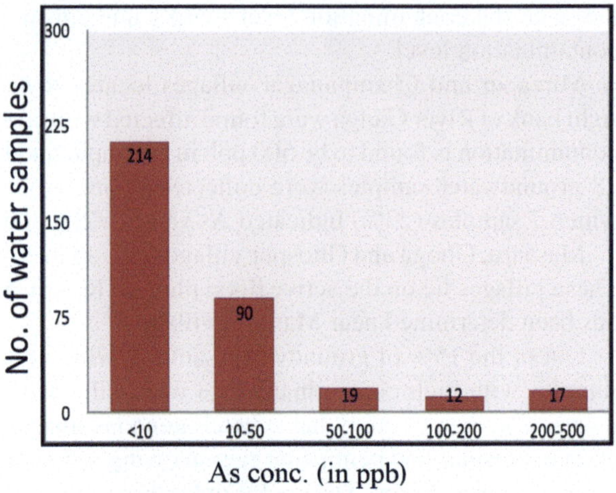

Fig. 3 Histogram showing the number of water samples falling towards right bank of river Ganga in different As conc. (in ppb) range

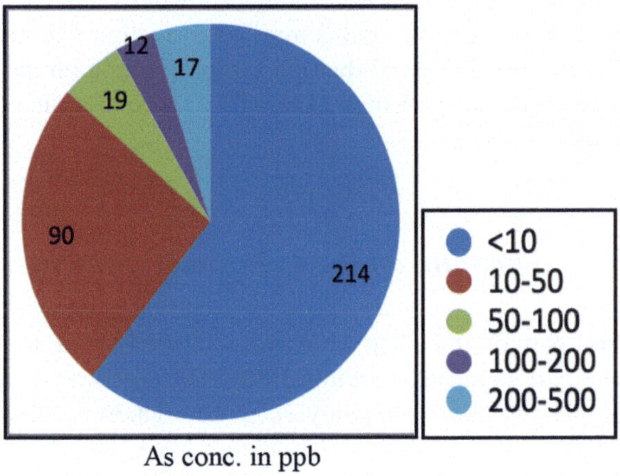

Fig. 4 Pie diagram showing the number of water samples falling towards right bank of River Ganga in different As conc. (in ppb) range

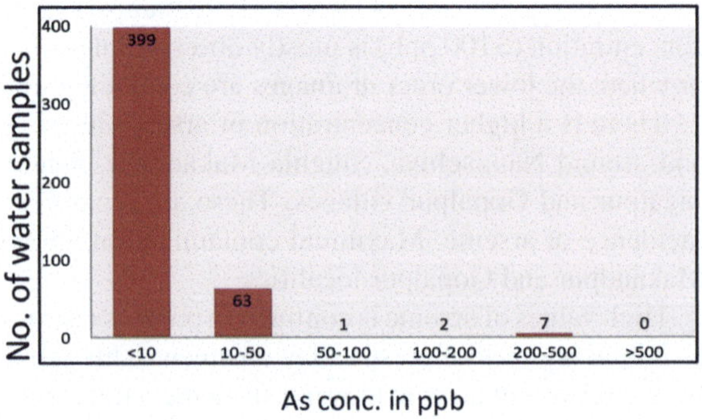

Fig. 5 Histogram showing the number of water samples falling towards left bank of River Ganga in different As conc. (in ppb) range

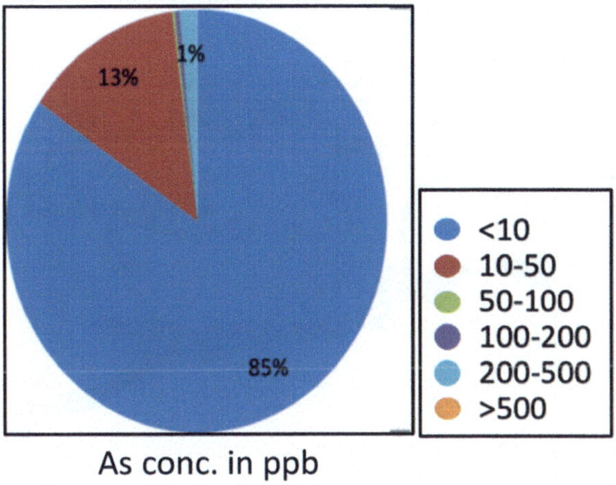

Fig. 6 Pie diagram showing the number of water samples falling towards north bank of River Ganga in different As conc. (in ppb) range

Arsenic distribution map superimposed on geological map of study area (Fig. 7) shows an overall spatial distribution of element that draws an analogy to the geological expression. The arsenic dispersion maps have been prepared using the Kriging interpolation method of the Spatial Analyst Tool in ArcGIS Software.

Kriging is a technique that fits the source data accurately and preserves local anomalies in the interpolation grid. Kriging interpolation assigns weights as a two-step process: first, the spatial covariance structure of the sampled points is determined by fitting a variogram; and second, weights derived from this covariance structure are used to interpolate values for unsampled points or blocks across the spatial field.

It is observed that the area falling towards the left bank of River Ganga has distinctly higher contamination of As compared to the right bank (Fig. 7). The left bank is drained only by Himalayan Rivers which is perceived to carry sediments containing arsenic. Various authors have already established that the store-house of arsenic is in the sediments of Siwalik groups of rock (e.g. Stanger 2005; Saunders et al. 2005) which act as secondary source while the primary source in Ganga basin within the foreland lies in the Indus Valley Suture Zone of Higher Himalaya. According to Mukherjee et al. 2014, active and palaeo-foreland basin of the Himalayan orogenic belt is considered as extensive arsenic enriched provenance globally. With conducive pH, redox condition, temperature, and solution composition, it gets released into groundwater, causing a severe impact on the population.

On the contrary, the rivers originating from Chotanagpur Gneissic Complex, Rajmahal and Kharagpur hills draining the southern part of the study area show occasional contamination. Even accentuated contamination is observed conspicuously along the southern bank of Ganga River within Present Day Formation and Belhar and fails to occur within the Sautadih Formation.

The pattern of contamination is thus pronounced on the northern bank and within the zone of oscillation of River Ganga (comprising levee portion of southern bank as well), resulting from the mixing of sediments due to River oscillation in north-south direction. Thus, the river flowing from west to east encounters several braided

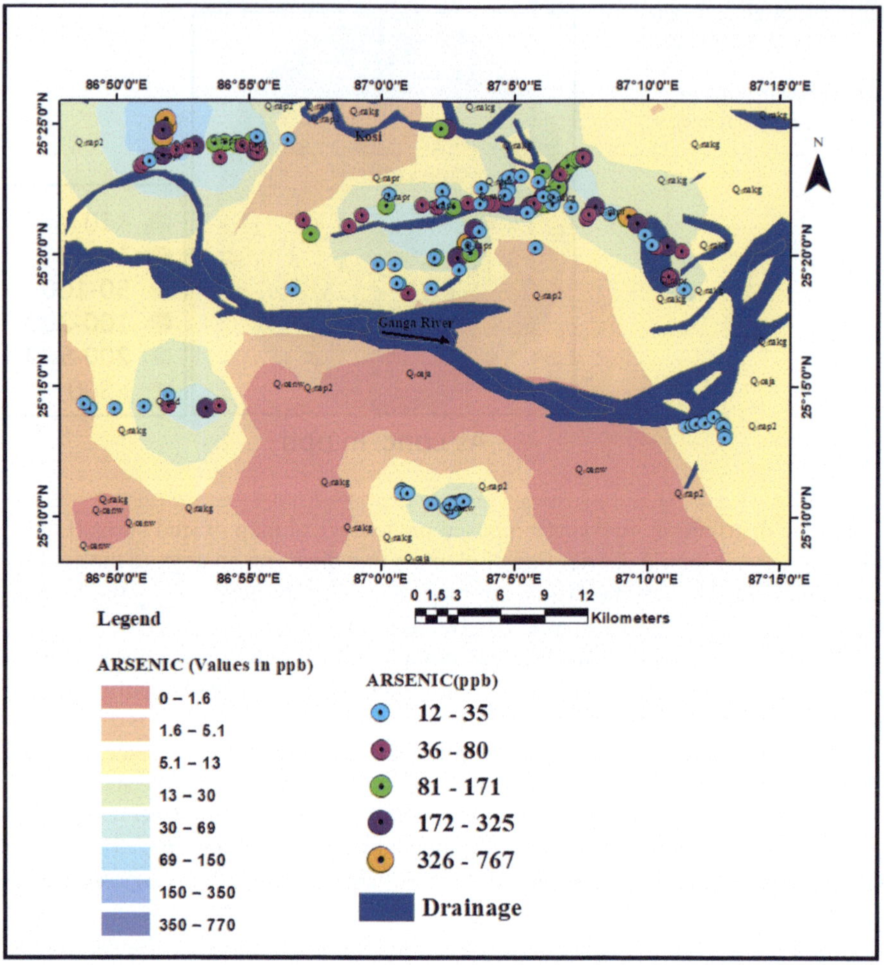

Fig. 7 Interpolated map of arsenic distribution of the study area

sections. The river width assumes appreciable width (~ 30 km) and distinct temporal change in flow geometry vouched by meanders, river islands, point bars and other geomorphic expressions. These pockets constitute the zone of miscibility of the inflow of sediments from contrasting sources.

Over the past few decades, the geomorphology of the anastomosing River Ganga (Holocene age) has been extensively modified. Temporal changes in the sinuosity of Ganga have been observed from Google Earth imagery (Fig 8). Abandoning earlier courses and acquiring new ones in selected sections has led to the formation of several abandoned channels, ox-bow lakes. From the Google Earth imagery, it is observed that aquifers falling in oscillation zone and the villages that are situated in or around the abandoned channels or present-day channel, back swamps, oxbow lakes, point bars are severely contaminated with arsenic. These appear to be the sites where the biomass gets accumulated due to water logging and the same is released in the aquifer during groundwater recharge. According to Harvey et al. 2002, Mc Arthur et. al. 2004, Wen et al. 2013, Gan et al. 2014, and Ghosh et al. 2021, microbially assisted reductive dissolution and reduction of oxidized organic matter utilizing Fe oxides/oxyhydroxides as electron acceptor helps in mobilization of As in groundwater (Fig. 9).

Arsenic Anomaly Along Natural Drainages Using Longitudinal Profile:

The traverse along the course of Gerua nadi and Katharia nala—Ghoga River (peninsular bound river) was taken to carry out close spaced sampling of water samples from dug wells of all the villages located immediately on the banks of these rivers. The arsenic contamination, if any, along the flow regime of rivers draining from the peninsular region would be a significant indicator to understand the agents of contamination and the depth control.

Gerua Nadi: This river enters the study area from the south near Dhobi village. It is a peninsular river flowing south to north and ultimately contrary to its east-west trend for few kilometres. The tubewells in around 06 villages located on either bank of Gerua nadi were sampled and the water tested for the presence of arsenic.

Accordingly, a L-profile (longitudinal) of the river (1:50,000 scale) was prepared with the X-axis representing the length of the river and the Y-axis representing the elevation in metres. Considering the fact, that the gradient along river is not appreciable, the scale along Y-axis has been kept at 1 cm = 1 m to pick up even minor variation in gradient fall. This L-section provides information on the gradient of the river and arsenic contamination along the banks (Fig. 10).

The villages falling on or close to the L-profile include Jagarnathpur, Chandpur, Shrimatpur, Milkitola, Shadopur and Amapur. The collected water samples were tested for the presence of arsenic using a field test kit. The results indicate that the villages on either side of the meandering course do not show any hazardous value. This could be because the rivers flowing from the peninsular region do not carry sediments rich in arsenic, or it may be that the aquifers present do not contain mobile/ dissolved (As), as they may not be supported by an environment favouring the release of As in water.

Katharia nadi and Ghoga Nala: Katharia nadi is a south to north aligned peninsular river which assumes conspicuous shift from west to east direction near Bhithi village and is rechristened as Ghoga nala before making confluence with River Ganga near Pannuchak. During field traverses and from imagery data, it is made out that the course of Ganga and most of the tributaries joining it are guided by a conspicuous west to east trending lineament/fracture(?) which prevails over the flow guiding it from west to east manner before making a confluence with the higher order drainage. It is also likely that this fracture has downthrown the active channel and upthrown the levee portion, the reason why drainages near confluence acquire west to east direction sympathetic to the flow of Ganga River.

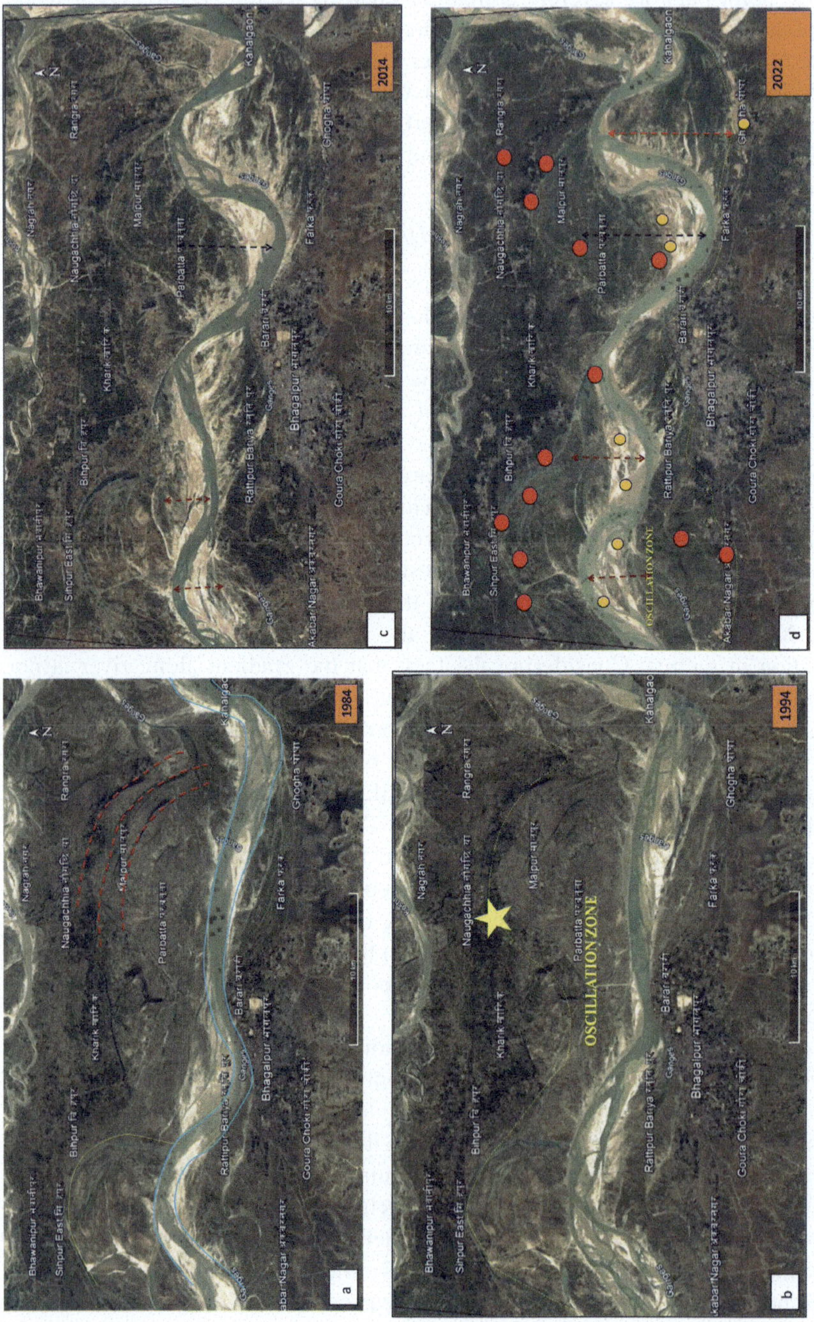

Fig. 8 Temporal changes in the sinuosity of Ganga have been observed from google Earth imagery. (**a**, **b**, **c**, **d**: The change in course during 1984–2022)

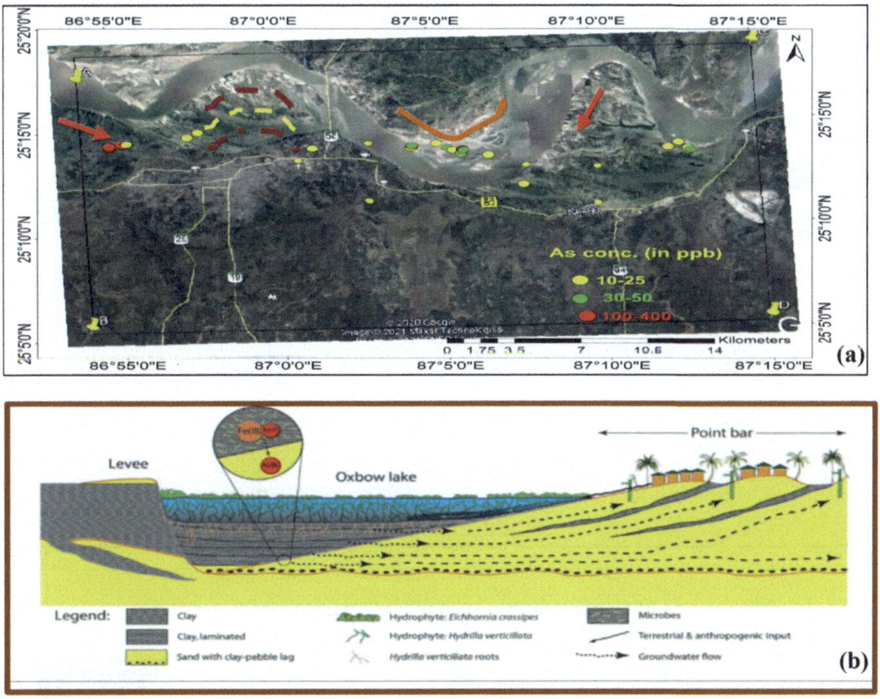

Fig. 9 **a** Temporal changes in geomorphic expression within oscillation zone of Ganga River and **b** mechanism showing release of As from stagnating clay horizon to adjoining sandy aquifer ultimately drafted by local population (Reproduced after Ghosh et al. 2021)

The juncture where Ganga trends towards north abruptly probably coincides with a fracture oblique to the west to east flow, resulting in intermittent change in flow direction within the parent median drainage. These could well be near surface or surface linear features. It is also observed that near the levee portion of the Ganga River, the northerly slope somewhat becomes southerly.

Similar to the Geruanadi, a L-profile section for the Katharia-Ghoga River was also prepared. A water sample was collected and subsequently tested for the presence of As from the villages on either bank of the rivers. It was observed that the villages lying on flood plain of Gerua channel do not show any alarming value of As.

The villages encountered were Sanhauli, Dhauri, Utsu and Machchipur. However, localities on either bank of the Ghoga River show high values, up to 100 ppb at some places. The villages with high contamination were English (25 ppb, 35 ppb), Masaru (100 ppb, 50 ppb, 30 ppb), Lailakh Mamlakha (20 ppb) and Shankarpur (15 ppb) (Fig. 11).

It is evident that the north-south channel (Katharia River) carries the sediments from the peninsular region and is represented by the Sautadih Formation. Sautadih Formation is of late Pleistocene, and the findings clearly show that water samples from the older formation occasionally show alarming contamination. However, the

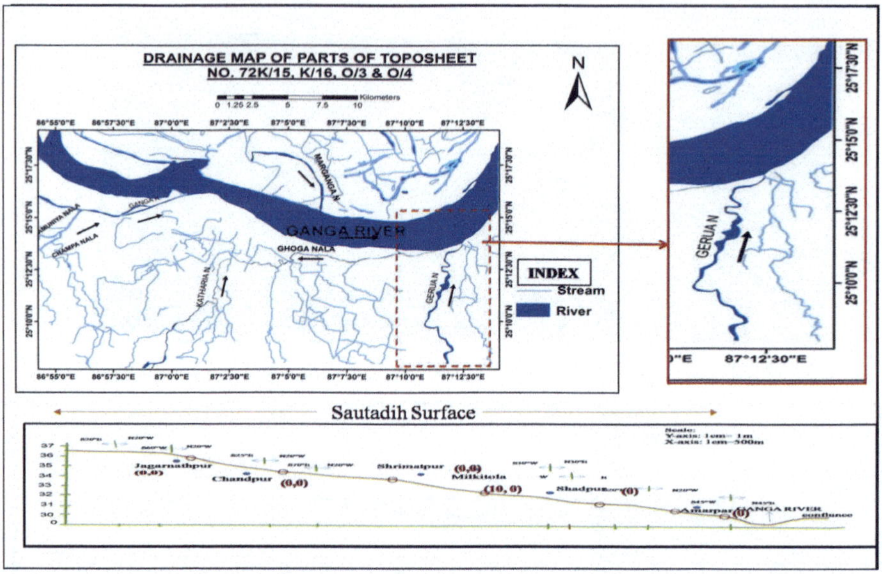

Fig. 10 L-profile section of Gerua Nadi (up to confluence with Ganga River)

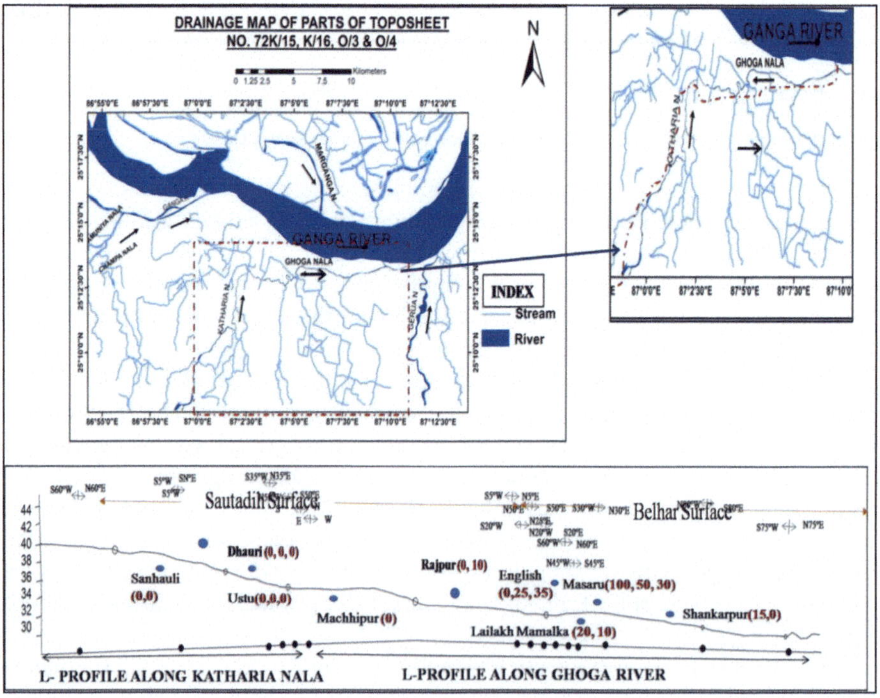

Fig. 11 L—Profile section of Kathria nala and Ghoga River (up to confluence with Ganga River) showing ascending values

Ghoga River trends east-west, flowing parallel to the Ganga River just before its confluence, and thus the villages located in this zone (Belhar formation of Holocene age) show high contamination. As explained earlier, this portion also constitutes a zone of oscillation of River Ganga where it not only flows from west to east but also oscillates in north-south direction thus constituting the zone of miscibility of inflow of sediments from contrasting sources. This sinuosity defines the active flood plain and older flood plain, which are supposedly contaminated with anomalous arsenic values greater than 10 ppb.

Section Showing Contamination Along the North and South Bank of the Ganga River:

Profiles cutting across the banks of the Ganga River have also been developed. This has been done to derive the contamination pattern across banks, as the profile line cuts across the River Ganga and various left-over channels and meander scrolls on the north bank of the Ganga. This approach gives a better representation of the varying depth of aquifers and their contamination range on either side of the river. Towards the left bank, alluvium representing Holocene and recent period is exposed while on the other bank of Ganga, alluvium representing Pleistocene to recent is present.

The cross section was prepared in SW-NE direction along profile line A-A'. This profile line intersects three borehole points. It was done to understand how the subsurface geology varied on either side of the Ganga. The first borehole point intersection falls south-west of the study area in Shakund village, followed by other drill point in Champanagar located on the right bank of River Ganga. The last drill point is located near Singhia Makandpur to the north of Ganga. Cross section along A-A' has been prepared with X-axis representing the length of the profile line (1 cm = 0.5 km) and Y-axis as the aquifer depth (1 cm = 1 m). All the villages lying along the profile line had been plotted with aquifer depth and arsenic concentration (Fig. 12).

It was observed that Shakund village (on right bank) is surrounded by isolated hillocks representing Munger-Kharagpur hill. Within the borehole, the only aquifer encountered was at a depth of 41.5 m. None of the water samples from this locality shows any hazardous value of As contamination. Despite the clay dominant column (which shows as enrichment on left bank), the mean As of core sample at this location is 3.9 ppm. This result indicates that sediments transported by Munger super group of rocks may not be containing arsenic enriched minerals. In addition to the above, other reason for very low presence of as in groundwater could be the environment in which aquifer formed and which may have been unfavourable for As release in water.

Moving further towards River Ganga, out of the two localities falling along the profile line, Purani Sarai and Champanagar, the latter shows as high as 400 ppb As value. Most of the high As contamination in this locality were present at the depth zone of 19.5–32 m, which forms the second tier of aquifer and falls within the Present-day formation. The deeper aquifer (third aquifer, 41.6–43 m deep) seems free from high arsenic contamination. Borehole drilled down to a depth of 60 m at this location shows high As value in the sediments also. The mean As value of 33.1 ppm has been observed at this drill point. From this section, it has been observed that towards the left side of the River Ganga, high arsenic contamination is observed

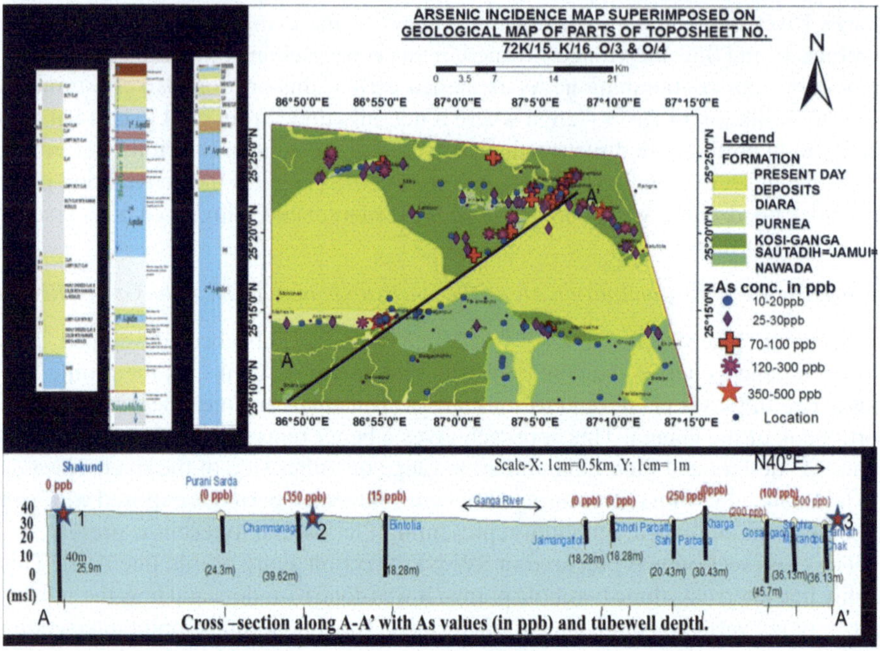

Fig. 12 Cross section along A-A' along with As value in ppb and tubewell depth

sporadically in pockets. The last borehole profile line shows an aquifer of colossal thickness with high As values.

Arsenic Distribution with Depth:

The observation depth range used for sampling was 5 ft. (1.5 m) to 350 ft. (106.68 m), and the average depth of the tubewells was 94 ft. (28.65 m). The median depth of the tubewell was 90 ft. (27.432 m), and the 95th percentile depth value was 150 ft. (45.72 m). Groundwater samples were collected from tubewells of different depths to represent the arsenic distribution of that particular aquifer (Fig. 14).

A total of 858 nos. of tubewells were sampled, their depth ranging from 5 ft. (1.5 m) to 200 ft. (106.68 m). The general depth of tubewells ranges between 50 and 150 ft. in the area. Out of 858 tubewells sampled, elevated concentration of arsenic is not confined to any specific depth zone. The samples from surface water and dug well water do not show aggravated As contamination. A bivariant plot of depth versus Arsenic was plotted.

R^2 value of linear equation shows very few to none of the variation in the y values being accounted for the x values. From these findings, it appears that maximum numbers of tubewell are present at the depth zone of 50–150 ft. This is the zone where As concentration ranges from 0 ppb to 500 ppb. However, deeper aquifers comprise lesser tubewell and hence appear to be occasional or less contaminated (Fig. 13).

Fig. 13 Bivariant plot of As with depth

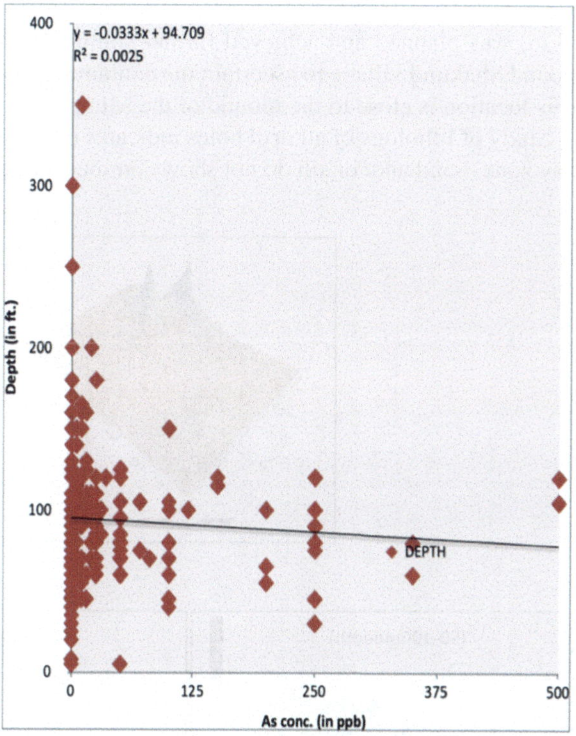

Subsurface Investigation:

During the study, drilling has been carried out at six locations of depth ranging from 35 to 60 m, in view of occurrence of arsenic within younger sediments vis-a-vis their comparison with older formations. This is done to decipher the morphostratigraphic control of As contamination within the subsurface sediments.

Considering texture and lithology, the sediments can be broadly categorized into clay, clayey silt, silty sand and sand. Grain size profiles characteristically fine upward in all cases indicating a fluvial environment of deposition.

Two drilling sites were selected along right bank of River Ganga in Champanagar village, west of Bhagalpur city and Masarhu village. Values of 500 ppb has been recorded in the tubewell near Champanagar. Drilling down to a depth of 60 m was achieved this location (BH1). Likewise (BH2; 60 m) at Masarhu village was achieved along the meandering course of River Ganga. Arsenic values up to 100 ppb in the tubewells have been recorded near this village and the area is occupied by younger alluvium sediments. Two boreholes were drilled in the interfluvial zone of Kosi and Ganga River at SinghiaMakandpur (BH4) and Narayanpur village (BH5). Arsenic values of 500–600 ppb have been observed from the tubewell near these villages. A borehole (BH3; right bank) was drilledat Ankita village, while the last borehole (BH6; 45 m) was planned and achieved further south-west of right bank of Ganga River around Shakund village to ascertain the contamination within peninsular sediments. This location is close to the mound of the Munger Group of rocks (Fig. 14).

Study of lithologs of all drill holes indicates that arsenic is mainly confined to the clay zone. Sand and/ or silt do not show commensurate alarming value of arsenic. In

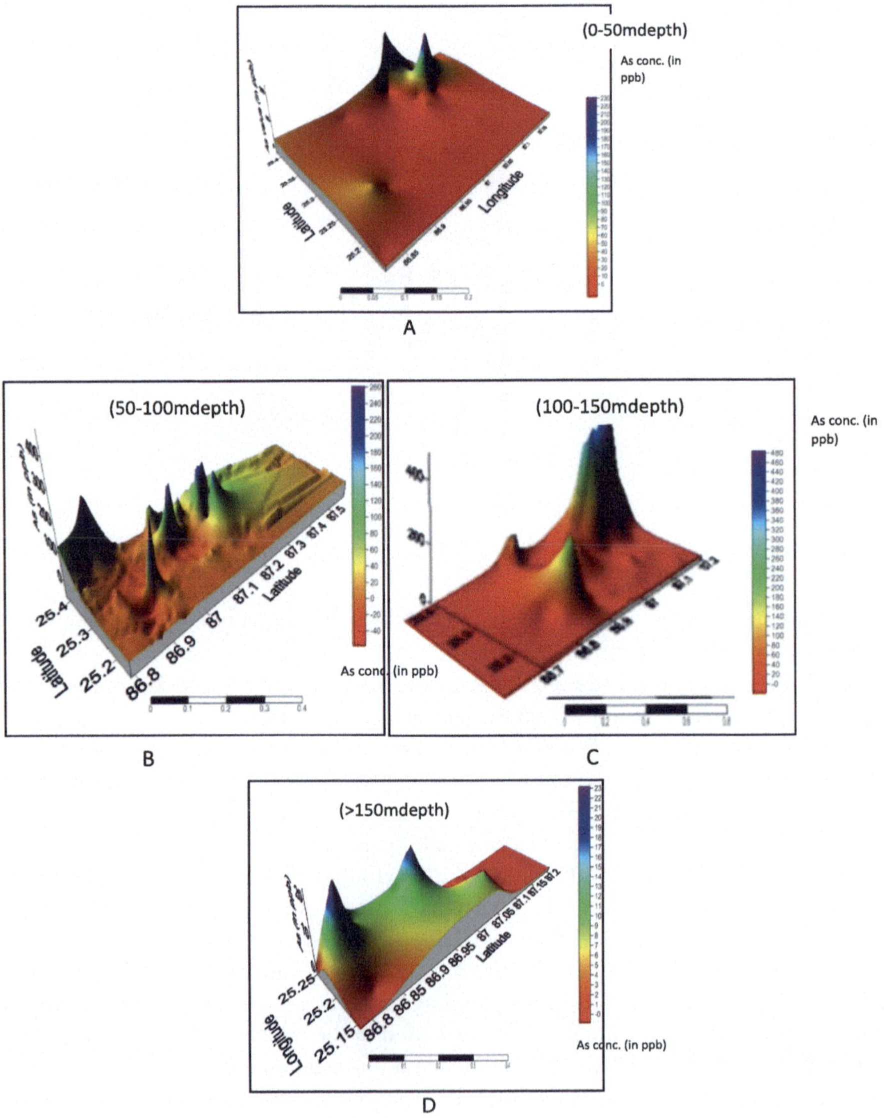

Fig. 14 Spatial distribution of incidence of As in 3D at different depths (A-D)

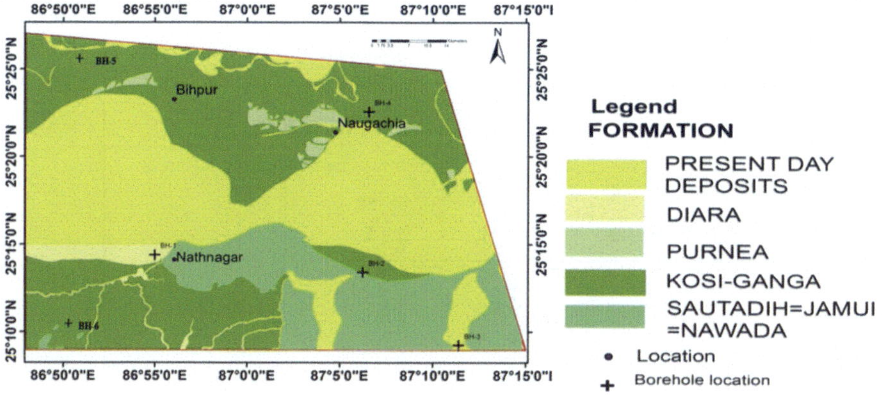

Fig. 15 Bore hole location superimposed on the geology of the study area

the northern part of the area, contrary to the high value of As in groundwater, core samples do not show anomalous values ranging from 4.39 to 1.67 ppm with a mean of 2.7 ppm. This may be due to dominance of medium to fine riverine sand within the drilled depth. The area lies in the interfluvial zone where sand horizon dominates.

The arsenic concentration of litho sequence within (BH1) ranges from 13 to 57 ppm with mean of 26 ppm. The higher values are confined mainly to the clay and silty clay layers. At Masarhu village (BH2), the value range between 8 and 34 ppm. In BH2, the values of arsenic within clay are higher within 2–6 m bgl, decrease in the depth range of 15–48 m and then spike between depth of 48–60 m.Comparative values are clearly lower in sand (Fig. 15).

In BH6 near Shahkund village, As within sediment ranges from 11 to 23 ppm with mean value of 3.9 ppm. Contrary to this, the value within core samples in BH3 is much higher and mean value of 33 ppm. Groundwater sample in this area does not show any alarming value (Fig. 16).

Soil Analysis:

Soil samples from the study area were analysed for the presence of trace elements. The result demonstrates six times higher arsenic concentration in the northern part of study area, i.e. the part which falls on the left bank of River Ganga compared to the right bank (Table 2). The northern part of the study area forms the interfluve portion between Ganga and Kosi and is occupied by the sediments of Himalayan orogeny, while the southern portion mainly derives its sediments from peninsular region. The mean of the As within sediments on left bank of Ganga is 25.657 ppm.

In addition, the trace elements of the soil samples were compared to Upper Continental Crust (UCC) values, showing high arsenic concentration. Results indicate that all other elements do not vary significantly with UCC on both the sides of Ganga River thus, except Arsenic (Table 3, Fig. 17).

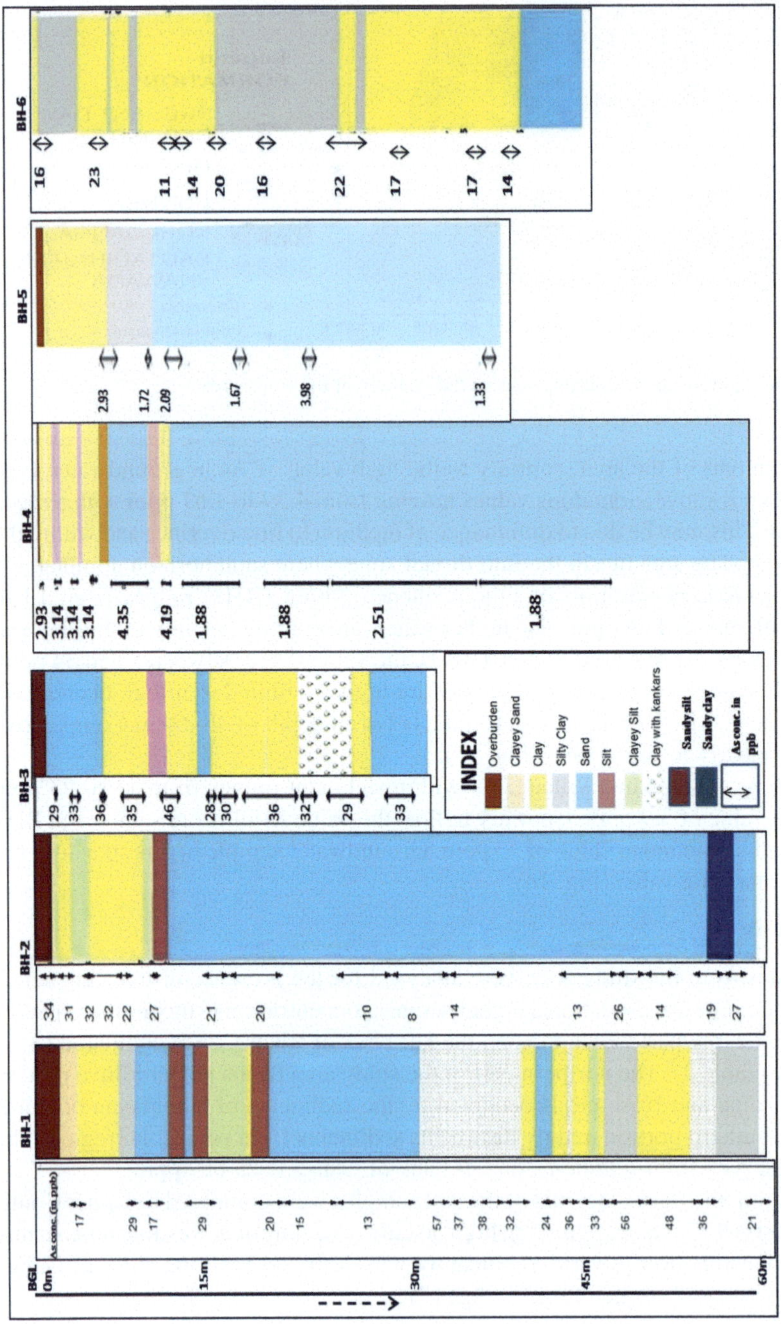

Fig. 16 Litholog of borehole (BH1 to 6) with commensurate As concentration in parts per billion

Table 2 Concentration of mean of trace elements in the soil samples from the study area

Elements (in ppm)	Zn	Cu	Pb	Co	Mn	Cr	As	Ni
Mean of the conc. Of element for the area falling on the left bank of Ganga	10.147	5.312	20.48	5	23.4	13.1	25.657	7.0909
Mean of the conc. Of element for the area falling on the right bank of Ganga	44.416	14.18	16.091	5.850	327	48.300	5.018	17.826

Table 3 Comparison between the mean of trace element from soil of the study area (both bank of River Ganga) with upper continental crust (UCC)

Elements (in ppm)	Zn	Cu	Pb	Co	Mn	Cr	As	Ni
UCC *(Taylor &McLenan, 1995)	71	25	20	10	600	35	1.5	20
Area falling in the left bank of Ganga	10.147	5.312	20.48	5	23.4	13.1	25.657	7.0909
Area falling on the right bank of Ganga	44.416	14.18	16.091	5.850	327	48.300	5.018	17.826
Area falling in the left bank of Ganga/UCC	0.143	0.212	1.024	0.500	0.039	0.374	17.105	0.355
Area falling bon the right bank of Ganga/UCC	0.626	0.567	0.805	0.585	0.545	1.380	3.345	0.891

*(Taylor & McLenan, 1995)

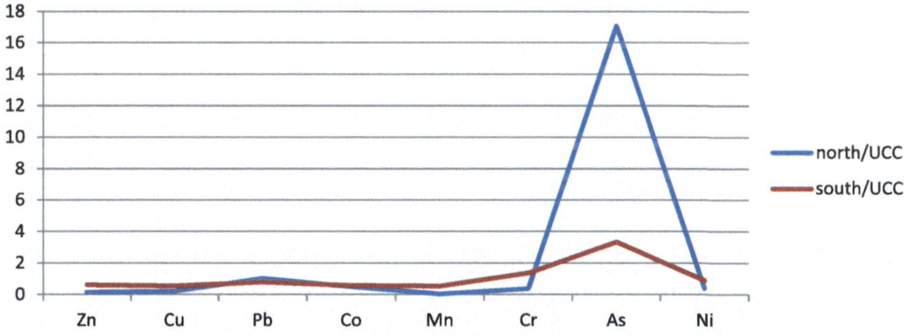

Fig. 17 Graph showing comparison between the mean of trace element from soil of the study area (both bank of River Ganga) with upper continental crust (UCC)

Geophysical Input: The Vertical Electrical Soundings (VES) using Schlumberger configuration have been carried out to map the subsurface lithological layers. The nature of sounding curves observed is H, HA, HK, KH and HKH types indicating 3–5 layered subsurface. The Vertical Electrical Soundings (VES) were conducted at 171 locations with a station interval of about 1.5–3.0 km with maximum current electrode separation (AB) 480 m.

The resistivity curve types A, Q, H and K and combinations were obtained. These curves are interpreted using the curve matching technique. The different basic types of curves are Q type (ρ 1>ρ2>ρ3), H type (ρ1 >ρ2<ρ3), K type (ρ1<ρ2>ρ3) and A type (ρ1<ρ2<ρ3). A combination of the basic curves can be obtained if the model is obtained for more than three layers. Then, the combinations possible can be as HA type (ρ1 >ρ2<ρ3<ρ4), AK type (ρ1<ρ2 <ρ3>ρ4), KH type (ρ1 <ρ2 >ρ3<ρ4), etc.

North Bank of the Ganga River

This part has been divided into four sections to describe the nature of curves dominating in the section. Details are given in table:

Eastern Part: The shape of the curve for the profile E55 is HKH type (Fig. 18). In respect to the layers pattern, first layer of resistivity around 26.6 Ohm-m and thickness around 4.16 m may represent top soil cover, second layer of resistivity around 6.6 Ohm-m and thickness around 3.56 m may represent sandy-silty clay and silty soil (wet to moist), third layer of resistivity 57.7 Ohm-m and thickness around 28.7 may represent silty sand, fourth layer of resistivity 19.2 Ohm-m and thickness 21.2 m may represent silty-sandy clay while fifth layer of resistivity 31.8 Ohm-m may represent silty sand.

Western Part: The shape of the curve for the profile W101 is HKH type (Fig. 19). In respect to the layers pattern, first layer of resistivity around 62.2 Ohm-m and thickness around 3.3 m may represent top soil cover, second layer of resistivity around 31.3 Ohm-m and thickness around 5.86 m may represent silty sand, third layer of resistivity 83.9 Ohm-m and thickness around 21.4 may represent fine to medium sand, fourth layer of resistivity 58.0 Ohm-m and thickness 44.9 m may represent silty sand while fifth layer of resistivity 76.2 Ohm-m may represent coarse sand.

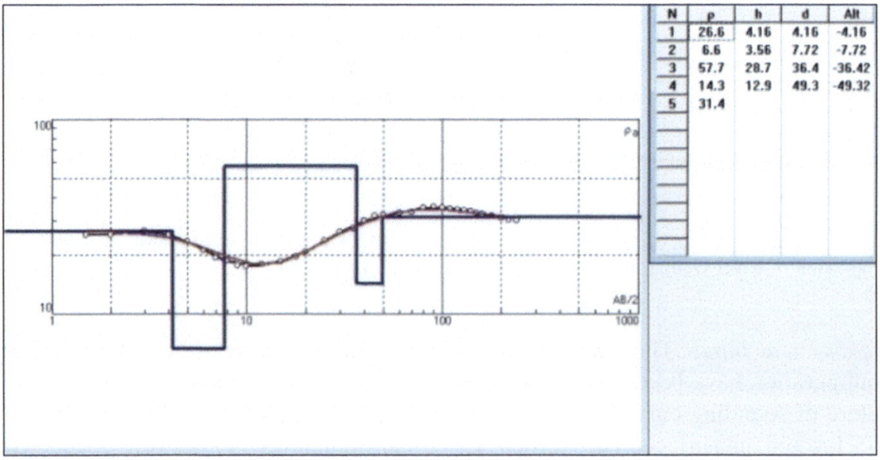

Fig. 18 VES curve of profile E55

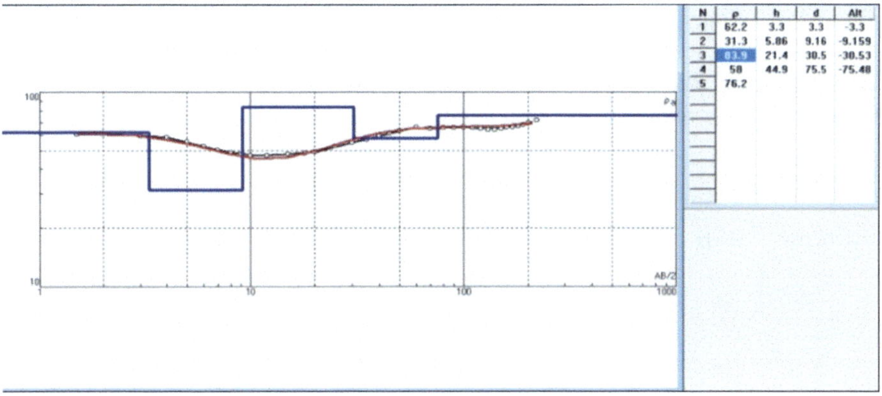

Fig. 19 VES curve of profile W101

Northern Part: The shape of the curve for the profile N59 is HKH type (Fig. 20). In respect to the layers pattern, first layer of resistivity around 37.2 Ohm-m and thickness around 3.01 m may represent top soil cover, second layer of resistivity around 15.3 Ohm-m and thickness around 6.66 m may represent sand, silt and clay, third layer of resistivity 44.5 Ohm-m and thickness around 3.72 may represent silty sand, fourth layer of resistivity 28.4 Ohm-m and thickness 18.1 m may represent silty sand, while fifth layer of resistivity 62.4 Ohm-m may represent medium to coarse sand.

Southern Part: The shape of the curve for the profile S22 is KH type (Fig. 21). In respect to the layers pattern, first layer of resistivity around 9.45 Ohm-m and thickness

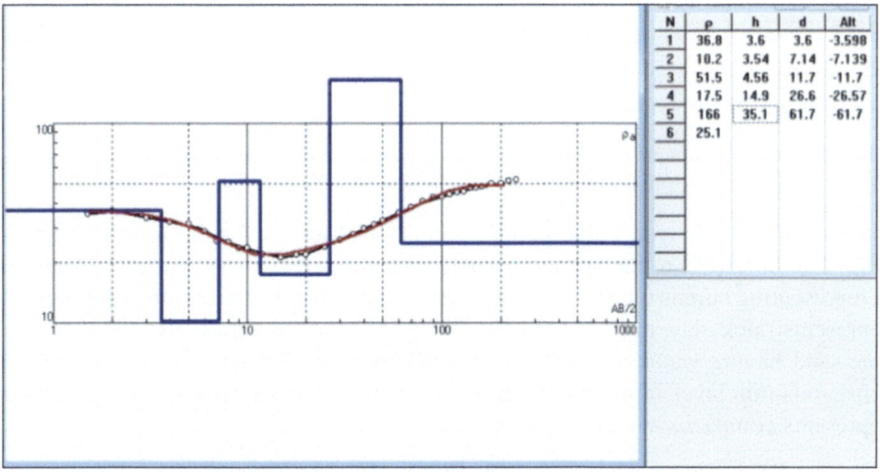

Fig. 20 VES curve of profile N59

around 2.91 m may represent top soil cover, second layer of resistivity around 19.8 Ohm-m and thickness around 4.07 m may represent silty clay, third layer of resistivity 2.15 Ohm-m and thickness around 25.3 may represent clay, fourth layer of resistivity 60.8 Ohm-m may represent medium to coarse sand.

Segments of survey	Type of resistivity curve	Layer	Resistivity (Ohm-m)	Thickness (m)	Lithology
Eastern part	HKH	1st	26.6	4.1	Top soil
		2nd	6.6	3.5	Sandy-silty clay and silty soil (wet to moist),
		3rd	57.7	28.7	Silty sand
		4th	19.2	21.2	Silty-sandy clay
		5th	31.8	–	Silty sand
Western part	HKH	1st	62.2	3.3	Top soil
		2nd	31.3	5.8	Silty sand
		3rd	83.9	21.4	Fine to medium sand
		4th	58.0	44.9	Silty sand
		5th	76.2	–	Coarse sand
Northern part	HKH	1st	37.2	3.0	Top soil
		2nd	15.3	6.6	Sand, silt and clay,
		3rd	44.5	3.7	Silty sand
		4th	28.4	18.1	Silty sand
		5th	62.4	–	Medium to coarse sand.
Southern part	KH	1st	9.5	2.9	Top soil
		2nd	19.8	4.0	Silty clay
		3rd	2.1	25.3	Clay
		4th	60.8	–	Medium to coarse sand.

South Bank of the Ganga River

This part also has been divided into four sections to describe the nature of curves dominating in the section. Details are given in table:

Eastern Part: HK-type curve pattern has been observed for the profile BP_S58 (Fig. 22). In respect to the layers, pattern first layer of resistivity around 16 Ohm-m is representing humus type top soil cover, second layer of resistivity around 4 Ohm-m represents thick clay, third layer of resistivity 12 Ohm-m represents homogeneous fine sand having water bearing zone, fourth layer of resistivity around 9 Ohm-m represents thin layer of old alluvium, while in fifth layer of resistivity 195 Ohm-m represents compact sand and hard formation.

Western Part: This section (BP_S23) shows HA-type curve pattern (Fig. 23). In respect to the layers, pattern first layer of resistivity around 16 Ohm-m represents humus type top soil cover, second layer of resistivity around 10 Ohm-m represents

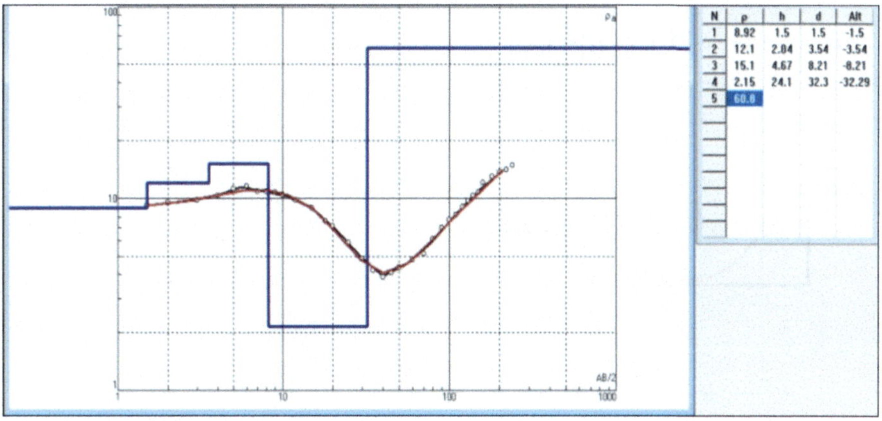

Fig. 21 VES curve of profile S22

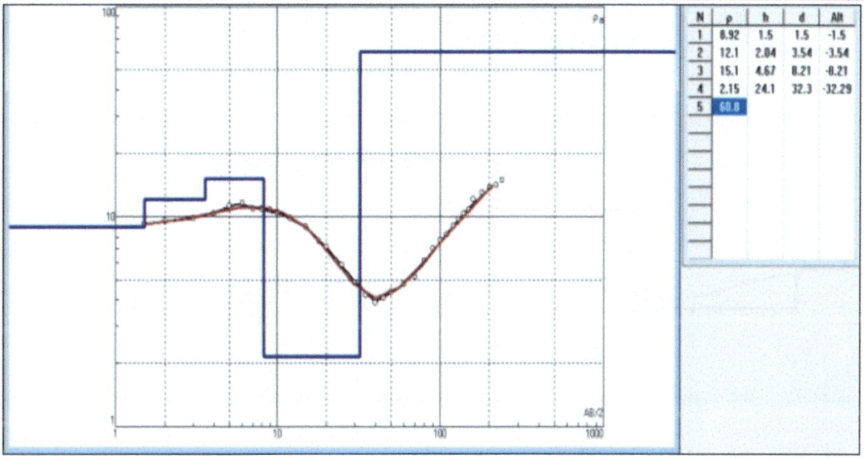

Fig. 22 VES curve of profile S22

clay, third layer of resistivity 33 Ohm-m represents homogeneous fine sand having water bearing zone and fourth layer of resistivity around 59 Ohm-m represents compact sand and hard rock formation.

Southern Part: In this section (BP_S61), H-type curve pattern has been observed (Fig. 24) in respect to the layer pattern first layer resistivity is around 29 Ohm-m representing humus type top soil cover, second layer of resistivity around 10 Ohm-m representing clay, third layer of resistivity 8 Ohm-m representing homogeneous fine sand having water bearing zone, fourth layer of resistivity around 61 Ohm-m representing thick layer of old alluvium while in fifth layer of resistivity 86 Ohm-m representing compact sand and hard formation.

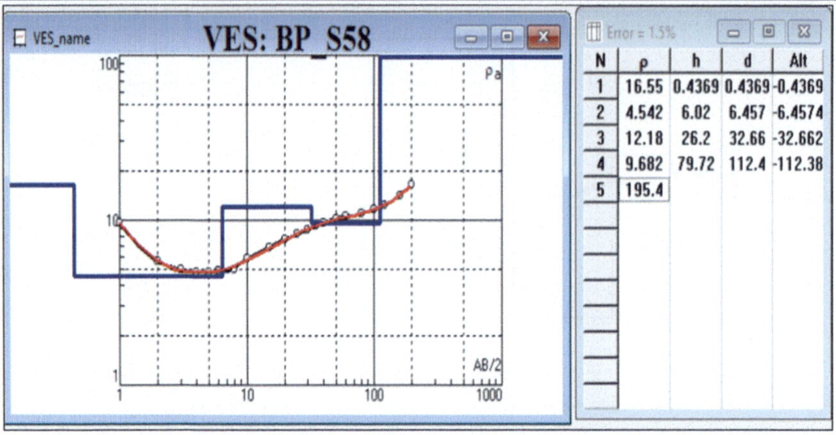

Fig. 23 VES curve of profile S28

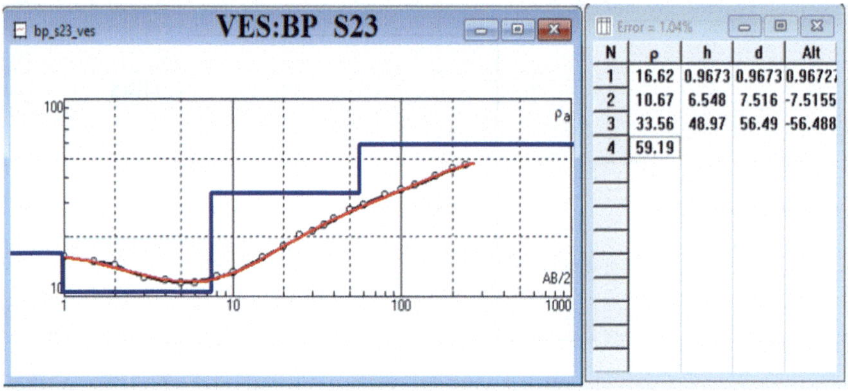

Fig. 24 VES curve of profile S23

Central Part: In this section (BP_S33), AH-type curve pattern has been shown (Fig. 25), in respect to the layer pattern first layer resistivity is around 41 Ohm-m representing humus type top soil cover, second layer of resistivity around 25 Ohm-m representing clay, third layer of resistivity 16 Ohm-m representing sand and silt type, fifth layer of resistivity around 8 Ohm-m representing homogeneous fine sand having water bearing zone.

Segments of survey	Type of resistivity curve	Layer	Resistivity (Ohm-m)	Thickness (m)	Lithology
Eastern part	KH	1st	16.5	0.4	Top soil cover
		2nd	4.5	6.0	Thick clay
		3rd	12.1	26.2	Homogeneous fine sand having water bearing zone
		4th	9.6	39.7	Thin layer of old alluvium
		5th	195.4	–	Compact sand and hard formation
Western part	HA	1st	16.6	0.9	Top soil cover
		2nd	10.6	6.5	Clay
		3rd	33.5	48.9	Homogeneous fine sand having water bearing zone
		4th	59.1	–	Compact sand and hard rock formation
Southern part	H	1st	29.2	0.7	Top soil cover
		2nd	10.5	7.6	Clay
		3rd	8.7	23.0	Homogeneous fine sand having water bearing zone
		4th	61.6	68.3	Thick layer of old alluvium
		5th	86	–	Compact sand and hard formation
Central part	AH	1st	41.7	1.0	Top soil cover
		2nd	25.6	3.9	Clay
		3rd	16.5	13.8	Sand and silt type
		4th	8.8	53.2	–

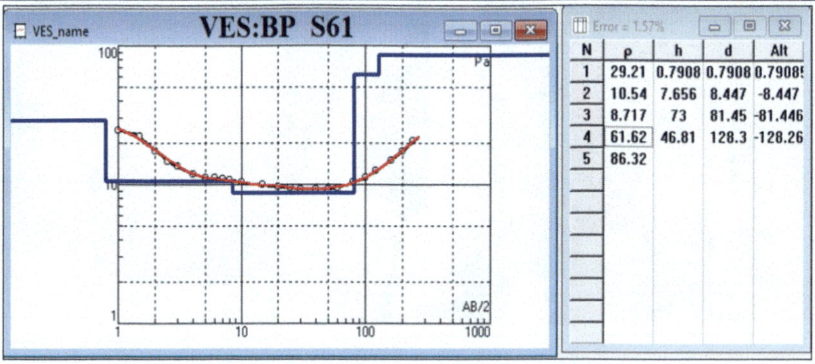

Fig. 25 VES curve of profile S61

Health Assessment:

Arsenic poisoning poses major health risks to the inhabitants of the Indian state of Bihar. People who have been drinking arsenic-tainted water for a long period are at high risk for major health problems. Arsenicosis symptoms, including skin manifestations, gastrointestinal illness, lung disease, cardiovascular disease, hormone imbalance, loss of appetite, poor immunity, bowel alterations, neurobehavioral abnormalities and cancer poisoning, have been seen in the exposed population (Chakraborti et al. 2003 & 2016; Kumar 2022[a], Kumar et al. 2022[b,c]; 2020[a,b]; 2021[a,b,c,d]; 2020; 2016; 2015).

In Bhagalpur district, the arsenic poisoning in groundwater has caused health-related problems in the exposed population. The village population in Champanagar and Nathnagar showed typical symptoms of arsenicosis such as keratosis, melanosis and rain drop pigmentation (Fig. 26).

The health survey conducted in an around Champanagar and Nathnagar area showed severe disease burden in the exposed population. The subjects reported with symptoms of general body weakness, constipation or diarrhoea, anaemia, breathlessness, mental disorders, fluctuations in blood pressure, diabetes, lumps and cancer.

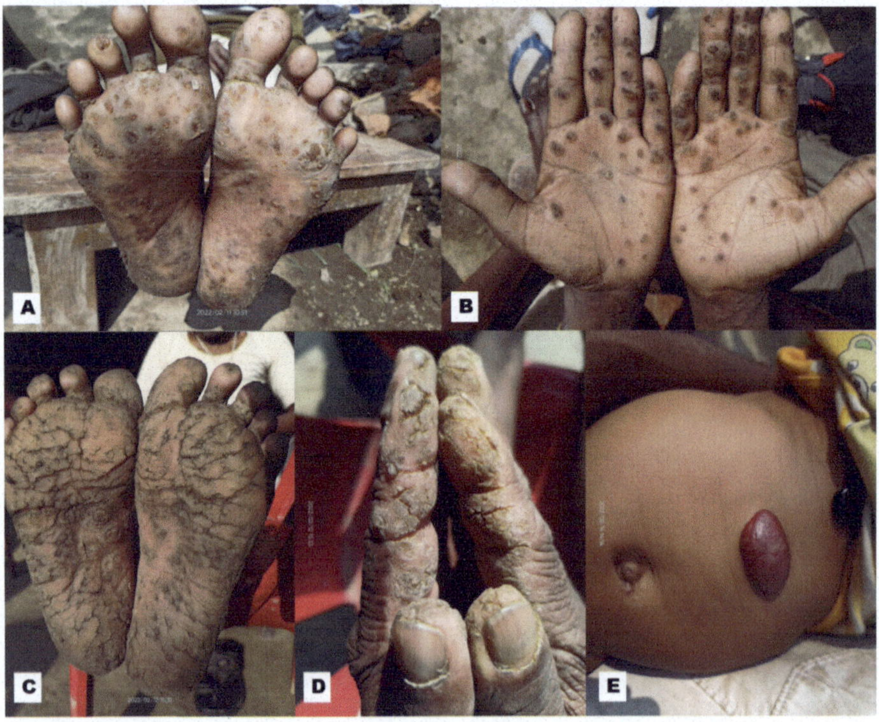

Fig. 26 Showing symptoms of arsenicosis **a**, **b** A 63 years old subject with keratosis in the sole and palm **c**, **d** A 57-year subject with keratosis in sole and fingers **e** A 4 months infant with tumour near his stomach

Table 4 Health assessment report of arsenic exposed population of the villages in Nathnagar block (including Champanagar village) (n = 1106 from households (n = 86)

Symptoms	Problems present in the population	Number of problems observed	Total cases	P-value
Keratosis in palms and sole	24 (2.2%)	1082 (97.8%)	1106	<0.01
Melanosis in sole, palm and trunk	08 (0.7%)	1098 (99.3%)	1106	<0.01
Other skin manifestations	368 (33.2%)	738 (66.8%)	1106	<0.01
Loss of appetite	297 (26.8%)	809 (73.2%)	1106	
Anaemia	44 (3.97%)	1062 (96.03%)	1106	<0.1
Constipation	284 (25.6%)	822 (74.4%)	1106	<0.001
Diarrhoea	27 (2.4%)	1079 (97.6%)	1106	<0.01
General body weakness	376 (33.9%)	730 (66.1%)	1106	<0.01
High or low blood pressure	89 (8.04%)	1017 (91.96%)	1106	<0.01
Breathlessness/ cough/asthma	106 (9.58%)	1000 (90.42%)	1106	< 0.001
Diabetes	44 (3.97%)	1062 (96.03%)	1106	<0.001
Mental disability	03 (0.27%)	1103 (99.73%)	1106	<0.01
Lumps in the body	08 (0.72%)	1098 (99.27%)	1106	<0.001
Cancer cases	06 (0.54%)	1100 (99.46%)	1106	<0.001
Other health problems	126 (11.39%)	980 (88.61%)	1106	<0.01

aAll data were analysed using One-way ANOVA ($p < 0.0001$ & $p < 0.001$)

Moreover, this area also reported very high values of As in groundwater (up to 500 ppb). As shown in Table 4, health assessment study of exposed population of village showed that 2.2% of people had the symptoms of keratosis, 0.7% had melanosis, 33.2% had other types of skin diseases, 26.8% had loss of appetite, 3.97% had anaemia, 25.6% of people were suffering from constipation, 2.4% had diarrhoea related problem, 33.9% had general body weakness, 9.58% had breathlessness/ cough our asthma problem, 3.97% had diabetes, 0.27% had mental disorder cases, 0.72% had some types of lumps in their body and 0.54% (n = 06) village population were reported with cancer disease such as Gallbladder cancer, liver cancer, breast cancer and skin cancer (Table 4). (Kumar et al. 2022[a]; 2021[a,b,c]; 2019[a]; Rahman et al. 2019[a];2019[b]).

4 Conclusions

From this study in Bhagalpur district, Bihar, it is observed that hazardous concentration of arsenic occurs in pockets and is influenced by geology and geomorphology, varying in concentration at various depth of aquifers. The dug well and surface water bodies are found to be generally free from hazardous incidence (>10 ppb) of arsenic.

Analysis of 858 nos. of water samples from surface as well as groundwater using Merck test kit from left and right bank of River Ganga in Bhagalpur district reveals that 29.15% of the samples are unsafe for domestic use containing more than 10 ppb arsenic. The area falling to the right bank of the River accounts for 15% (66 out of 490) of contaminated samples with ~2% (~10 samples) showing >100 ppb As value. The left bank constitutes 39% of samples above 10 ppb (145 out of 368) with 8% (~30 samples) of samples having concentration > 100 ppb.

The area lying to the left bank of River Ganga shows pervasive and higher values of arsenic as compared to the right bank. Along the left bank, groundwater being virtually stagnant due to frequent flooding and poor drainage facilities easily releases arsenic accumulated in the trapped sediments and thus contaminates groundwater. Secondary enrichment is observed in favourable locales. Southern part of the area is mostly free from Arsenic contamination. Younger Alluvium is showing higher values of arsenic concentration as compared to older alluvium in the study area.

The source of arsenic contamination in the groundwater appears to be associated with the sediments of the Himalayan provenance by drainages emanating from Extra Peninsular (Himalayan) region constituting the Quaternary succession. Higher As concentration (>100 ppb) is observed in pockets along the paleochannels, favourable locales of secondary enrichment or receptacle points of higher order drainages.

The study almost conclusively indicates an enhanced enrichment pattern to the northern bank of Ganga River which voices for a possible extrapeninsular source. As the deeper aquifers are largely free of Arsenic, it is unlikely that the underlying lithology beneath colossal cover of alluvium would have a role to play in contaminating the younger holocene sediments. The threshold point for release of As from sediments has also been defined globally and any value less than 20 ppm in soil (Proceedings of the expert meeting on critical limits for heavy metals and methods for their application, Berlin: Umweltbundesamt; 2003. pp. 29–78) considered within acceptable limits of caution for their release into alluvium dominated aquifers. The health study concludes that the village population of Champanagar/ Nathnagar are suffering from serious health problems and health interventions are urgently required to combat the problem.

These observations somewhat indicate a likely distant source of As in the Himalayas where drainage geomorphology, wet climatic spikes and carrying capacity of the Rivers (drainages) may have carried the sediments laden with As into the foreland sedimentary basin.

Acknowledgments The authors are indebted to Dr. Rajesh Asthana, Dy. Director Generals (Retd.), SU: Bihar, GSI, Patna, Shri S.K. Dutta, Dy. Director Generals (Retd.), SU: Bihar, GSI, Patna under whose direction and guidance the study was undertaken.

The authors also acknowledge the Chemical Division, SU: Bihar, GSI, Patna; Chemical Division and Eastern Region, Kolkata for facilitating the support for the chemical analyses of samples.

The authors are also thankful to the analytical facilities available at Mahavir Cancer Institute and Research Centre, Patna. The authors are thankful to all their colleagues and individuals, whose assistance was received for the work.

References

Acharyya SK, Shah BA (2004) Risk of arsenic contamination in groundwater affecting Ganga Alluvial Plain, India? Environ Health Perspect112:A19–A20

Ahmed, K.M., Bhattacharya, P., Hasan, M.A., Akhter, S.H., Alam, M.A., Bhuyian, H., Imam, M.B., Khan, A.A., Sracek, O., (2004). Arsenic enrichment in groundwater of the alluvial aquifers in Bangladesh: an overview. Appl. Geochem. 19, 181–200

Bhattacharya, P., Chatterjee, D., Jacks, G., 1997. Occurrence of arsenic contamination of groundwater in alluvial aquifers from Delta Plain, Eastern India: option for safe drinking supply. Int. J. Water Res. Dev. 13, 79–92.

Bhattacharya, P., Hasan, M.A., Sracek, O., Smith, E., Ahmed, K.M., von Brömssen, M., Haq, S.M.I., Naidu, R., 2009. Groundwater chemistry and arsenic mobilization in the Holocene flood plains in south-central Bangladesh. Environ. Geochem. Health 31, 23–44.

Bhattacharya, P., Jacks, G., Ahmed, K.M., Khan, A.A., Routh, J., 2002. Arsenic in Groundwater of the Bengal Delta Plain aquifers in Bangladesh. Bull. Environ. Cont. Toxicol. 69 (4), 538–545

Chaturvedi, R., Thakur, L.K., & Prasad, S.C., Final report on incidence of arsenic in groundwater in parts of Saran and Vaishali districts, Bihar, Unpublished Report Geological Survey of India; F.S. 2007-08 and2008-09.

Gan, Y., Wang, Y., Duan, Y., Deng, Y., Guo, X., Ding, X., 2014. Hydrogeochemistry and arsenic contamination of groundwater in the Jianghan Plain, central China. J. Geochem Explor. 138 (2014), 81–93.

Ghosh, D., Kumar, S., Donseelar, M.E., Corrote, C., Ghosh, A.K., 2021. Organic carbon transport model of abandoned River channels—A motif of floodplain geomorphology influencing biogeochemical swaying of arsenic. Science of total environment 762 (2021) 144400.

Goodbred, S.L., Kuehl, S.A., 2000. The significance of large sediment supply, active tectonism, and eustasy on margin sequence development: late Quaternary stratigraphy and evolution of the Ganges–Brahmaputra delta. Sedimentary Geology 133, 227–248.

Haque M.W. and Naskar B., Incidence of Arsenic in Groundwater in parts of Bhojpur District, Bihar, Unpublished Report Geological Survey of India; F.S.2003-04.

Harvey, C., Swartz, C.H., Badruzzaman, A.B.M., Keon-Blute, N.E., Yu, W., Ashraf Ali, M., Jay, J., Beckie, R., Niedam, V., Brabander, D.J., Oates, P.M., Ashfaque, K.N., Islam, S., Hemond, H.F., Ahmed, M.F., 2002. Arsenic mobility and groundwater extraction in Bangladesh. Science 298, 1602–1606.

Hasan, M.A., von Brömssen, M., Bhattacharya, P., Ahmed, K.M., Sikder, A.M., Jacks, G., Sracek, O., 2009. Geochemistry and mineralogy of shallow alluvial aquifers in Daudkandi, Environmental Geology 57, 499–511

Kumar A, Ali M, Kumar R, Rahman R.S., Srivastava A., Chayal N.K., Sagar V., Kumari R., Parween S., Kumar R., Niraj P.K., Anand G., Singh S. K., Ghosh A.K. (2018), High Arsenic Concentration in Blood Samples of People of Village GyaspurMahaji, Patna, Bihar Drinking Arsenic-Contaminated Water Exposure and Health, 12, 131–140

Kumar, A., Ali, M., Raj V, Kumari A, Rachamalla M, Niyogi S, Kumar D, Sharma A, Saxena A, Panjawani G, Jain P, Vidyarthi A, Kumar N, Kumar M, Niraj PK, Rahman MS, Bishwapriya A, Kumar R, Sakamoto M, Kumar S, Singh M, Ghosh AK., 2023. Arsenic causing gallbladder cancer disease in Bihar. Scientific reports, 13(1), 4259.

Kumar, A., Rahman, M. S., Ali, M., Salaun, P., Gourain, A., Kumar, S., Kumar, R., Niraj, P. K., Kumar, M., Kumar, D., Bishwapriya, A., Singh, S., Murti, K., Dhingra, S., Sakamoto, M., & Ghosh, A. K., 2022a[a]. Assessment of disease burden in the arsenic exposed population of Chapar village of Samastipur district, Bihar, India, and related mitigation initiative. Environmental science and pollution research international, 29(18), 27443–27459.

Kumar, A., Ravi, C., Dhingra, S., Krishna Murti, M. A., & Ghosh, A. K., 2022b[b]. Arsenic Causing Gallbladder Cancer Disease near the Himalayan bound Rivers in Bihar: A Case study of Gallbladder Cancer. Journal of Cancer Science and Clinical Therapeutics, 6, 388–391.

Kumar, A., Raj, V., Srivastava, A., Ali, M., Ghosh, A. K., Rachamalla, M., & Kumar, D.,2022[c]. Autophagy in arsenic exposed population and cancer patients. In Autophagy and Metabolism 141–161. Academic Press.

Kumar, A., Ali, M., Kumar, R., Kumar, M., Sagar, P., Pandey, R. K., Akhouri, V., Kumar, V., Anand, G., Niraj, P. K., Rani, R., Kumar, S., Kumar, D., Bishwapriya, A., & Ghosh, A. K., 2021a[a]. Arsenic exposure in Indo Gangetic plains of Bihar causing increased cancer risk. Scientific reports, 11(1), 2376.

Kumar, A., Kumar, R., Rahman, M. S., Ali, M., Kumar, R., Nupur, N., Gaurav, A., Raj, V., Anand, G., Niraj, P. K., Kumar, N., Srivastava, A., Biswapriya, A., Chand, G. B., Kumar, D., Rashmi, T., Kumar, S., Sakamoto, M., & Ghosh, A. K., 2021b[b]. Assessment of arsenic exposure in the population of Sabalpur village of Saran District of Bihar with mitigation approach. Environmental science and pollution research international, https://doi.org/10.1007/s11356-021-13521-5. Advance online publication.

Kumar, A., Rahman, M.S., Ali, M., Kumar, R., Niraj, P.K., Akhouri, V., Singh, S.K., Kumar, D., Rashmi, T., Bishwapriya, A., Chand G.B., Sakamoto, M., Ghosh, A.K., (2021c). Assessment of arsenic exposure and its mitigation intervention in severely exposed population of Buxar district of Bihar, India. Toxicol. Environ. Health Sci. 13, 287–297

Kumar A., Ghosh A.K.,2021. Assessment of Arsenic Contamination in Groundwater and Affected Population of Bihar. In: Kumar N. (eds) Arsenic Toxicity: Challenges and Solutions. Springer, Singapore.

Kumar A, Ali M, Kumar R, Rahman MS, Srivastava A, Chayal NK, Sagar V, Kumari R, Parween S, Kumar R, Niraj PK, Anand G, Singh SK, Ghosh AK., 2020. High Arsenic Concentration in Blood Samples of People of Village GyaspurMahaji, Patna, Bihar Drinking Arsenic-Contaminated Water. Springer Nature Journal Exposure and Health, 12, 131–140 (published print version 2020).

Kumar, M., Kumari, K., Ramanathan, A.L., Saxena, R., 2007. A comparative evaluation of groundwater suitability for irrigation and drinking purposes in two agriculture dominated districts of Punjab, India. Environ. Geol. 53, 553–574.

Kumar, M., Ramanathan, A.L., Rao, M.S., Kumar, B., 2006. Identification and evaluation of hydrogeochemical processes in the groundwater environment of Delhi, India Environ. Geol. 50, 1025–1039.

Lone, S.A., Jeelani, G., Deshpande, R.D., Mukherjee, A., 2019. Stable isotope (δ18O and δD) dynamics of precipitation in a high-altitude Himalayan cold desert and its surroundings in the Indus River basin, Ladakh. Atmos. Res

McArthur, J.M., Banerjee, D.M., Hudson-Edwards, K.A., Mishra, R., Purohit, R., Ravenscroft, P., Cronin, A., Howarth, R.J., Chatterjee, A., Talukder, T., Lowry, D., Houghton, S., Chadha, D.K., 2004. Natural organic matter in sedimentary basins and its relation to arsenic in anoxic ground water: the example of West Bengal and its worldwide implications. Appl. Geochem. 19 (8), 1255–1293.

McArthur, J.M., Ravenscroft, P., Safiullah, S., Thirlwall, M.F., 2001. Arsenic in groundwater: testing pollution mechanisms for sedimentary aquifers in Bangladesh. Water Resour. Res. 37 (1), 109–117.

Mishra, S.; Dwivedi, S.; Kumar, A.; Chauhan, R.; Awasthi, S.; Tripathi, R.; Mattusch, J. Current status of ground water arsenic contamination in India and recent advancements in removal techniques from drinking water. Int. J. Plant Environ. 2016, 2, 1–2.

Mitra, I., Shareef, N.M, Sethi, P.S and Prasad, S.C. Report on assessment of incidence of arsenic in groundwater in parts of Buxar district, Bihar, Unpublished Report Geological Survey of India; F.S.2004-05.

Mukherjee, A., Gupta, S., Coomar, P., Fryar, A.E., Guillot, S., Verma, S., Bhattacharya, P., Bundschuh, J., Charlet, L., 2019. Plate tectonics influence on geogenic arsenic cycling: from primary sources to global groundwater enrichment. Sci. Total environ. 683, 793–807.

Mukherjee A., Verma S., Gupta S., Henke, K. R., Bhattacharya P., 2014. Influence of tectonics, sedimentation and aqueous flow cycles on the origin of global groundwater arsenic: Paradigms from three continents; Journal of Hydrology 518.284–299

Nickson, R.T., McArthur, J.M., Burgess, W.G., Ahmed, K.M., Ravenscroft, P., Rahman, M., 1998. Arsenic poisoning of Bangladesh groundwater. Nature 395, 338.

Nickson, R.T., McArthur, J.M., Ravenscroft, P., Burgess, W.G., Ahmed, K.M., 2000. Mechanism of arsenic release to groundwater, Bangladesh and West Bengal. Appl. Geochem. 15, 403–413.

Proceedings of the expert meeting on critical limits for heavy metals and methods for their application, Berlin, 2–4 December 2002 Held under the UN/ECE Convention on Long—Range Trans boundary Air Pollution UBA-Texte 47/03. Berlin: Umweltbundesamt; 2003. pp. 29–78.

Saunders, J.A., Lee, M.K., Mohammad, S., 2005. Natural arsenic contamination of Holocene alluvial aquifers by linked tectonic, weathering, and microbial processes. Geochemistry, Geophysics, Geosystems 6 (4), Q04006.

Stanger, G., 2005. A palaeo-hydrogeological model for arsenic contamination in southern and south-east Asia. Environmental Geochemistry and Health 27, 359–367.

Taylor SR, Mclennan SM., 1995. The geochemical evolution of the continental-crust. Rev Geophys 33, 241–265

Umitsu, M., 1993. Late Quaternary sedimentary environments and landforms in the Ganges Delta. Sed. Geol. 83, 177–186.

UNICEF Report on "Water in India: Situation and Prospects", 2013.

Wen, D., Zhang, F., Zhang, E., Wang, C., Han, S., Zheng, Y., 2013. Arsenic, fluoride and iodine in groundwater of China. J. Geochem. Explor. 135, 1–21.

Part II
Climate Changes in Recent

Chapter 3
The Geodetic Mass Balance of the High Mountain Asia Glaciers: A Critical Analysis of the Estimates

Debmita Bandyopadhyay

1 Introduction

Mass balance estimation is one of the most effective parameters for understanding the glacier dynamics. To calculate mass balance, two main processes; ablation and accumulation are considered. Ablation involves mass loss processes like melting, sublimation of glacier ice, wind erosion, avalanche, ice flow into the sea, etc., whereas accumulation encompasses all processes that add to the mass of snow or glacier ice like snowfall, precipitation, freezing of ice water, and even avalanche is considered. The relationship between the two processes gives an idea of whether the mass balance is positive or negative. Such studies are performed for a glaciological year (time taken by the glacier to undergo one cycle of accumulation and ablation). Mathematically, the annual mass balance is expressed as:

$$B_a = \int b_a dA, \tag{1}$$

where B_a is the annual mass balance for the entire glacier, b_a is the mass loss at a point which is sometimes also referred to as point mass balance and A is the glacier area. While conducting ground survey for mass balance measurements, point mass balance is measured and extrapolated to the mass balance of the entire glacier. The unit of measurement is meter water equivalent (m.w.eq.). Water equivalent is the volume of water one would obtain by melting the measured glacier ice or snow. Meter water equivalent is the volume divided by the area which is the measure of how much the glacier would thicken or be thinner if the mass accumulation or ablation occurs over the entire surface. Another unit that is commonly used to express

D. Bandyopadhyay (✉)
Ministry of Earth Sciences, New Delhi, India
e-mail: b.debmita@gmail.com

N. Khare (ed.), *Changing Climate over India*,
https://doi.org/10.1007/978-981-92-1098-5_3

mass balance is Gt which is numerically equivalent to 1 km^3 w.eq.. As one performs continuous monitoring, the mass balance is treated as a rate of change per year and expressed as m.w.eq. a^{-1}. Furthermore, to convert the mass loss to sea-level rise equivalent, the mass loss is calculated in terms of kg m^{-2} knowing that (1000 kg m^{-2})/ ρ is 1 m.w.eq., where ρ is the density of water. This when divided by the area of the ocean i.e. 362.5 x 10^{12} m^2, gives the sea-level rise in meters (Cogley (2012)). A positive mass balance accounts for higher accumulation than ablation for the particular glacial period, whereas a negative balance suggests a lower accumulation which, if repeatedly occurs, might lead to the retreat of the particular glacier.

Mass balance calculations require elevation change information which is done using a variety of parameters like Accumulation Area Ratio (AAR), Equilibrium Line altitude (ELA), Firn Line altitude (FLA) estimation and even using Digital Elevation Models (DEMs). There are several methods by which the glaciers can be monitored viz. direct methods which include glaciological measurements and hydrological method, and indirect methods like AAR, ELA method or the geodetic method (using DEMs). Undoubtedly, the direct measurements give a better understanding of the glacial melt processes however, the glaciers in certain regions like that of the Himalayas, have a steep and rugged topography, rendering it impossible to do large-scale monitoring. In fact, out of over 10,000 glaciers in Indian terrain only 38 glaciers (Azam 2026) have been monitored for their mass balance rates.

Of all the above-mentioned methods for continuous monitoring, geodetic method has been the most promising one. Hence, long-term studies (multi-decade), over a difficult terrain (like the Himalayas) and on a global scale using this very method is a preferable option and has been performed by several research groups (Bandyopadhyay et al. 2019; Vijay and Braun 2016; Brun et al. 2017; Dehecq et al., 2019; Fan et al. 2023; Kaab et al. 2012; Rankl and Braun 2016; Shean et al., 2020), especially over High Mountain Asia. Even though the large-scale glacier mass balance estimates obtained from satellite imagery give us an idea about the glacier dynamics, glacier responses to climate changes, the methodology needs to be applied with caution. The main reason is that the spatial heterogeneity of geodetic mass balance at both regional and local scales use regional average approximations to estimate mass balance or sea-level rise contributions for the full glacierized region and may not always be appropriate.

Furthermore, the method of disaggregating decadal geodetic glacier mass balance to an annual scale using remote-sensing proxies has been developed, providing a new approach to determining point surface mass balances on glaciers from remote sensing measurements. However, the uncertainty of the estimated geodetic mass balance needs careful consideration especially when proposing policy related solutions on a catchment or even more local scale. These gaps of geodetic glacier mass balance indicate the need for further studies to improve the understanding of glacier response to climate change and its implications for sea-level rise. In this chapter, the various large-scale studies reported in the recent past using the geodetic method are referred to, a critical analysis of the reason behind the uncertainty is mentioned and aims to provide a holistic approach to the existing issue.

2 Characteristics of High Mountain Asia and Datasets Used for the Geodetic Method

2.1 Study Area

High Mountain Asia (HMA) is a region with unique characteristics that significantly influence the climate and water resources of the surrounding areas. The study area encompasses a vast and complex terrain, including the Tibetan Plateau, Tien Shan Mountains, Himalayas, Hindu Kush, Karakoram, Pamir, Qilian, and Hengduan mountains (Fig. 1). The geographical range stretches from about 25°N to 46°N and 64°E to 106°E, with over 50% of the area being at or above 4000 m above sea level. Glaciers in the study area are a mixture of debris covered and clean glaciers. The region experiences abundant snowfall and a long period of snow cover annually, making it a crucial source of water for the headwaters of many prominent Asian Rivers. Studies have shown that the headwaters of several major Rivers in Asia depend in part on HMA meltwater. The distribution of snowfall in HMA exhibits spatial variability, with high values for observed amounts and days of snowfall mainly distributed in the western Tienshan Mountains and southwest and southeast of the Tibetan Plateau (Sun et al. 2023).

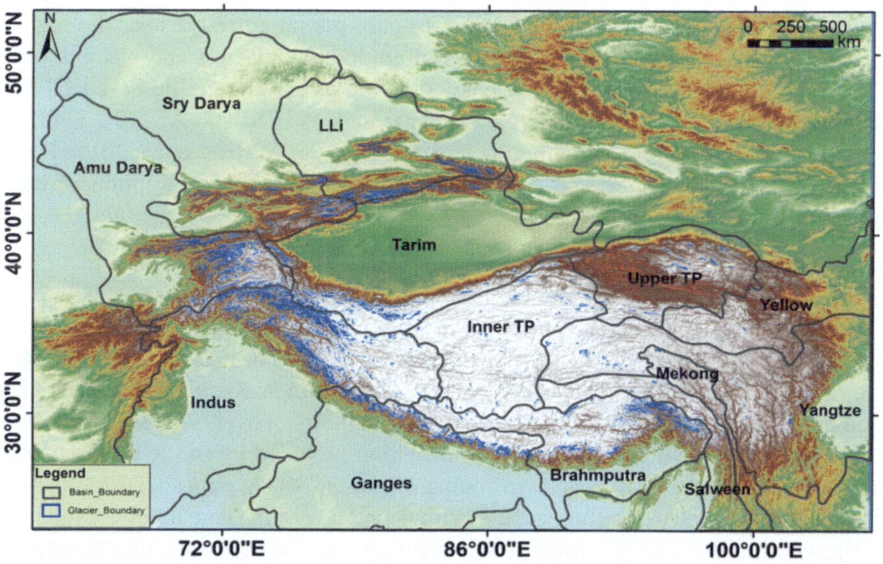

Fig. 1 High Mountain Asia (HMA) extent with grey boundaries highlighting the different basin boundaries and the blue color depicts the glaciated regions of HMA

2.2 Data Sets

Photogrammetry: To derive the DEMs for the glaciated terrain, the photogrammetric approach is the most conventional one. In the 1970's, when ground survey was the only means to map the glacier developing topographic sheets, photogrammetry was a useful tool. Recent years have seen a resurgence in photogrammetry for very high-resolution data which can be utilized for the generation of DEMs, including the development of structure-from-motion (SfM) photogrammetry (Fugazza et al. 2018). This method utilizes airborne photography which gives high resolution images, a huge bonus when interpretation of small individual glaciers is necessary. To validate the findings, the images are compared to historical data or the map-based inventories. However, a large number of well distributed GCPs are necessary for the interpretation of the area of interest. This requires in-situ GPS observations as well, which limits its accuracy to no more than 10 m (Fischer 2011) owing to the manual collection of data with induced errors from the ground measurements. In addition, a high contrast in the image is needed to estimate the elevation of the glaciers (Bamber and Rivera 2007). In absence of such conditions in the image, generation of DEMs is difficult. To facilitate the process, high resolution and better coverage are needed which escalates the cost, making the entire process highly uneconomical. Further, high mountain environments such as the Himalayas have mostly tropical climate (cloud cover mostly in the ablation period i.e. August to September) and restricted areas (sharing boundaries with neighboring countries), leading to only partial coverage over the region of interest. Hence, this method is restricted to individual glaciers and glaciers with favorable geometry. This limitation has led to the usage of satellite based DEMs generated from optical stereo-photogrammetry such as IRS 1-C, ASTER, SPOT 5 HRVIR and Landsat ETM (Bahuguna and Kulkarni 2005; Brun et al. 2017; Kaab et al., 2005; Quincey et al. 2005), with a compromise on the resolution of data collected.

Altimetry: The second category of geodetic data is where the height information is collected in spots with a formidable radius using altimeter from laser (satellite or airborne) or a radar altimeter. Gardelle, Berthier, Arnaud, and Kaab (2013); Gardner et al. (2013); Kaab et al. (2012) demonstrated the potential of ICESat laser altimetry and estimated the elevation and mass changes in the Hindu-Kush Himalaya (HKH) region. However, it is argued that there is a fair possibility of over or under-estimation of these changes owing to the sparse information provided by the ICESat data. In fact, data is interpolated to be able to compare with the SRTM dataset, DEM derived from photogrammetry or airborne data. With the release of the new ICESat-2 data (in 2019) with its six-beam geometry, it has 150 times more repeat-track measurements than ICESat was capable of. Hence, it is expected that new elevation change studies utilizing this data series shall be able to minimize the data void, thereby avoiding under- or over-estimation of the glacier elevation change. Alternatively, Cryosat-2 can be utilized in the high mountain Asia (HMA) region, but only a limited number of studies have been possible (Dehecq et al. 2013; Jakob et al. 2020) owing to the complex topography of the region. Elevation changes studies using Cryosat-2 have been specifically done over ice-sheets (Simonsen and Sørensen 2017), but it has often been debated that, due to the switching between the two modes (LRM and SAR- InM), the observations are quite noisy and difficult to interpret. Hence, for short-term and individual glacier systems, this dataset might be very effective owing to the high resolution and accuracy (horizontal and vertical), but large-scale studies are still better with the complementary technique of Interferometric SAR (InSAR).

Gravimetry: Gravimetry, on the other hand, measures the mass loss of glaciers and ice sheets by essentially" weighing" Earth's ice through the gravitational pull on a pair of satellites. This method directly measures the mass loss of glaciers and ice sheets and is used to understand how Earth's ice contributes to sea level rise. The main tools for this measurement have been the Gravity Recovery and Climate Experiment Follow On (GRACE-FO) satellites—a joint mission launched in 2018 by NASA and the German Research Center for Geosciences (GFZ)—and GRACE, a predecessor mission between NASA and the German Aerospace Center (DLR) from 2002 to 2017 (Taking a Measure of Sea Level Rise: Gravimetry n.d.).

Interferometric Synthetic Aperture Radar: Elevation change, or height information can be estimated using InSAR. A schematic representation of the geometry of a repeat InSAR system is depicted in Fig. 2(a). The Fig.shows S_1 and S_2 as the locations of the satellite with baseline as B, α as the inclination angle, h as the altitude and θ as the look angle of the SAR satellite antenna. R_1 and R_2 are the range distance for the same target z(y), denoting the surface elevation. With the advent of remote sensing technology various optical stereo pairs or InSAR based DEMs have been used to monitor the steep terrains like that of the Indian Himalayas. Optical data like ASTER (Bhattacharya et al. 2016, Bolch et al. 2017) or SPOT (Berthier et al. 2007, Gardelle et al. 2012, Gardelle et al. 2013, Vincent et al. 2013) or Cartosat (Agarwal et al. 2017, Bhushan et al. 2018) have also been used to understand the mass changes in various selective regions in the Indian Himalayas. However, to have a full coverage of the Himalayas, global DEM is a priority, which recently has been provided by the German Space Agency, DLR in September 2018.

To test the potential of interferometry based DEMs, Pandey and Venkataraman (2013) have generated TanDEM-X DEM for the year 2011 in the Chandra-Bhaga basin, Himachal Pradesh, India and compared it with the SRTM to estimate the elevation change of the glacier, volume change and mass balance for a 11-year period. Bandyopadhyay et al. (2023) utilized the recently available global Tandem-X/ TerraSAR-X DEM and estimated the mass balance over the entire Indian Himalaya. However, large-scale studies using InSAR technology remain an arduous task and, hence, an ongoing research domain for the scientific community.

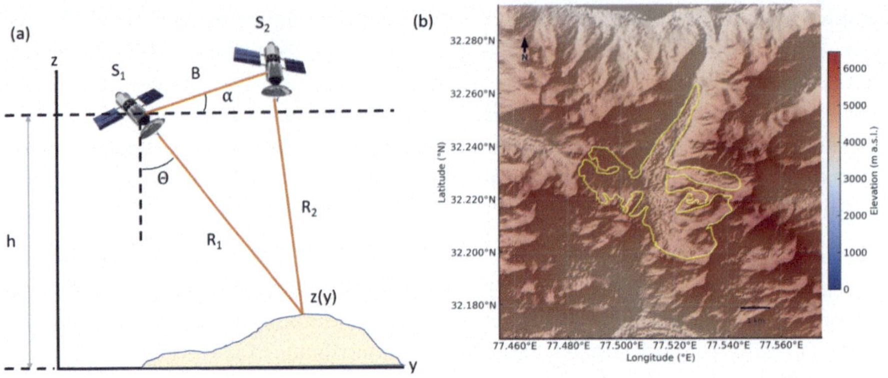

Fig. 2 **a** Schematic diagram showing the geometry for interferometric synthetic aperture radar (InSAR) in a repeat or single pass measurement **b** DEM generated using InSAR technique over Chota Shigri glacier, Himachal Pradesh

3 Methodology

Various datasets are used when performing the geodetic method (as mentioned in the previous section). However, the prime step in the geodetic method requires two-time period DEMs. For the year (T_1 and T_2), optical/SAR DEM is used (Fig. 4). A pictorial representation of the same is in Fig. 3. To maintain planimetric consistency, all the DEMs are co-registered at the same spatial reference system with the same spatial resolution. This process facilitates the removal of horizontal and vertical offsets in the two DEMs that are being compared. In the regions where DEMs have been generated, the accuracy assessment is performed using the methodology by Deo et al. (2013).

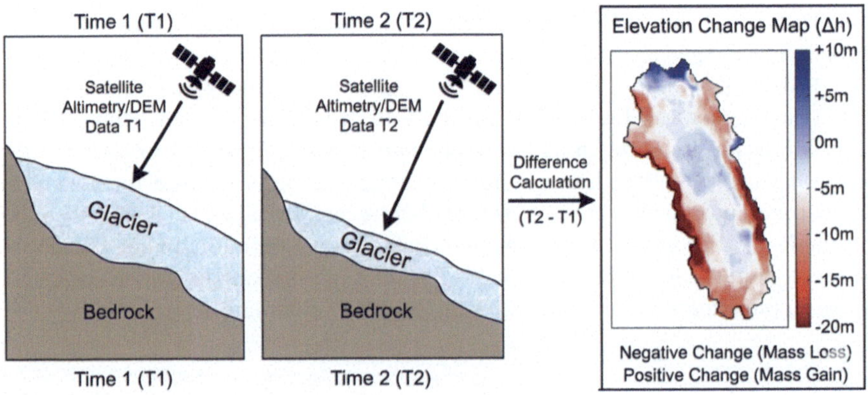

Fig. 3 Geodetic mass balance calculation for dates T1 and T2 with elevation change as Δh which is inclusive of the internal melting processes

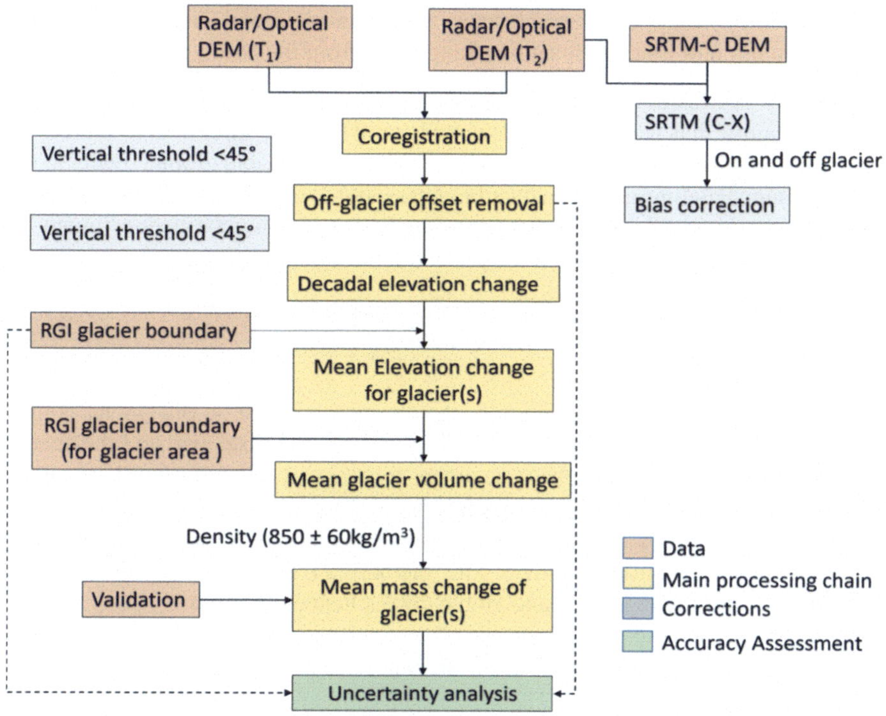

Fig. 4 Workflow for mass change calculation where the light-colored boxes are the dataset used in the process and the green colored box implied the last but prominent step of uncertainty analysis

methodology used by Deo et al. (2013). For penetration bias correction, SRTM-C and SRTM-X DEMs are used for glaciated and non-glaciated terrain. Upon implementation of this correction, the decadal elevation changes (such as from 2000–2014) is obtained if utilizing TerraSAR-X/TanDEM-X (composite year: 2014) and SRTM-C (year: 2000) bands. Modified RGI 6.0 glacier outlines (i.e. manually correcting the RGI inventory based glacier outlines) is used to extract elevation change on glaciers and this is further converted into volume change by multiplying with the area of the glacier under consideration. Accuracy assessment of the elevation change results for each catchment is performed as discussed in the following subsections. Finally, the mean volume change (for each catchment) is converted into a mass budget using a constant density conversion factor of 850 ± 60 kg m^{-3} for ice/firn for the entire Himalayan glaciers in India and nearby regions.

RGI inventory The glacier boundaries from Randolph Glacier Inventory (RGI) 6.0 (RGI Consortium 2017) are used to calculate the elevation change of the glaciers from the DEM difference image. However, this inventory was updated in the period 2003–2011, and in this study, we monitor changes from the year 2000 to 2014. Hence, the glacier boundaries were manually updated using cloud free, False Color composite (FCC) Landsat images (scenes from Landsat 5 and 7) of the year 2000 (since calculation of glacier elevation changes are from the base year, 2000).

Uncertainty Assessment The measurements taken on surface have an accompanied uncertainty that needs to be accounted for. There are three major sources of uncertainty in mass change: 1) fluctuation in elevation on stable ground owing to the DEM, 2) manual delineation in glacier boundary and 3) the glacier volume to glacier ice-mass conversion factor. To account for the first uncertainty or error estimate for high resolution DEM data used in Uttarakhand, the following equation is used:

$$\text{Uncertainty} = \frac{\Delta E_h}{\sqrt{N_{\text{eff}}}}, \tag{2}$$

Where N is the standard deviation of the elevation change on off-glacier points, and N_{eff} is given by the total number of observations, the pixel size and the spatial auto-correlation. This approach of accuracy assessment has been used in earlier studies (Gardner et al. 2013, Vijay and Braun, 2018).

For large-scale study in the Himalaya, standard method used is the normalized median absolute deviation (NMAD). This method is robust for a distribution with large number of outliers (Hohle and Hohle 2009), hence preferred.

$$NMAD = 1.4826 \times \text{median}_i(|\Delta h_i - m_{\Delta h}|) \tag{3}$$

where Δh_i is the sum of individual errors resulting from each slope interval on the stable (off-glacier) terrain and $m_{\Delta h}$ is the median of the errors. This method of calculating error has proven to be more resilient to the calculation of standard deviation especially with outliers in the dataset.

The next uncertainty occurs due to the delineation of glacier boundary which is accounted for using the formula used by Braun et al. (2019).

$$\delta A = \frac{r_{P/A}}{R_{P/A}} \times 0.03 \tag{4}$$

where δA denotes the uncertainty that results from glacier boundary, $r_{P/A}$ is the ratio of perimeter and area of the glacier and $R_{P/A}$ was calculated previously for alpine region as 5.03 km^{-1} by Paul et al. (2013).

The uncertainty that accompanies volume to mass conversion is performed using the ± 60 kg m^{-3} value, following Huss (2013). This methodology has also been adopted by Berthier et al. (2007), Brun et al. (2017) and Braun et al. (2019) over various test sites in glaciated terrains of Alaska, High Mountain Asia, South Patagonia respectively.

4 The Uncertainty in MB Estimates

Recent geodetic studies have provided varying estimates of glacier mass loss across the Himalayas, utilizing different sensor technologies and image processing methodologies. Bandyopadhyay et al. (2023) estimated a total glacier-ice mass loss of 153 ± 16 Gt for the Himalayan mountains between 2000 and 2014. Their study employed the InSAR method, comparing the SRTM-C DEM from 2000 with the TanDEM-X global DEM (primarily from 2014), both offering 90 m resolution and extensive coverage. They reported regional heterogeneity: mass loss was prevalent in the west and south of the Jammu-Kashmir (JK) region, while the Karakoram glaciers showed mass gain. Overall, the region experienced an average thickness decrease of 3.1 ± 1.4 m during this period. The most significant thinning occurred in Himachal Pradesh (HP) at a rate of 8.3 ± 2.9 m over the 14 years. These contrasting trends between JK and HP are attributed to differing dominant atmospheric circulation patterns i.e. the Westerlies and the Indian Summer Monsoon (ISM), respectively. Further east in Uttarakhand, thickness loss was less pronounced, partially due to mass gains in glaciers with high accumulation area ratios (AAR) in the Alaknanda catchment (Bandyopadhyay et al. 2019). However, studies relying on radar interferometry (InSAR) like this face specific uncertainty challenges; notably, radar signal penetration into snow and firn can introduce vertical biases that vary with wavelength (C-band vs. X-band) and surface conditions, potentially complicating the separation of seasonal signals from long-term trends.

Global assessments have sought to unify these regional estimates. Hugonnet et al. (2021) utilized the 20-year archive of ASTER stereo images to derive a globally consistent estimate of glacier mass change, finding that between 2000 and 2019, glaciers lost 267 ± 16 Gt /yr. This equates to 21.3% of observed sea-level rise, with an accelerated loss rate of 48 ± 16 Gt/ yr. per decade. Their approach represents a "time series" methodology, which effectively utilizes dense stacks of optical DEMs to mitigate random errors and improve robust trend detection compared to simple DEM differencing. However, optical sensors like ASTER are inherently limited by cloud cover and sensor saturation over bright snow, which can result in data voids in accumulation zones that necessitate interpolation, thereby adding to the uncertainty budget.

Focusing specifically on High Mountain Asia (HMA), Shean et al. (2020) combined ASTER data with high-resolution commercial stereo imagery (WorldView/GeoEye) to estimate a mass balance of −19.0 ± 2.5 Gt/yr. from 2000 to 2018. This study highlighted the complexity of uncertainty propagation, noting that aggregating mass balance uncertainty for large regions is not a straightforward summation due to spatial correlations. Standard approaches often aggregate statistics by elevation bins or scaling by hypsometry, but they frequently overlook longer-range spatial autocorrelation (e.g., tens of kilometers). To address this, Shean et al. (2020) aggregated uncertainty using spatially correlated samples (hex cells) to derive robust regional error estimates. Despite these advancements, calculating a "true" uncertainty remains a significant challenge for the glaciological community. A recent inter-comparison experiment (Piermattei et al. 2024) revealed that even when using identical input data, mass balance results from different research groups can diverge by more than 0.5 m/yr., a spread often exceeding the reported uncertainty limit. This divergence is largely driven by methodological choices in the image processing workflow, specifically in DEM generation, co-registration, and DEM void filling, rather than just sensor precision. For instance, co-registration corrections alone can alter results by magnitudes larger than the long-term glacier change rates. Furthermore, handling systematic errors, such as those arising from varying elevation precision based on terrain slope and quality as discussed by Hugonnet et al. (2022), varies significantly between studies. While some groups employ rigorous probabilistic frameworks or spatial inference from stable terrain, there is currently no consensus on best practices for uncertainty assessment, leading to inconsistent error reporting across regional studies. Consequently, while ensemble estimates may help reduce random errors, systematic offsets related to processing workflows and radar penetration remain critical sources of uncertainty in geodetic mass balance estimates.

5 Summary and Outlook

The steep topography and rugged terrains of the High Mountain Asia glaciers lead to sparse ground measurements of the glaciers' dynamics. Remote sensing technology offers the luxury of making observations without much discontinuity on a multi-decade scale. Still, utilization of apt technology for seamless and accurate estimates is of prime importance. The geodetic method is one of the most preferred methods of measuring the mass balance of a glacier by comparing changes in surface elevation over time using satellite or airborne remote sensing data, as it accounts for all spatial variability in the estimated glacier mass balance, assuming that the elevations are accurate everywhere, and references a stable geographic datum. However, one of the major drawbacks of this method is that the errors in the geodetic balance can result from seasonal adjustments, density adjustments, and image control and quality. Further, the relative error of the digital elevation models (DEMs) is more important than their absolute error when calculating geodetic mass balance, and differentiate wavelength penetration may make glacier-level measurements over bedrock less accurate than those over ice especially when using radar-based images. It is clear from the discussion above that no one approach can currently provide the necessary spatial and temporal coverage to allow a complete assessment of the mass balance of the High Mountain Asia glaciers. Hence one must consider the importance of point measurements (from glaciological observations on the ground), which would help validate and rectify the high uncertainty that the geodetic method offers at the cost of spatial coverage and continuity of measurements.

Acknowledgments DB would like to thank Prof G. Singh, Centre for Studies in Resources Engineering, IIT Bombay, for his technical input and support, as this analysis was conceptualized from the thesis work conducted under his supervision, while DB pursued PhD at the said institute.

References

Agarwal, V., Bolch, T., Syed, T. H., Pieczonka, T., Strozzi, T., & Nagaich, R. (2017). Area and mass changes of Siachen Glacier (East Karakoram). *Journal of Glaciology, 63* (237), 148–163.

Azam, M. F., Wagnon, P., Berthier, E., Vincent, C., Fujita, K., & Kergel, J. S. (2018). Review of the status and mass changes of Himalayan-Karakoram glaciers. *Journal of Glaciology*, 1–14.

Azam MF. Fifty years of Himalayan glacier mass-balance monitoring: Recommendations in honour of IYGP 2025. Journal of Glaciology. 2026;72:e43. https://doi.org/10.1017/jog.2026.10151.

Bahuguna, I., & Kulkarni, A. (2005). Application of digital elevation model and orthoimages derived from irs-1c pan stereo data in monitoring variations in glacial dimensions. *Journal of the Indian Society of Remote Sensing, 33* (1), 107–112.

Bamber, J. L., & Rivera, A. (2007). A review of remote sensing methods for glacier mass balance determination. *Global and Planetary Change, 59* (1–4), 138–148.

Bandyopadhyay, D., Mukherjee, S., Singh, G., & Coomes, D. (2023). The rapid vegetation line shift in response to glacial dynamics and climate variability in Himalaya between 2000 and 2014. *Environmental Monitoring and Assessment, 195* (1), 70. https://doi.org/10.1007/s10661-022-10577-9

Bandyopadhyay, D., Singh, G., & Kulkarni, A. V. (2019). Spatial distribution of decadal ice-thickness changes and glacier stored water loss in the Upper Ganga basin, India during 2000–2014. *Scientific Reports, 9* (1), 16730.

Berthier, E., Arnaud, Y., Kumar, R., Ahmad, S., Wagnon, P., & Chevallier, P. (2007). Remote sensing estimates of glacier mass balances in the Himachal Pradesh (Western Himalaya, India) *Remote Sensing of Environment, 108*, 327–338.

Bhattacharya, A., Bolch, T., Mukherjee, K., Pieczonka, T., Kropacek, J., & Buchroithner, M. F. (2016). Overall recession and mass budget of Gangotri Glacier, Garhwal Himalayas, from 1965 to 2015 using remote sensing data. *Journal of Glaciology, 62* (236), 1115–1133.

Bhushan, S., Syed, T. H., Arendt, A. A., Kulkarni, A. V., & Sinha, D. (2018). Assessing controls on mass budget and surface velocity variations of glaciers in Western Himalaya. *Scientific Reports, 8* (1), 1–11.

Bolch, T., Pieczonka, T., Mukherjee, K., & Shea, J. (2017). Brief communication: Glaciers in the Hunza catchment (Karakoram) have been nearly in balance since the 1970s. *Cryosphere, 11* (1), 531–539.

Braun, A. M. H., Malz, P., Sommer, C., & Barahona, D. F. (2019). Constraining glacier elevation and mass changes in South America. *Nature Climate Change letters, 9* (region 03), 130–136.

Brun, F., Berthier, E., Wagnon, P., Kääb, A., & Treichler, D. (2017). A spatially resolved estimate of High Mountain Asia glacier mass balances from 2000 to 2016. *Nature Geoscience, 10* (9), 668–673. https://doi.org/10.1038/ngeo2999.

Cogley, J. G. (2012). Area of the ocean. *Marine Geodesy, 35* (4), 379–388. Dehecq, A., Gourmelen, N., Gardner, A. S., Brun, F., Goldberg, D., Nienow,

Dehecq, A., Gourmelen, N., Shepherd, A., Cullen, R., & Trouve, E. (2013). Evaluation of cryosat-2 for height retrieval over the himalayan range. In *Cryosat-2 third user workshop.*

Dehecq, A., Gourmelen, N., Gardner, A.S. et al. Twenty-first century glacier slowdown driven by mass loss in High Mountain Asia. Nature Geosci 12, 22–27 (2019). https://doi.org/10.1038/s41561-018-0271-9.

R. Deo, S. Manickam, Y. S. Rao and S. S. Gedam, Evaluation of interferometric SAR DEMs generated using TanDEM-X data, 2013 IEEE International Geoscience and Remote Sensing Symposium - IGARSS, Melbourne, VIC, Australia, 2013, pp. 2079–2082. https://doi.org/10.1109/IGARSS.2013.6723221.

Fan, Y., Ke, C.-Q., Zhou, X., Shen, X., Yu, X., & Lhakpa, D. (2023). Glacier mass-balance estimates over high mountain asia from 2000 to 2021 based on icesat-2 and nasadem. *Journal of Glaciology, 69* (275), 500–512.

Fischer, A. (2011). Comparison of direct and geodetic mass balances on a multi-annual time scale. *Cryosphere, 5* (1), 107–124.

Fugazza, D., Scaioni, M., Corti, M., D'Agata, C., Azzoni, R. S., Cernuschi, M.,... Diolaiuti, G. A. (2018). Combination of uav and terrestrial photogrammetry to assess rapid glacier evolution and map glacier hazards. *Natural Hazards and Earth System Sciences, 18* (4), 1055–1071.

Gardelle, J., Berthier, E., & Arnaud, Y. (2012). Slight mass gain of Karakoram glaciers in the early twenty-first century. *Nature Geoscience, 5* (5), 322–325.

Gardelle, J., Berthier, E., Arnaud, Y., & K¨a¨ab, A. (2013). Region-wide glacier mass balances over the Pamir-Karakoram-Himalaya during 1999–2011. *Cryosphere, 7* (4), 1263–1286.

Gardner, A. S., Moholdt, G., Cogley, J. G., Wouters, B., Arendt, A. A., Wahr, J., Paul, F. (2013). A reconciled estimate of glacier contributions to sea level rise: 2003 to 2009. *Science (New York, N.Y.), 340* (6134), 852–857.

H¨ohle, J., & H¨ohle, M. (2009). Accuracy assessment of digital elevation models by means of robust statistical methods. *ISPRS Journal of Photogrammetry and Remote Sensing, 64* (4), 398–406.

Hugonnet, R., Brun, F., Berthier, E., Dehecq, A., Mannerfelt, E. S., Eckert, N., & Farinotti, D. (2022). Uncertainty analysis of digital elevation models by spatial inference from stable terrain. *IEEE Journal of Selected Topics in Applied Earth Observations and Remote Sensing, 15*, 6456-6472.

Hugonnet, R., McNabb, R., Berthier, E., Menounos, B., Nuth, C., Girod, L., (2021). Accelerated global glacier mass loss in early twenty-first century. *Nature*, 592 (7856), 726–731.

Huss, M. (2013). Density assumptions for converting geodetic glacier volume change to mass change. *The Cryosphere, 7* (4), 877–887.

Jakob, L., Gourmelen, N., Ewart, M., & Plummer, S. (2020). Ice loss in high mountain asia and the gulf of alaska observed by cryosat-2 swath altimetry between 2010 and 2019. *The Cryosphere Discussions*, 1–29.

Kaab, A., Berthier, E., Nuth, C., Gardelle, J., & Arnaud, Y. (2012). Contrasting patterns of early twenty-first-century glacier mass change in the Himalayas. *Nature, 488* (7412), 495–498.

Kaab, A., Huggel, C., Fischer, L., Guex, S., Paul, F., Roer, I., (2005). Remote sensing of glacier-and permafrost-related hazards in high mountains: an overview. *Natural Hazards and Earth System Sciences, 5* (4), 527–554.

Pandey, P., & Venkataraman, G. (2013). Comparison of dems derived from tandem-x and srtm-c for himalayan terrain. *2013 IEEE International Geo- science and Remote Sensing Symposium - IGARSS*, 322-325.

Paul, F., Berrand, N., Baumann, S., Berthier, E., Bolch, T., Casey, K., Winsvold, S. (2013). On the accuracy of glacier outlines derived from remote- sensing data. *Annals of Glaciology, 54* (63), 171–182.

Piermattei, L., Zemp, M., Sommer, C., Brun, F., Braun, M. H., Andreassen, L. M., Belart, J. M. C., Berthier, E., Bhattacharya, A., Boehm Vock, L., Bolch, T., Dehecq, A., Dussaillant, I., Falaschi, D., Florentine, C., Floricioiu, D., Ginzler, C., Guillet, G., Hugonnet, R.,... Yang, R. (2024). Observing glacier elevation changes from spaceborne optical and radar sensors—an inter-comparison experiment using ASTER and TanDEM-X data. *The Cryosphere, 18*, 3195–3230.

Quincey, D. J., Lucas, R. M., Richardson, S., Glasser, N. F., Hambrey, M., & Reynolds, J. M. (2005). Optical remote sensing techniques in high-mountain environments: application to glacial hazards. *Progress in Physical Geography, 29* (4), 475–505.

Rankl, M., & Braun, M. (2016). Glacier elevation and mass changes over the central Karakoram region estimated from TanDEM-X and SRTM/X- SAR digital elevation models. *Annals of Glaciology, 57* (71), 273–281.

RGI Consortium. (2017). *Randolph Glacier Inventory—A Dataset of Global Glacier Outlines: Version 6.0* (Tech. Rep. No. July). Colorado, USA: Global Land Ice Measurements from Space, Boulder, Colorado, USA. Digital Me- dia.

Shean, D. E., Bhushan, S., Montesano, P., Rounce, D. R., Arendt, A., & Os- manoglu, B. (2020). A systematic, regional assessment of high mountain asia glacier mass balance. *Frontiers in Earth Science, 7*, 363.

Simonsen, S. B., & Sørensen, L. S. (2017). Implications of changing scattering properties on Greenland ice sheet volume change from cryosat-2 altimetry. *Remote Sensing of Environment, 190*, 207–216.

Sun, F., Li, Y., Chen, Y., Duan, W., Li, B., Fang, G., & Li, Z. (2023). Evaluation of multiple gridded snowfall datasets using gauge observations over high mountain Asia. *Journal of Hydrology*, 130346.

Taking a measure of sea level rise: Gravimetry. (n.d.). https:// earthobservatory.nasa.gov/images/ 147437/taking-a-measure-of-sea-level-rise-gravimetry.

Vijay, S., & Braun, M. (2016). Elevation change rates of glaciers in the Lahaul-Spiti (Western Himalaya, India) during 2000–2012 and 2012–2013. Remote Sensing, 8(12), 1038.

Vijay, S., & Braun, M. (2018). Early 21st century spatially detailed elevation changes of Jammu and Kashmir glaciers (Karakoram–Himalaya). *Global and Planetary Change, 165* (April), 137–146.

Vincent, C., Ramanathan, A., Wagnon, P., Dobhal, D. P., Linda, A., Berthier, E., Gardelle, J. (2013). Balanced conditions or slight mass gain of glaciers in the Lahaul and Spiti region (northern India, Himalaya) during the nineties preceded recent mass loss. *The Cryosphere, 7* (2), 569–582.

Chapter 4
Ecological Risk Assessment from the Distribution of Recent Benthic Foraminifera from Palar Estuarine Sediments, Tamil Nadu, India

N. Suresh and M. Suresh Gandhi

1 Introduction

Benthic foraminifera have been shown to be useful environmental markers (Alve et al. 2009; Dolven et al. 2013). In recent years, many studies on benthic foraminiferal assemblages have been conducted in various regions of the world. Benthic foraminifera have great potential as "pollutant markers" and are among the most sensitive and cost-effective indicators of environmental stress in both naturally and anthropogenically stressed sites. Studies have shown that environmental contamination, which can lead to intracellular pathological processes, can lead to abnormal (teratological) testing of benthic foraminifera, as well as changes in foraminiferal abundance and taxonomic composition has been shown to be an essential component that determines size differences and conformational changes.

Assessment of the environmental status of the Palar Estuary in Tamil Nadu, India using foraminiferal bioindicators is of great importance with the increasing anthropogenic encroachment in coastal areas. Therefore, the aim of this study is to examine the recent distribution patterns of benthic foraminifera and their relationships with various physicochemical parameters and substrate attributes to explore the ecological potential of benthic foraminifera in these estuaries record the risk assessment.

Study Area

Palar Estuary is located in Kancheepuram District, Tamil Nadu, South India. The Palar River in South India ($12^0 28' 04'' N$, $80^0 09' 16'' E$) rises in the Nandi Durg Hills in the Kolar District of Karnataka, passes through Karnataka, Andhra Pradesh, Tamil Nadu and ends at Vayalur, Bay of Bengal south of Chennai (Fig. 1). The Palar Estuary

N. Suresh · M. S. Gandhi (✉)
Department of Geology, University of Madras Guindy Campus, Chennai, India
e-mail: msureshgandhi@gmail.com

N. Khare (ed.), *Changing Climate over India*,
https://doi.org/10.1007/978-981-92-1098-5_4

is a timber estuary, so it is cut off from the open ocean to the northeast, especially during the monsoon season (January to October). The presence of several longitudinal islands divides the actual estuary. The estuary is 1.5 km wide and 10.5 km long, with water depths ranging from 0.5 m to 2 m. The channel originating from Lake Madhurantakam joins the Palar Estuary at higher elevations, and during monsoons fresh water flows into the estuary.

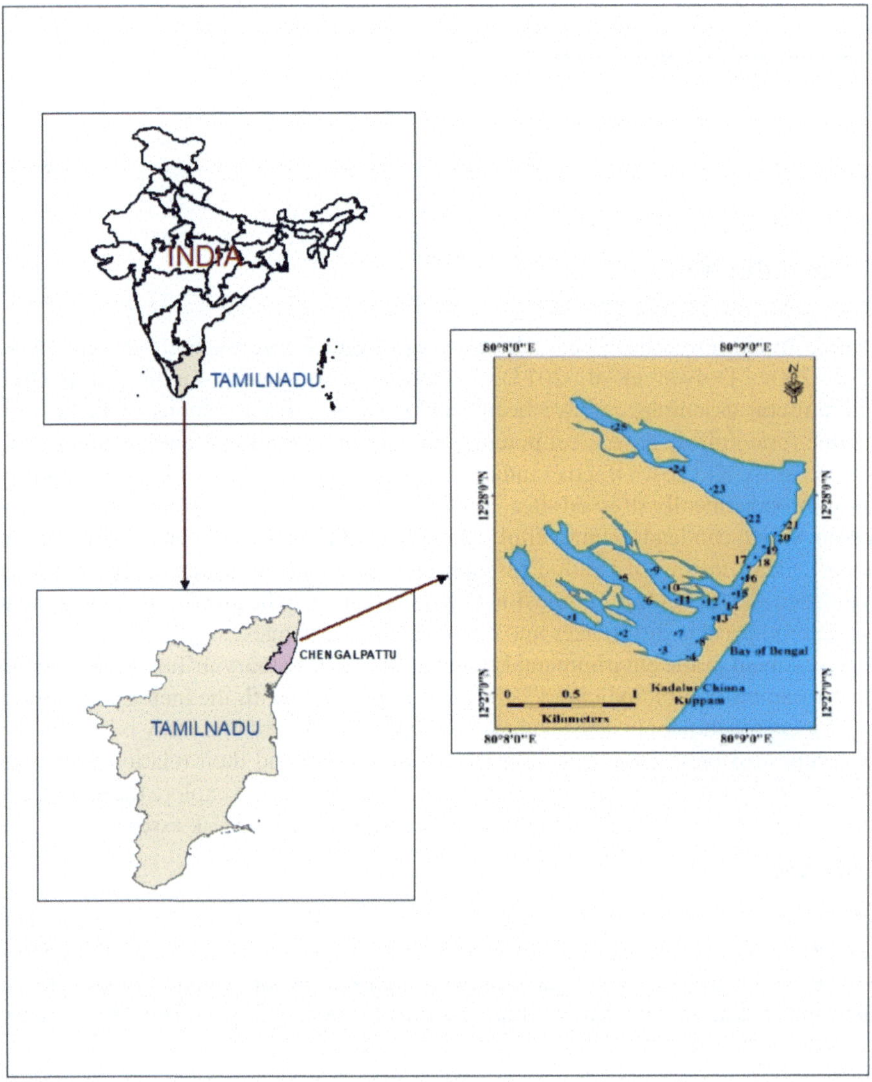

Fig. 1 Location map of Palar Estuary with sample locations

Methodology

A systematic collection of bottom sediment and bottom water was carried out at each point in both the study area, using a fisherman boat, mud grab and water sampler. Sediment and water samples were collected from a total of 25 locations in Palar Estuary. A total of 25 sediment and water samples were collected from the study areas in two seasons. (Table 1).

In the laboratory, the identification of the foraminiferal tests was carried out using Loeblich and Tappan (1988), Brady (1884), foraminifera and gallery-illustrated catalogue (*www.foraminifera.eu*), World Modern Foraminifera Database (*www.marinespecies.org/foraminiferal/*), and Paleontologia Electronica Webmaster 1998–2015. The pipette technique is used to analyze the fine fraction (silt and clay)

Table 1 Sample collection station latitude/longitude and water depth for Palar Estuary

Stations	Latitude	Longitude	Water depth (m)	
			Pre-monsoon	Post-monsoon
1	12°27′23"N	80° 08′15″E	1.5	1.5
2	12°27′18"N	80°09′07″E	1.0	1.0
3	12°27′13"N	80° 08′38″E	0.8	0.8
4	12°27′35"N	80° 08′28″E	0.5	0.5
5	12°27′28"N	80° 08′34″E	1.5	1.5
6	12°27′18"N	80° 08′42″E	1.0	1.0
7	12°27′11"N	80° 08′45″E	0.5	0.5
8	12°27′15"N	80° 08′49″E	0.9	0.9
9	12°27′17"N	80° 08′52″E	0.5	0.5
10	12°27′24"N	80° 08′55″E	0.5	0.5
11	12°27′40"N	80° 09′09″E	0.5	0.5
12	12°27′40"N	80°08′09″E	2.0	2.0
13	12°27′48"N	80° 09′11″E	1.5	1.5
14	12°27′51"N	80° 09′10″E	1.0	1.0
15	12°27′46"N	80° 09′07″E	0.8	0.8
16	12°27′53"N	80° 09′00″E	0.5	0.5
17	12°28′02"N	80° 08′51″E	1.5	1.5
18	12°28′08"N	80° 08′41″E	1.0	1.0
19	12°28′21"N	80° 08′26″E	0.5	0.5
20	12°27′28"N	80° 08′57″E	0.9	0.9
21	12°27′28"N	80° 08′48″E	0.5	0.5
22	12°27′32"N	80° 08′40″E	0.5	0.5
23	12°27′26"N	80° 08′45″E	0.5	0.5
24	12°27′20"N	80° 08′54″E	2.0	2.0
25	12°27′16"N	80° 08′53″E	2.0	2.0

in the washings, following Krumbein and Pettijohn, (1938). The sand, silt, and clay fractions' individual weights are converted to weight percentages and displayed on a trilinear diagram. The sediments in this investigation were described using Trefethen's (1950) textural nomenclature.

Calcium Carbonate: Calcium carbonate is measured using a fast titration technique (Piper, 1947). Any dissolved oxygen present, at the time of sampling, rapidly oxidizes an equal quantity of divalent manganese to basic hydroxide of higher valency states, as described by Winkler's technique (Strickland and Parsons, 1968). As a result, the diversity indices, 1) Simpson index (D), 2) Shannon-Weiner index (H′), and 3) Equitability (J), are determined in this study in order to more clearly describe the difference in taxonomical diversity of each sample.

Result and Discussion

In Palar Estuary a total of 40 foraminiferal species belonging to 15 genera was observed during the Pre-Monsoon period whereas in the Post-Monsoon period 53 species belonging to 24 genera was observed.

Living Population

The living foraminiferal specimens in Palar Estuary sediments are 14 out of the total 25, i.e., (56%) sampling stations during the Pre-Monsoon whereas, 12 out of the total 25 (48%) sampling stations in the Post-Monsoon. The total living population/100 ml. of wet-sediment is ranging from 0 to 14 and 0 to 65 during Pre- and Post-Monsoon, respectively. The Pre-Monsoon sediments comprise of 9 diversified foraminiferal species, out of 55 living species population, while in the Post-Monsoon samples 5 different foraminiferal species in the living population was 198. The results outlined that during the Pre-Monsoon and Post-Monsoon period considered for this study, that the maximum living foraminiferal population were recorded in the Northern part of the Palar Estuary (Pre-Monsoon: Stations 20 and 23 and Post-Monsoon: Stations 22, 24, and 25 that lies in the mouth of the estuary (Fig. 2). The investigation at stations; 1–2, 5–10, and 13 (Pre-Monsoon); 1–3, 5–9, and 13 (Post-Monsoon) insights that the living foraminiferal population was minimum in the above-mentioned stations that lied in the southern part of the estuary. Thus, the investigation infers that there is a gradual decline in living foraminifera toward the upstream side where, the flushing by the sea water drastically decreases.

Total Population of Foraminifera

In the Palar Estuary, total foraminiferal specimens were experiential in 24 out of the total 25 (96%) sampling stations during the Pre-Monsoon whereas, in the Post-Monsoon total foraminifera specimens of 18 out of the total 25 stations (72%). The sum of the total foraminifera population/100 ml. of wet-sediment is ranging from 0 to 294 and 0 to 151 during Pre-Monsoon and Post-Monsoon, respectively. In the Pre-Monsoon samples, the total (Dead+Live) population was 1503, comprising 40 different foraminiferal species, whereas in the Post-Monsoon samples, the total living population was 877, comprising 53 different foraminiferal species.

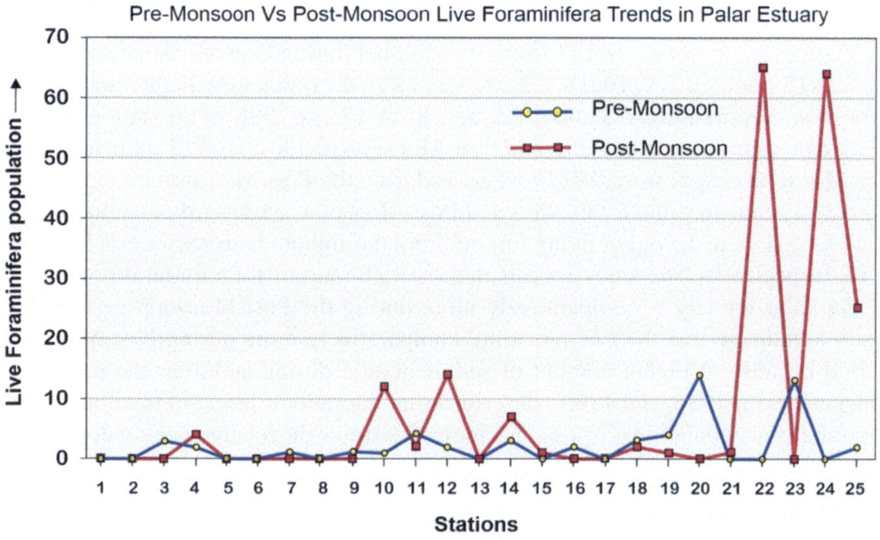

Fig. 2 Living foraminiferal trends between Pre-Monsoon and Post Sediments of Palar Estuary

Table 2 Population density of foraminifera

Category	Pre-monsoon		Post-monsoon
Feebly populated: <50	1, 5, 6, 7, 8, 9, 11, 12, 13, 15, 16, 17, 18, 19, 21, 22, 25 (68%)		1, 2, 3, 4, 5, 6, 7, 8, 9, 10, 11, 14, 15, 16, 17, 18, 19, 20, 21, 23 (80%)
Moderately populated: 50–150	3, 4, 10, 14, 20, 23 (24%)		12, 13, 22, 24 (16%)
Densely populated: >150	2, 24 (8%)		25 (4%)

The sampled stations were divided into three groups based on the population density of foraminifera (Table 2).

1.1 Living-Total Population Ratios—Palar

In the present study, an attempt has been made to examine the relationship between living-total population ratios of foraminifera and the rates of sedimentation. The L/T ratios for Pre-Monsoon and Post-Monsoon period are presented in Table 3. "0" value indicates the absence of living population in that particular station. The comparative L/T ratio for the Pre-Monsoon and Post-Monsoon period is represented in (Fig. 3).The study revealed that in 11 stations (Station: 4, 10, 11, 12, 14, 15, 19, 21, 22, 24, 25) out of 25 stations, i.e., 44%, the L/T ratios were higher during Post-Monsoon, while in 7 stations (Station: 3, 7, 9, 16, 18, 20, 23), i.e., 28%, the ratios were higher during Pre-Monsoon. Seven stations (Station: 1, 2, 5, 6, 8, 13, 17), i.e., 28% of the stations devoid of living foraminifera in both Pre-and Post-Monsoon periods. The L/T ratio during the Pre-Monsoon ranges from 0–0.19 whereas during the Post-Monsoon it ranges from 0–0.69. The mean value for the Pre- and Post-Monsoon seasons (after excluding the stations that were devoid of living foraminifera throughout both seasons) is 0.06 and 0.17, respectively. The study reveals that the L/T ratio of foraminiferal population of the Palar Estuary is comparatively more during the Post-Monsoon period. The study adjudicate that the L/T ratios are comparatively more during Post-Monsoon period because of higher amount of sedimentation during and after the monsoon. Phleger (1955, 1956, and 1960), described that the ratio of living to total benthonic foraminiferal population is a possible clue to determine the relative rates of deposition of sediment in which tshey occur.

Sand-Silt-Clay-Ratio

The textural nomenclature of Trefethen (1950) was used to characterize the sediment types in this investigation. Trefethen's figure depicts the proportions of sand, silt, and clay in all samples taken from 25 sites throughout the Pre-Monsoon and Post-Monsoon periods (Table 4, Figs. 3 and 4).

As shown in Table 4, the sediment types identified in Palar Estuary sample sites during Pre-Monsoon and Post-Monsoon fall into three sediment categories: Silty sand, clayey sand, and sand. The Sand-Silt Clay ratio of the substrate during the Post-Monsoon period indicated that sand should be the sediment type in 68 percent of the stations, while silty sand should be in 24 percent of the stations, and clayey sand should be in 8% of the stations. A comparative study of the ternary plots (Trefethen's,1950) textural nomenclature for the Pre-Monsoon and Post-Monsoon periods of Palar Estuary reveals that the substrate became coarser during the Post-Monsoon period. The relative rate of sedimentation is higher when the sediment size is coarser, thus, it is inferred that the rate of sedimentation was higher during the Post-Monsoon period as compared to Pre-Monsoon period. These results are well corroborated with the L/T ratio elucidated in the preceding chapter which indicated that the rate of sedimentation was higher during the Post-Monsoon period. Total foraminifera population reduced drastically during the Post-Monsoon period when compared to the Pre-Monsoon period in Palar Estuary, thus validating the fact that finer the sediments are ideal for foraminiferal population.

Calcium Carbonate

In the present study in Palar Estuary, during the Pre-Monsoon, the Calcium Carbonate content varies from 1% to 5.5% and the mean Calcium Carbonate content is 3.1%. During the Post-Monsoon the Calcium Carbonate content varies from 1.2% to 4.9% and the mean Calcium Carbonate content is 2.72%. A comparative reading of the Calcium Carbonate concentration in the 25 stations during Pre- and Post-Monsoon (Table 4) indicating that the Calcium Carbonate percentage in 17 stations, i.e., 68%

Table 3 L/T ratio for Pre-Monsoon and Post-Monsoon season for Palar Estuary (in numbers)

Stations	Pre monsoon			Post monsoon		
	Live forams (1)	Total forams (1)	L/T Ratio	Live forams (2)	Total forams (2)	L/T ratio
1	0	30	0	0	1	0
2	0	252	0	0	0	0
3	3	54	0.06	0	0	0
4	2	110	0.02	4	47	0.09
5	0	5	0	0	0	0
6	0	31	0	0	0	0
7	1	18	0.06	0	0	0
8	0	9	0	0	0	0
9	1	49	0.02	0	0	0
10	1	68	0.01	12	31	0.39
11	4	26	0.15	2	11	0.18
12	2	37	0.05	14	141	0.10
13	0	28	0	0	65	0
14	3	64	0.05	7	40	0.18
15	0	24	0	1	25	0.04
16	2	46	0.04	0	42	0
17	0	48	0	0	27	0
18	3	37	0.08	2	30	0.07
19	4	21	0.19	1	2	0.5
20	14	78	0.18	0	26	0
21	0	32	0	1	22	0.05
22	0	0	0	65	97	0.67
23	13	93	0.14	0	26	0
24	0	294	0	64	93	0.69
25	2	49	0.04	25	151	0.17

of the total stations is higher than the Post-Monsoon concentration. In 3 stations (give the station numbers), i.e., 12% of the stations the $CaCO_3$ values during the Pre-Monsoon and Post-Monsoon are same, whereas in 5 stations, i.e., 20% of the stations the $CaCO_3$ values are lesser than the Post-Monsoon values.

Organic Matter Content

In the study area, during the Pre-Monsoon period, the organic matter content varies from 1.17% to 3.95% whereas during the Post-Monsoon it varies from 0.9% to 3.90% (Table 4). The organic matter content is higher in all the stations during the Pre-Monsoon when compared with the Post-Monsoon period. The mean organic content is 1.88% and 1.80% for the Pre-Monsoon and Post-Monsoon period, respectively.

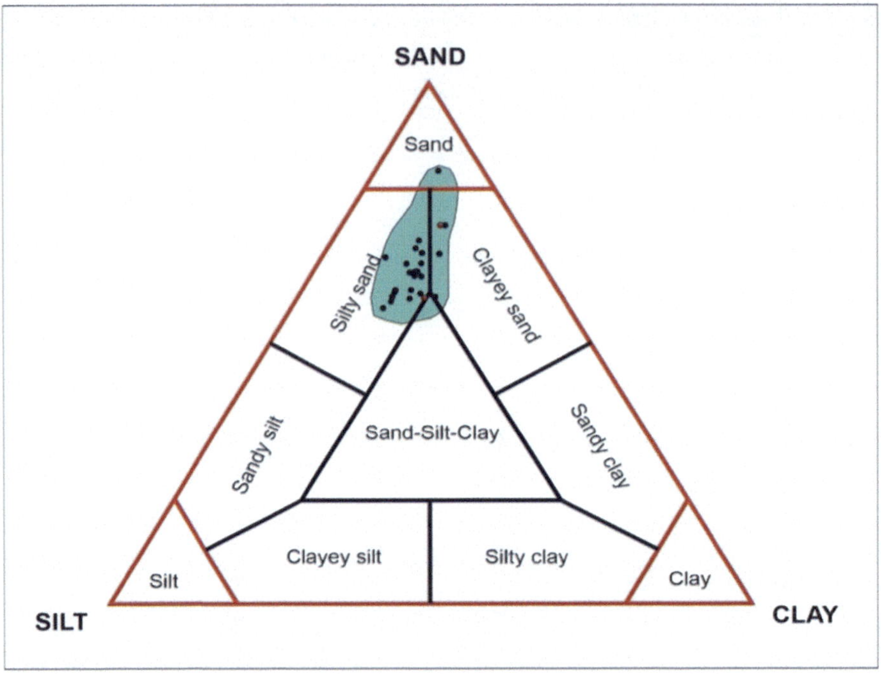

Fig. 3 Sand-silt-clay plot for Pre-Monsoon sediments of Palar Estuary

The overall reduction in the mean organic matter content is attributed to the total coarsening of the substrate sediments from Pre- to Post-Monsoon period. The station 11, that has clayey sand as substrate during the Pre- and Post-Monsoon periods (as compared to the total coarsening of sediments in Palar Estuary from Pre-Monsoon to Post-Monsoon as observed in this chapter) recorded the highest organic content of 3.95% and 3.90% during Pre- and Post-Monsoon, respectively.

Salinity

Table 5 shows the salinity levels of the bottom waters of the sample sites in the Palar Estuary for the Pre- and Post-Monsoon periods. The range of salinity measured in the 25 stations during Pre- and Post-Monsoon is as follows: Pre-Monsoon: Maximum: 27.1; Minimum: 24.9; Mean: 25.8 and in Post-Monsoon: Maximum: 26.5; Minimum: 21.1 °C; Mean: 24.5 is recorded. The marginal reduction in salinity is having a bearing on the total foraminiferal population during the Pre- and Post-Monsoon period, as substantial reduction in total foraminiferal (TF) population between Pre-Monsoon (TF: 1503) and Post-Monsoon period (TF: 877).

Statistical Analysis

The diversity indices simpson Index (D), Shannon-Weiner index (H′), and Pielou's Equitability Index (J) for Pre-Monsoon and Post-Monsoon seasons in Palar are

Table 4 Ecological parameters of the study area

Station	Pre-monsoon					Type	Post-monsoon					Type
	Sand %	Silt %	Clay %	$CaCO_3$%	OM %		Sand %	Silt %	Clay %	$CaCO_3$%	OM %	
1	s	11.9	15.0	3.0	1.42	Clayey sand	86.6	6.9	6.5	2.2	1.4	Sand
2	67.2	17.5	15.3	1.5	1.17	Silty sand	75.2	11.7	13.1	1.8	1.1	Clayey sand
3	59.6	25.7	14.7	1.0	2.73	Silty sand	68.6	19.5	11.9	1.2	2.2	Silty sand
4	56.8	29.0	14.2	2.5	2.05	Silty sand	73.5	16.3	10.2	2.6	1.5	Silty sand
5	83.0	7.3	9.7	4.0	1.2	Sand	87.1	4.3	8.6	3.4	0.9	Sand
6	57.7	27.0	15.3	3.5	1.85	Silty sand	77.6	13.2	9.2	3.5	1.8	Silty sand
7	72.3	11.3	16.4	1.0	2.88	Silty sand	81.7	7.6	10.7	1.7	2.8	Sand
8	65.1	18.6	16.3	4.0	1.71	Silty sand	80.4	6.2	13.4	3.9	1.6	Sand
9	64.9	20.9	14.2	3.5	2.05	Silty sand	84.6	6.3	9.1	2.4	2.0	Sand
10	62.7	21.1	16.2	2.5	1.51	Silty sand	88.6	3.8	7.6	3.1	1.4	Sand
11	66.5	15.3	18.2	4.0	3.95	Clayey sand	74.1	10.8	15.1	3.4	3.9	Clayey sand
12	58.4	26.6	15.0	5.0	1.56	Silty sand	79.3	12.2	8.5	3.9	1.5	Sand
13	67.9	18.1	14.0	3.0	1.42	Silty sand	89.0	6.6	4.4	2.8	1.3	Sand
14	59.7	25.3	15.0	5.5	1.46	Silty sand	79.6	10.4	10	4.9	1.4	Sand
15	69.0	17.0	14.0	2.0	2.54	Silty sand	89.6	2.8	7.6	1.7	2.5	Sand
16	66.6	23.2	10.2	3.0	1.27	Silty sand	83.1	11.7	5.2	2.6	1.2	Sand
17	59.3	21.6	19.1	2.0	1.61	Silty sand	93.2	2.5	4.3	2.0	1.5	Sand
18	58.4	23.6	18.0	3.0	1.37	Silty sand	90.1	1.3	8.6	3.0	1.3	Sand
19	63.8	20.0	16.2	3.5	1.17	Silty sand	93.4	1.3	5.3	2.9	1.1	Sand
20	58.4	21.1	20.5	2.5	1.81	Silty sand	81.4	6.2	12.4	1.6	1.7	Sand

(continued)

Table 4 (continued)

Station	Pre-monsoon					Type	Post-monsoon					Type
	Sand %	Silt %	Clay %	$CaCO_3$%	OM %		Sand %	Silt %	Clay %	$CaCO_3$%	OM %	
21	62.5	21.1	16.4	3.5	1.61	Silty sand	68.6	17.5	13.9	2.3	1.5	Silty sand
22	63.4	20.2	16.4	2.0	2.34	Silty sand	66.9	18.9	14.2	1.7	2.3	Silty sand
23	59.5	23.1	17.4	5.5	2.05	Silty sand	94.2	1.5	4.3	4.3	2.0	Sand
24	57.8	20.5	21.7	1.5	2.44	Clayey sand	84.6	8.9	6.5	1.9	2.4	Sand
25	61.9	20.4	17.7	4.0	1.9	Silty sand	75.6	15.2	9.2	3.3	1.8	Silty sand

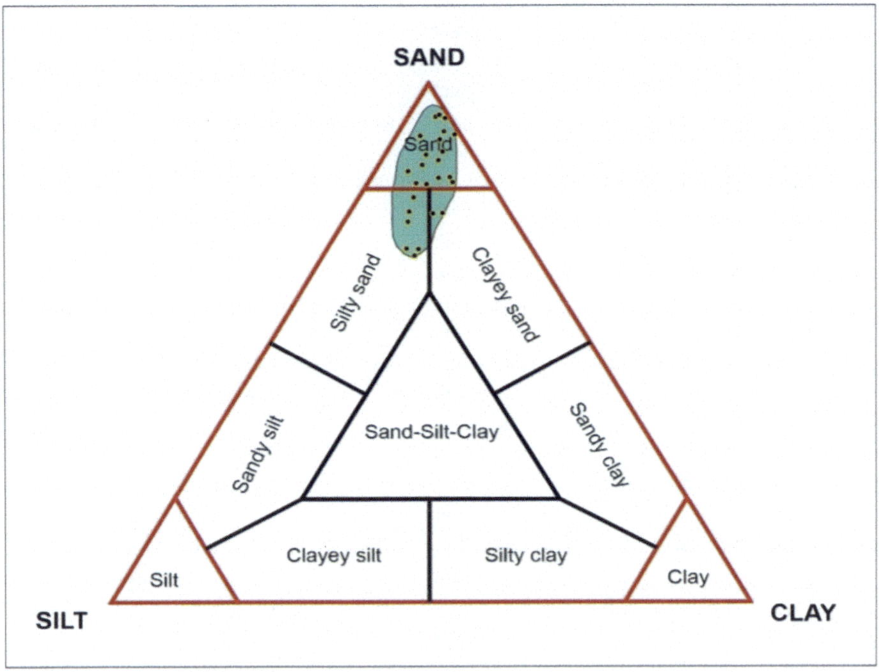

Fig. 4 Sand-silt-clay plot for Post-Monsoon sediments of Palar Estuary

depicted in Tables 6 and 7. The limited diversity in the Palar Estuary may be due to the environment being severely stressed during the Pre-Monsoon and Post-Monsoon seasons. This might be due to a number of factors, including: 1) the closing of the estuarine mouth, which restricted sea water circulation in the estuary (during Pre-Monsoon) and 2) the fast regional change of biological characteristics (during

Table 5 Salinity (ppm) values for Pre-Monsoon and Post-Monsoon Seasons of Palar Estuary

Stations	Salinity (Pre-Monsoon)	Salinity (Post-Monsoon)
1	26.3	25.2
2	27.1	26.5
3	26.3	25.9
4	26.9	26.0
5	26.6	25.8
6	25.6	25.6
7	26.0	23.0
8	25.8	23.9
9	25.4	24.3
10	24.9	23.1
11	25.0	22.0
12	24.9	22.8
13	26.3	24.5
14	27.1	25.5
15	26.3	25.5
16	26.9	26.3
17	26.6	26.1
18	25.6	25.0
19	25.4	23.3
20	25.8	24.4
21	25.4	24.5
22	24.9	23.8
23	25.6	21.1
24	24.9	24.6
25	24.9	24.1

Post-Monsoon). In addition, anthropological activities such as fish and crab farming and dredging of estuary bottoms for edible shells contributed to ecological instability.

Statistical Approach to the Foraminiferal Distribution-Cluster Analysis

Cluster analysis is a statistical technique for assigning a collection of biotopes such that objects in the same biotope are more comparable to each other than to other biotopes. (Kafecioglu, 1975; Reddy and Reddy, 1989). Cluster analysis is extensively used in ecological studies because it is simple, logical and allows for a pair-by-pair comparing of samples or variables. (Halfar and Ingle, 2003; Elshanawany et al., 2011, Nagendra, et al., 2015).

There are two methods for clustering: "R" mode and "Q" mode. In the "R" mode, clustering prioritizes taxonomic groups that should be completed in rows. You can also find groupings of variables or correlations by putting taxa in columns, which is

Table 6 Diversity indices D, H′ and J for Pre-Monsoon in Palar Estuary (SR = Species Richness, D=Simpson Index, H′ = Shannon-Weiner index, J = Pielou's Equitability Index)

Stations	Palar Pre Monsoon			
	SR	D	H'	J
1	9	0.79	1.7	0.79
2	4	0.61	1.1	0.77
3	13	0.87	2.2	0.85
4	3	0.48	0.7	0.66
5	2	0.00	0.5	0.72
6	10	0.85	2.0	0.86
7	6	0.76	1.5	0.81
8	3	0.67	1.0	0.88
9	7	1.00	1.2	0.62
10	3	0.09	0.2	0.19
11	3	0.34	0.6	0.53
12	9	0.82	1.9	0.86
13	6	0.81	1.6	0.91
14	12	0.85	2.0	0.82
15	4	0.71	1.3	0.95
16	7	0.50	1.1	0.58
17	5	0.57	1.0	0.61
18	11	0.87	2.1	0.88
19	5	0.60	1.2	0.73
20	17	0.81	2.1	0.74
21	5	0.34	0.7	0.44
22	0	1.00	0.0	0
23	11	0.73	1.7	0.69
24	5	0.53	0.9	0.55
25	10	0.77	1.7	0.72

known as Q mode clustering. The easiest method of computing distances between objects in a multi-dimensional space is to use Euclidean distances. Though there are numerous clustering approaches and algorithms to use in ecological research, there is no one way to determine which the best is. As a result, outcomes are subjective and may vary depending on the methodologies utilized.

The Q mode hierarchical clustering method was used in this research because it offers a hierarchy for site similarities. A dendrogram, or clustering tree, depicts the clustering findings. The dendrogram illustrates the level at which clusters were connected, as well as the locations inside each cluster.

In the present study, only the total (dead + live) foraminifera were considered for cluster analysis because the living taxa are sparse and absent in many stations in the

Table 7 Diversity indices D, H′ and J for Post-Monsoon in Palar Estuary (SR = Species Richness, D=Simpson Index, H′ = Shannon-Weiner index, J = Pielou's Equitability Index)

Stations	Palar post monsoon			
	SR	D	H'	J
1	1		0	
2	0		0	
3	0		0	
4	3	0.55	0.8	0.70
5	0		0	
6	0		0	
7	0		0	
8	0		0	
9	0		0	
10	3	0.53	0.7	0.67
11	8	0.11	1.9	0.91
12	18	0.25	2.0	0.70
13	19	0.09	2.5	0.86
14	5	0.41	1.0	0.65
15	11	0.15	2.0	0.84
16	9	0.21	1.8	0.80
17	11	0.14	2.1	0.86
18	10	0.11	2.2	0.94
19	1	1	0	
20	1	1	0	
21	8	0.17	1.8	0.87
22	4	0.86	0.3	0.23
23	7	0.26	1.5	0.77
24	2	0.88	0.2	0.35
25	5	0.70	0.6	0.36

Palar Estuary. For each station, the number of species and total number of individuals were calculated from the data collected from Palar Estuary (25 stations) during Pre-Monsoon and Post-Monsoon period. The data was compared among the sampled stations for each estuary and each season of sampling. The stations showing closeness in linkage distance were considered to have similarity in faunal characteristics. Based on the similarity in faunal abundance, the stations were clustered as elucidated in the following section.

The single linkage approach was adopted and the Euclidean distance was calculated for any two stations by making use of the PAST Software. The calculations were done using desktop computer.

Pre-Monsoon Season (Palar Estuary)

The dendrogram for the Pre-Monsoon season (Palar Estuary) is depicted in Fig. 5. Six clusters are recognized and are classified as Clusters 1–6. Cluster 3 has 3 sub-clusters and Cluster 5 has 2 sub-clusters. All the clusters are grouped at a closer ecological distance.

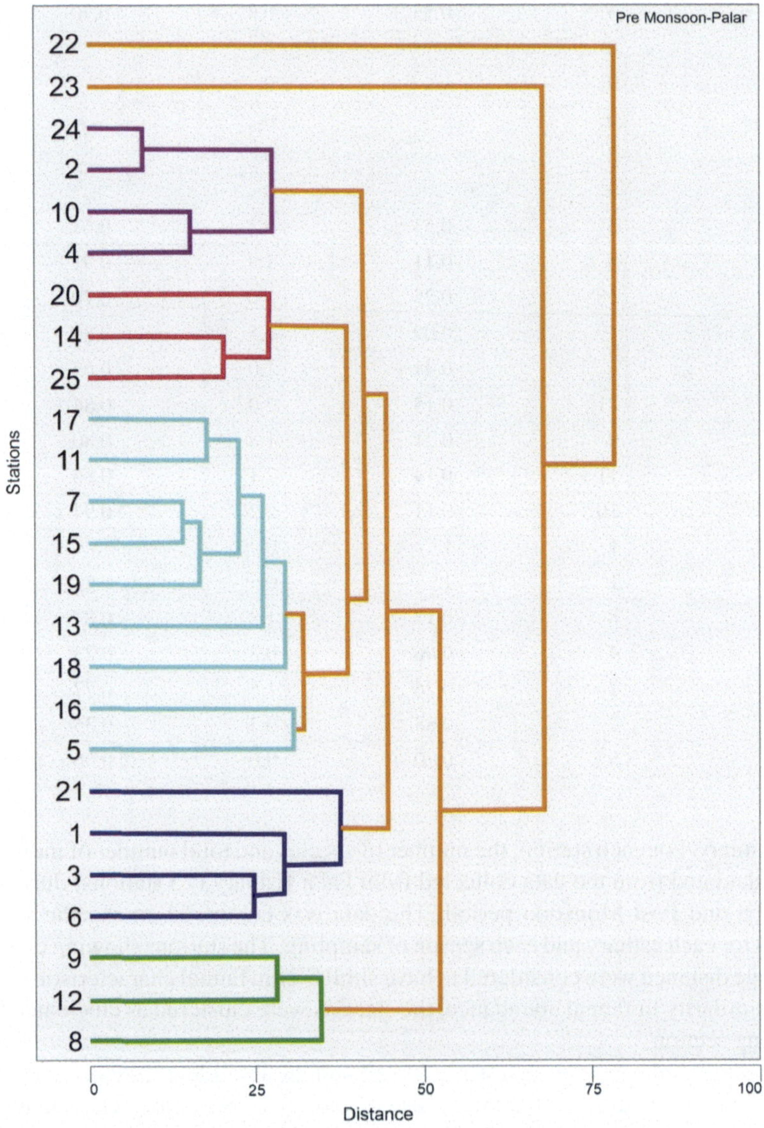

Fig. 5 Hierarchical dendrogram for the Pre-Monsoon season, Palar Estuary

Cluster 1: It consists of stations 8, 9, and 12, which are mainly located in the middle Kayal. The substrate is dominantly represented by silty sand. The salinity ranges from 24.9‰ at stn.12 to 25.8‰ at stn.8. The Dissolved Oxygen (DO) ranges from 7.0 mL/L at stn.12 to 9.2 mL/L at stn.9. The foraminiferal assemblage represented dominantly by *Quinqueloculin astelligera* (38% at stn 12 and 61% at stn 9) and *Pyrgo depressa* (11% at stn 12 and 44% at stn 8). Species richness (S) is low ranging from 3 at stn.8 to 9 at stn.12. The H'diversity index range from 1 at stn.8 to 1.9 at stn.12, and species dominance (D) is 0 at stn 9 and 0.33 at stn 8. Pielou's Equitability Index (J) ranges from 0.62 to 0.88. The low species richness (S), total foraminifer number (TFN), Shannon-Weiner index H′), and Simpson dominance (D) suggest that the Cluster 1 is in poor ecological status during the Pre-Monsoon.

Cluster 2: This cluster consists of 4 stations, 1, 3, 6, and 21. The salinity varies from 25.4‰ at stn.21 to 26.3‰ at stns.1 and 3, while DO fluctuates from 7.5 mL/L at stn.1 to 9.2 mL/L at stn.6 and stn 21. The substrate is predominantly silty sand with OM content ranges from 1.42 to 2.73. The faunal assemblage is dominated by milioilds, *Spiroloculina depressa* (40%, 9.3%, and 13% at stns.1, 3, 6, respectively), *S. communis* (6.7%, 20%, 32%, and 3% at stns.1, 3, 6, and 21, respectively) and *Triloculina rupertiana* (81% at stn.21). The S varying from 5 at stn.21 to a high of 13 at stn.3, while TFN is low at 30 at stn.1 to 54 at stn.3. The D ranges from 0.13 to 0.66, J (Pielou's equitability) value ranges from 0.44 to 0.86. H′ ranges from 0.7 to 2.2. Overall, the values of S, TFN and diversity indices infer that this cluster is in poor ecological status.

Cluster 3: This cluster is further sub-divided into 3 sub-clusters (3a, 3b, and 3c).

Sub-cluster 3a: It contains stns 5 and 16. The TFN is 5 and 46 while the S is 2 and 7 at stn 5 and 16, respectively. The H'diversity is too low at 0.5 at stn.5 and 1.1 at stn.16 and dominance is 1 at stn 5 and 0.5 at stn.16. The evenness index J ranges from 0.58–072, indicating better evenness. The foraminifer assemblage is represented by *Ammonia tepida* (80% at stn.5 and 13% at stn.16), *Quinqueloculina laevigata*, 4% at stn.16 and *Triloculina terquemiana*, 70% at stn 16. The salinity varies from 26.60‰ at stn.5 to 26.90‰ at stn.16. The DO ranges from 7.8 mL/L to 8.3 mL/L at stn 16 and stn 5, respectively. Low S, TFN and low diversity indices amply suggests that this sub-cluster is in poor ecological status.

Sub-cluster 3b: This sub-cluster comprises of stns.7, 13, 15, 18, and 19. *Ammonia parkinsoniana* and *A. tepida* are the dominant and constant foraminifer assemblage in this cluster. *A. parkinsoniana* has highest of 61.9% at stn. 19 and lowest of 30% at stn.18, while *A. tepida* is 33% at stn.7 and 5.4% at stn.18. The S ranges from minimum of 4 at stn.15 and a high of 11 at stn.18. The TFN varies from 18 at stn. 7 and 37 at stn.18. The diversity H′ is more than 1 in this sub-cluster. It varies from 1.2 at stn.19 to 2.1 at stn. 18. Species dominance (D) is low, ranging from 0.13 at stn.18 to 0.4 at stn.19. The evenness index J, ranges from 0.73 at stn. 19 to 0.95 at stn. 15, indicating better evenness of the species. The salinity changes from 25.4‰ at stn.19 to 26.3‰ at stns.13 and15. The DO changes from 7.8 mL/L at stn.13 to 9.5 mL/L at stns.19. The low S and TFN and the dominance of *Ammonia*, an opportunistic and stress tolerant taxon suggests that the sub-cluster 3b is ecologically in poor status.

Sub-cluster 3c: This cluster comprises stns.11 and 17. The foraminifer assemblage dominated by *Ammonia parkinsoniana* and *A. tepida* (81% and 15.4% at stn.11, 56.3% and 35.4% at stn.17, respectively). The salinity, DO and pH are 25‰, 10.1 mL/L, 7.5 at stn.11, and 26.6‰, 8.1 mL/L, 7.5 at stn.17, respectively. The S and TFN are 3 and 26 at stn.11 and 5 and 48 at stn.17, respectively. The diversity H′ varies from 0.6 at stn.11 to 1.0 at stn.17 and similarly D varies from 0.43 at stn 17 to 0.66 at stn 11. J varies from 0.53 at stn 11 to 0.61 at stn 17. The dominance is 0.43 at stn 17 and 0.66 at stn 11. The very low S and TFN and H′ advocate that prevalence of poor ecological conditions in cluster 3c.

Cluster 4: It consists of stns.14, 20, and 25. The substrate is mainly silty sand. The salinity, DO and pH show values of 27.1‰, 8.1 mL/L, 7.4 at stn.14; 25.8‰, 8.9 mL/L, 7.6 at stn.20; 24.9‰, 6.0 mL/L, 7.3 at stn. 25. The dominant foraminifer assemblage in this cluster includes *Ammonia parkinsoniana, A. tepida, Pararotalia nipponica,* and *Quinqueloculina lata. A. parkinsoniana* represents 23.4%, 18%, and 20%; *A. tepida* is 20%, 38%, 31%; *Q. lata* 16%, nil, 33% at stns.14, 20, 25, respectively. *P. nipponica* present at stn.14 with 19%. The S and TFN are relatively high. S ranges from 10–17 and TFN from 49–78. The diversity indices H′ is >1 in this cluster, while dominance D is low and J index shows more evenness. The relatively higher S, TFN, and H′ values in this cluster imply that moderate ecological status.

Cluster 5: This cluster is divided into 2 sub-clusters 5a and 5b.

Sub-cluster 5a: This sub-cluster comprises of stns.4 and 10. The silty sand is the main substrate type in this cluster. The S is low at 3 for both stns and TFN vary from 110 at stn.4 to 68 at stn.10. H′ is relatively higher value of 0.70 at stn.4 and lower value of 0.20 at stn.10; The dominance is 0.52 at stn 4 and 0.91 at stn.10; J varies from 0.19 to 0.66 at stns.104 and 4, respectively. The foraminifera assemblage is predominantly characterized by *Ammonia* at >95%. However, faunal assemblage corresponds to *Ammonia parkinsoniana* and *A. tepida* at stn. 4 and *A. parkinsoniana and Nonion grateloupi* at stn. 10. The salinity and DO vary from 24.9–26.9‰ and 8.5–9.4 mL/L at stns.4 and 10, respectively. The pH for both the stations is 7.6. The low S, H′, and predominance of *Ammonia* abundantly suggest that poor ecological status of sub-cluster 5a.

Sub-cluster 5b: It includes widely apart stations 2 and 24. The S is very low with 4 and 5 and high TFN of 252 and 294 at stns.2 and 24, respectively. The foraminifer assemblage is characterized by *A. parkinsoniana* (stn.2—48%; stn. 24—60%), *A. tepida* (stn.2—40%; stn.24—34%), *Q. lata* (stn.2—9%; stn.24—5%), and *Trochammina rotaliformis* (stn.2—4%; stn.24—0.7%). The H′ varies from 1.1 to 0.9, while D is low at 0.39 and 0.47; J fluctuates from 0.77 to 0.55 at stns.2 and 24, respectively. Bottom water Salinity varies from 27.12‰ to 24.9‰; DO ranges from 7.8 mL/L to 5.5 mL/L and pH fluctuates from 7.4 to 7.3 at stns.2 and 24, respectively. The low S, H′, and D values and predominance of species of *Ammonia* indicate that this sub-cluster is in poor ecological status.

Cluster 6: This cluster consists of solitary station, i.e., stn.23. *Ammonia dentata* (24%), *Asterorotalia pulchella* (45%), *Elphidium poeyanum* (7.5%), and *Lobatula lobatula* (4.3%) are the characteristic foraminifer assemblage in this cluster. Salinity 25.6‰, DO 10.1 mL/L, and pH 7.5, H′ at 1.7, D is low at 0.27 and J, 0.69. The substrate is dominantly silty. The occurrence of *A. pulchella* makes this cluster a unique and different from other clusters.

Post-Monsoon Season

The dendrogram for the Post-Monsoon season (Palar Estuary) is depicted in Fig. 6. In this season, six clusters are recognized and are classified as Clusters 1–6. Cluster 4 has 3 sub-clusters and Cluster 6 is the grouping done for all stations where no foraminifer was retrieved.

Cluster 1: It consists of two stations 19 and 21. The substrate is represented by sand at stn. 19 and silty sand at stn. 21. The foraminiferal assemblage represented dominantly by *Ammonia parkinsioniana,* 100% at stn 19 and 36.4% at stn. 21. Station 21 further has *Nonion soaphum (18.2%), Elphidium poeyanum (13.6%),* and *Elphidium williamsoni (9.1%).* The TFN and S ranges from 2 to 22 and 1 to 8, respectively. Salinity, DO and pH ranges from 23.3‰ to 24.5‰; 9.6 mL/L to 10.3 mL/L; 7.5–7.6 respectively. The H'diversity index ranges from 0 to 1.8, while the species dominance (D) ranges from 0.17 to 1. The Pielou's Equitability Index (J) ranges from 0 to 0.87. The low Species richness (S), total foraminifer number (TFN), Shannon-Weiner index (H′), and Simpson dominance (D) suggest that the Cluster 1 is in poor ecological status during the Post-Monsoon season.

Cluster 2: This cluster consists of four stations, 4, 10, 14, and 20. The substrate is represented by silty sand at stn. 4 and sand at other stations. The foraminiferal assemblage is represented by *Ammonia parkinsioniana* at stations 4, 10, 14, 20 at 72.3%, 68%, 38%, and 100%, respectively. The other dominant species is *Ammonia tepida* that occurs at stn. 4 (10.6%), 10 (295), and 14 (53%). The TFN and S ranges from 26 to 47 and 1 to 5, respectively. Salinity, DO and pH ranges from 23.1‰ to 26.0‰;8.1 mL/L to 9.8 mL/L; 7.2–7.6, respectively. The H'diversity index ranges from 0 to 1, while the species dominance (D) ranges from 0.41 to 1. The Pielou's Equitability Index (J) ranges from 0 to 0.7. Overall, the values of S, TFN and diversity indices infer that this cluster is in poor ecological status.

Cluster 3: This cluster consists of three stations, 22, 24, and 25. The substrate is silty sand to sandy in nature. The foraminiferal assemblage is represented by *Ammonia parkinsioniana and Ammonia tepida* that occurs at all stations ranging from 83% to 94% and 5.2% to 14%. The TFN and S ranges from 93 to 151 and 2 to 5, respectively. Salinity, DO and pH ranges from 23.8‰ to 24.6‰; 6.1 mL/L to 9.8 mL/L; 7.3–7.4, respectively. The H'diversity index ranges from 0.7 to 0.88, while the species dominance (D) ranges from 0.2 to 0.6. The Pielou's Equitability Index (J) ranges from

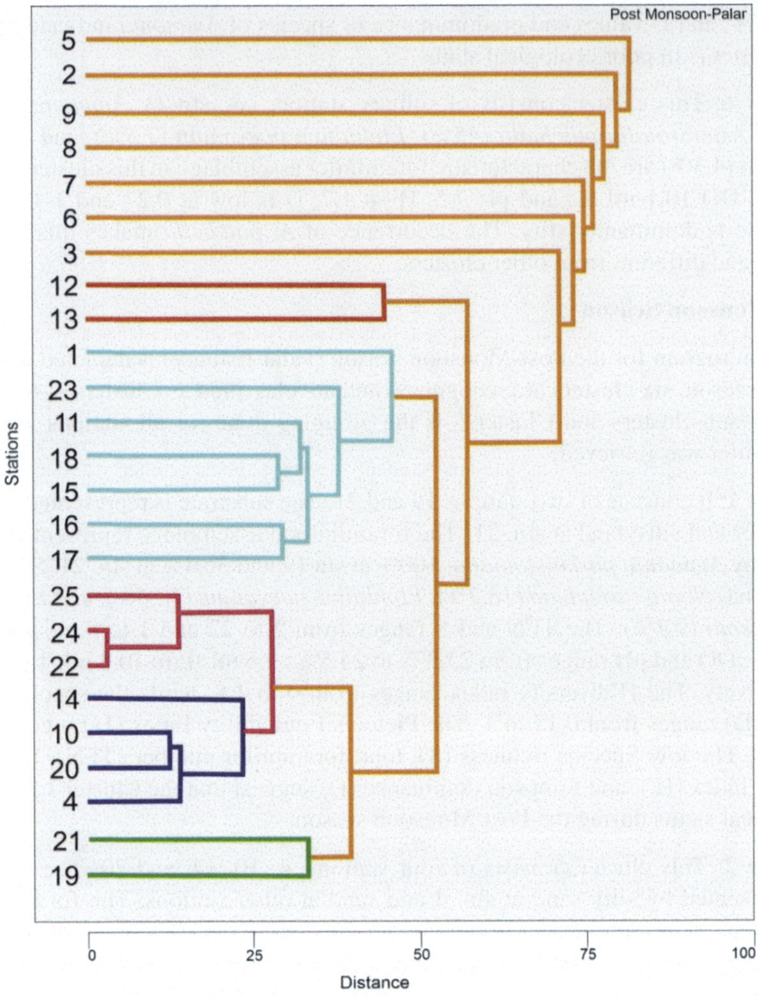

Fig. 6 Hierarchical dendrogram for the Post-Monsoon season, Palar Estuary

0.23 to 0.36. Low S, and low diversity indices amply suggests that this sub-cluster is in poor ecological status.

Cluster 4: This cluster is further sub-divided into 3 sub-clusters (4a, 4b, and 4c).

Sub-cluster 4a: This sub-cluster comprises of two stations 16 and 17. The substrate is sandy. *Ammonia beccarii* and *A. tepida* are the dominant species found in the two stations ranging 11.1% to 11.9% and 33.3% to 40.5%, respectively. *Elphidium crispum* (16.7%) is found only at station 16 and *Quinquelonculina ata* (14.8%) at stn. 17. The TFN and S ranges from 27 to 42 and 9 to 11, respectively. Salinity and pH ranges from 26.1‰ to 26.3‰ and 7.5–7.6, respectively. The DO is 8.9 mL/L at both the stations. The H'diversity index ranges from 1.8 to 2.1, while the species dominance (D) ranges from 0.14 to 0.21. The Pielou's Equitability Index (J) ranges from 0.80 to 0.86. The S values, marginally moderate H′ and evenness tending toward 1 indicates that this cluster is having moderate ecological status.

Sub-cluster 4b: This cluster comprises of three stations 11, 15, and 18. The substrate is clayey sand to sandy. The foraminifer assemblage *Ammonia tepida*and*Pararotalianipponica* occurs at all stations ranging from 26.7% to 36% and 3.32% to 28%, respectively. The TFN and S ranges from 11 to 30 and 8 to 11, respectively. Salinity, DO and pH ranges from 22‰ to 25.5‰; 9.5 mL/L to 10.5 mL/L; 7.3–7.5, respectively. The H'diversity index ranges from 1.9 to 2.1, while the species dominance (D) ranges from 0.11 to 0.15. The Pielou's Equitability Index (J) ranges from 0.84 to 0.94. The S values, moderate H′ and evenness tending toward 1 indicates that this cluster is having moderate ecological status.

Sub-cluster 4c: It consists of two stations 1 and 23. The substrate is sandy. The foraminifera *Ammonia tepida* is the only species recorded in stn 1. In station 23, *Spiroloculina depressula, Triloculina oblonga, Ammonia tepida,* and *Spiroloculina communis* occur at 42.3%, 30.8%, 7.7%, and 7.7%, respectively. The TFN and S ranges from 1 to 26 and 1 to 7, respectively. Salinity and DO ranges from 21.1‰ to 25.2‰ and 8.1 mL/L to 10.5 mL/L, respectively. The pH in both the stations is 7.5. The H'diversity index ranges from 0 to 1.5, while the species dominance (D) ranges from 0 to 0.26. The Pielou's Equitability Index (J) ranges from 0 to 0.77. The wide ranges in the diversity indices indicate that this cluster has poor ecological status.

Cluster 5: This cluster comprises of two stations 12 and 13. The substrate is sandy. The foraminifera that is common to both the stations is *Elphedium advenum* and *E. poeyanum* occurring 2.8% and 3.1%, 7.8% and 3.1% at station 12 and 13, respectively. The dominant species in stn. 12 is *Asterorotalia pulchella* (48%) while in stn. 13, *Ammonia beccarii* and *Peneroplis pertusus* dominates with 18.5% each. The TFN and S ranges from 65 to 141 and 18 to 19, respectively. Salinity, DO and pH ranges from 22.8‰ to 24.5‰; 7.1 mL/L to 8.1 mL/L; 7.4–7.6, respectively. The H'diversity index ranges from 2 to 2.5, while the species dominance (D) ranges from 0.09 to 0.25. The Pielou's Equitability Index (J) ranges from 0.7 to 0.86. The relatively high S and TFN with moderate H′ value, poor dominance and evenness tending toward 1 indicate that this cluster has moderate ecological status.

Cluster 6: This cluster consists of seven stations 2, 3, 5, 6, 7, 8, and 9 where no foraminifera was retrieved. The substrate is represented by silty sand in stn. 5 followed by sand at all other stations. Salinity, DO and pH ranges from 23‰ to 26.5‰; 8.1 mL/L to 10.3 mL/L; 7.3–7.6, respectively. The diversity indices could not be calculated as no foraminifers were retrieved.

2 Conclusions

In the Palar estuary, a total of 40 foraminiferal species belonging to 15 genera were observed during the Pre-Monsoon period, whereas in the Post-Monsoon period, 53 species belonging to 24 genera were observed. A positive correlation between the Mean CaCO3 values during the Pre-Monsoon and Post-Monsoon total foraminifera was observed in Palar. The diversity indices D, H′, and J for Pre-Monsoon and Post-Monsoon seasons in Palar Estuary indicate limited diversity is due to the environment being severely stressed during the Pre-Monsoon and Post-Monsoon seasons. The Palar Estuary had 53 benthic species, although many of them were sporadic. *Ammonia beccarii* and *Ammonia parkinsoniana* were the two most prevalent species, accounting for more than 80 percent of the foraminiferal abundance. The faunal assemblage was dominated by the most tolerant species, *A. beccarii* and *A. parkinsoniana*, which thrived and reproduced in a stressed environment. The single linkage cluster analysis using Euclidean distance in this study reveals the differences in the surface faunal assemblages and defines each cluster characterized by a group of environmental variables. All the clusters are grouped at close ecological distance. The apparently unsystematic grouping of stations as disclosed by cluster analysis for total foraminifera suggests stressed ecological conditions. The substrate has an impact on faunal distribution and relative abundance. The lack of connection between the estuary and the open sea has a direct impact on environmental factors and, as a result, foraminiferal abundance and geographical distribution.

Acknowledgement The authors are thankful to Head of the Department, Department of Geology, University of Madras for providing necessary lab facilities to carry out the work.

References

Alve, E., Lepland, A., Magnusson, J., and Backer-Owe, K. (2009). Monitoring strategies for re-establishment of ecological reference conditions: possibilities and imitations. Marine Pollution Bulletin, vol. 59(8-12), pp. 297-310.

Brady, H. B. (1884). *Report on the Foraminifera collected by HMS Challenger during the years 1873-1876*. HM Stationery Office.

Elshanawany, R., Ibrahim, M. I., Milker, Y., Schmiedl, G., Badr, N., Kholeif, S. E., & Zonneveld, K. A. (2011). Anthropogenic impact on benthic foraminifera, Abu-Qir Bay, Alexandria, Egypt. *The Journal of Foraminiferal Research*, *41*(4), 326-348.

Halfar, J., & Ingle Jr, J. C. (2003). Modern warm-temperate and subtropical shallow-water benthic foraminifera of the southern Gulf of California, Mexico. *The Journal of Foraminiferal Research*, *33*(4), 309-329.

Kafecioglu, I.A. (1975) Quantitative distribution of foraminifera on the continental shelf and uppermost slope off Massachusetts. Micropaleontology, vol.21, pp.261-305.

Krumbein, W. C., & Pettijohn, F. J. (1938). Manual of sedimentary petrography.. Appleton-Century Co., Inc. XIV+549 pages

Loeblich and Tappan (1988). Foraminiferal evolution, diversification, and extinction. *Journal of Paleontology*, 695-714.

Nagendra, R., Sathiyamoorthy, P., Reddy, A. N., & Ramachandran, A. (2015). Spatial distribution of benthic foraminifera in the Palar estuary, Tamil Nadu. *Journal of the Geological Society of India*, *86*(3), 305-316.

Phleger, F. B. (1955). Ecology of foraminifera in southeastern Mississippi Delta area. *AAPG Bulletin*, *39*(5), 712-752.

Phleger, F. B. (1956). Significance of Living Foraminiferal Populations Along the Central Texas Coast. *Journal of Sedimentary Research*, *26*(2).

Phleger, F. B. (1960). Ecology and Distribution of Recent Foraminifera. The John Hopkins Press, Baltimore, 297 p.

Piper, O. A. (1947). I John and the Didache of the Primitive Church. *Journal of Biblical Literature*, 437-451.

Reddy, A. N., & Reddy, K. R. (1989). Cluster analysis and distribution of recent foraminifera, Araniar River estuary Tamil Nadu, India. *Journal of the Geological Society of India*, *33*(1), 76-81.

Strickland, J.D.H. and Parsons, T.R. (1968) A Practical Handbook of Seawater Analyses. Bulletin Fisheries Research Board of Canada, n. 167, Fisheries Research Board of Canada, Ottawa, 311 p.

Trefethen, J.M.1950, Classification of sediments, Am. Jour. Soc., v.248pp.55-62.

Part III
Climate Changes on Short Scale

Chapter 5
Morphometric and Morphotectonic Characterization of the Kopili River Basin, Northeast India: Integrating Geospatial and Remote Sensing Techniques

Ibitmeki Chadong, Shikhar Kumar, Anamika Upadhaya, and Devesh Walia

1 Introduction

The Kopili River Basin in Northeast India represents a vital region for conducting morphometric and morphotectonic studies. These analyses provide significant insights into drainage networks, fluvial geomorphology, and tectonic activity, which are essential for understanding the area's geological and hydrological evolution. Morphometry analysis helps evaluate the hydrological behavior and geomorphological progression of a basin, highlighting the influence of lithology and tectonics on fluvial landforms (Pophare and Balpande, 2014; Prakash et al., 2016). Drainage systems exhibit high sensitivity to even minor surface morphology changes, often reflecting underlying structural deformations (Leeder and Jackson, 1993; Rai et al., 2017; Kale and Shejwalkar, 2008; Singh et al., 2020; Prakash et al., 2017).

Advancements in Remote Sensing and Geographic Information Systems (GIS) have revolutionized the study of geomorphometry by enabling the efficient and accurate extraction of morphometric parameters from Digital Elevation Models (DEMs). These technologies facilitate systematic observation and rapid analysis of geomorphometric indices and drainage characteristics (Yunus et al. 2014; Bhat et al., 2013; Avinash et al., 2014). The ability to establish geometric relationships between drainage basin properties, stream networks, and channel geometry provides a framework for predicting channel shape parameters and assessing basin dynamics, further

I. Chadong · S. Kumar · D. Walia (✉)
Department of Geology, North-Eastern Hill University, Shillong, India
e-mail: drdeveshwalia@gmail.com

A. Upadhaya
Department of Environmental Studies, North-Eastern Hill University, Shillong, India

N. Khare (ed.), *Changing Climate over India*,
https://doi.org/10.1007/978-981-92-1098-5_5

enhancing our understanding of geomorphic processes (Doornkamp and Cuchlaine, 1971; Strahler, 1957; Flint, 1974; Schumm, 1956; Leopold and Maddock, 1953; Horton, 1945, 1932).

Morphotectonic analysis serves as a crucial tool for identifying active tectonics and evaluating the impact of geomorphological, structural, and neotectonic processes on basin evolution (Bhatt et al., 2007). These studies are indispensable for understanding the role of tectonics in shaping drainage basins and contribute significantly to geo-hazard identification, land-use planning, and civil engineering projects. By integrating morphometric parameters with morphotectonic indices, this study aims to examine the geomorphic evolution and active tectonism within the Kopili River Basin, providing vital information for both scientific research and practical applications.

1.1 Study Area

The Kopili River Basin, located in Northeast India, spans an area of 1,35,600 hectares between longitude 91°–93° E and latitude 25°–27° N. Originating from the south-western slope of Shillong Peak, the River flows northward into the Brahmaputra River, serving as an interstate River between Meghalaya and Assam. The basin exhibits a sub-dendritic drainage pattern and includes major tributaries such as the Myntang, Kharkor, Mynriang, Saipung, and Amring Rivers, among others. These Rivers primarily flow from south to north, contributing to the hydrological dynamics of the region.

The climatic conditions of the basin range from moderate to sub-humid, with temperatures varying significantly across seasons. Winter temperatures drop as low as 4 °C, while summer temperatures can reach as high as 44 °C. The area experiences annual rainfall ranging from 980 mm to 1700 mm, with the highest precipitation occurring during the monsoon months of June to September. More than 80% of the annual runoff, which totals around 600 mm, occurs during this season. The flat floodplain of the Kopili River often overflows due to recurrent floods caused by heavy rainfall, resulting in water accumulation across the basin (Figs.1 and 2).

Geomorphologically, the basin is characterized by a diverse range of features, including dissected plateaus, hills, valleys, alluvial plains, floodplains, and piedmont slopes. These geomorphic units encompass both highly and moderately dissected hills and plateaus, with the highly dissected plateaus predominantly found in the western and eastern portions, while moderately and low dissected plateaus are scattered across the northern, southern, western, and eastern areas. The basin also features pediment peneplain complexes, dissected valleys, and water bodies. Along the highest stream order of the basin, identified as the eighth order, dams, reservoirs, and floodplains are present, contributing to water management and hydrological complexities (Fig. 3).

Geologically, the basin includes formations from the Shillong, Jaintia, and Barail groups, alongside Quaternary-age sediments, adding to the region's rich geological

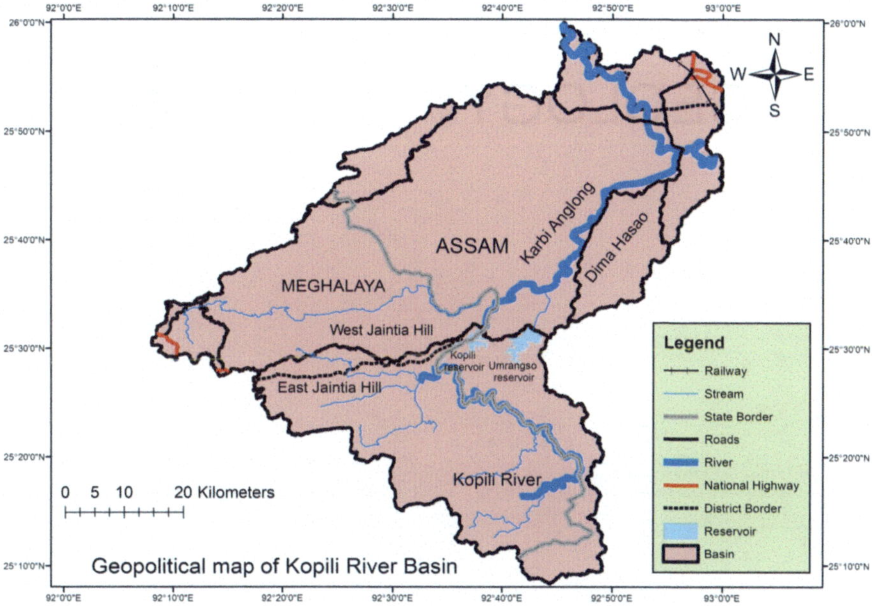

Fig. 1 Map of the study area

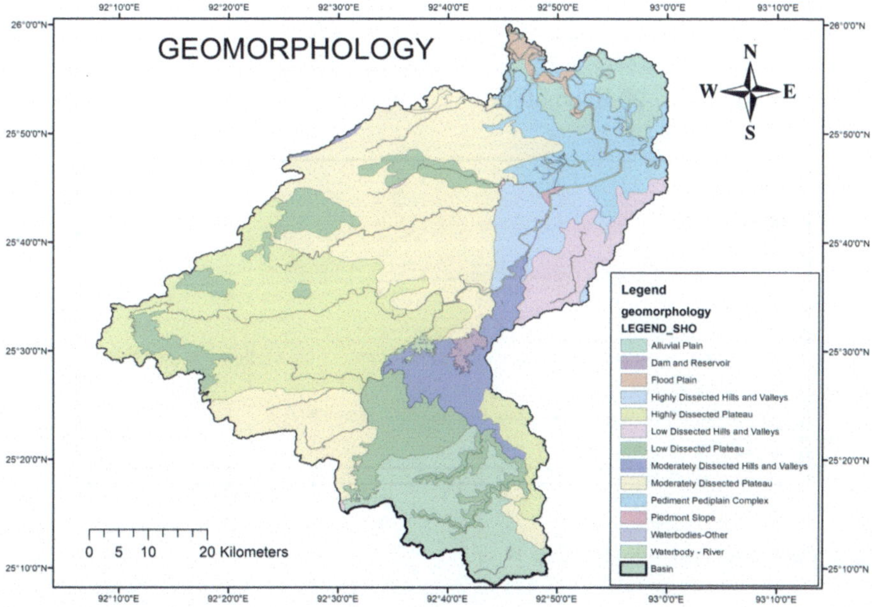

Fig. 2 Geomorphological map of Kopili River Basin

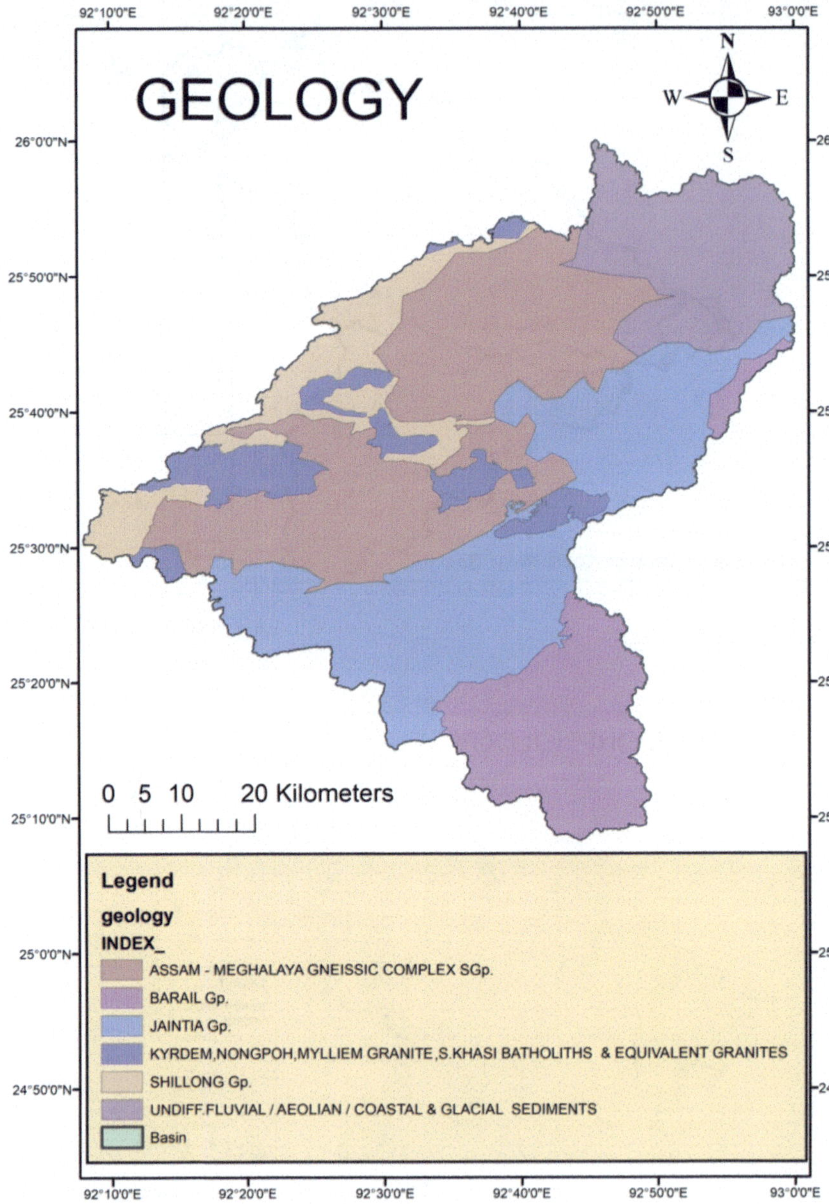

Fig. 3 Geological map of Kopili River Basin

tapestry (Walia and Mipun, 2010). The diverse topography creates varying climatic zones across the study area, influencing ecological and hydrological patterns.

2 Methodology

2.1 Data Collection

Topographic and elevation data were essential for the analysis. Topographic maps were obtained from Survey of India toposheets at a scale of 1:50,000 to ensure high-resolution and accurate spatial representation. Additionally, SRTM Digital Elevation Model (DEM) images with a resolution of 30 meters were sourced from the USGS/EROS database. The DEM was utilized for elevation profiling and slope analysis. The data were processed using ArcGIS 10.4.1, where georeferencing of maps and digitization of drainage networks were carried out. This ensured that the spatial data aligned accurately and could be further analyzed using GIS-based techniques (Fig. 4).

2.2 Stream Network Analysis

The stream network was analyzed using Horton's and Strahler's methods of stream ordering. Horton's method categorized streams based on their hierarchical arrangement, while Strahler's technique refined the classification by considering the intersection of tributaries. Key morphometric parameters were calculated to understand the hydrological behavior of the basin:

- Stream Order: The hierarchical organization of streams within the basin.
- Stream Length: Measurement of individual stream lengths to analyze flow characteristics.
- Bifurcation Ratio: A parameter assessing the branching pattern and geomorphic maturity of the basin.
- Drainage Density: The total stream length divided by the basin area, indicative of surface runoff efficiency.
- Form Factor: A shape-based parameter that determines the flow concentration within the basin.

These parameters were computed using standard mathematical formulae and GIS tools for better accuracy.

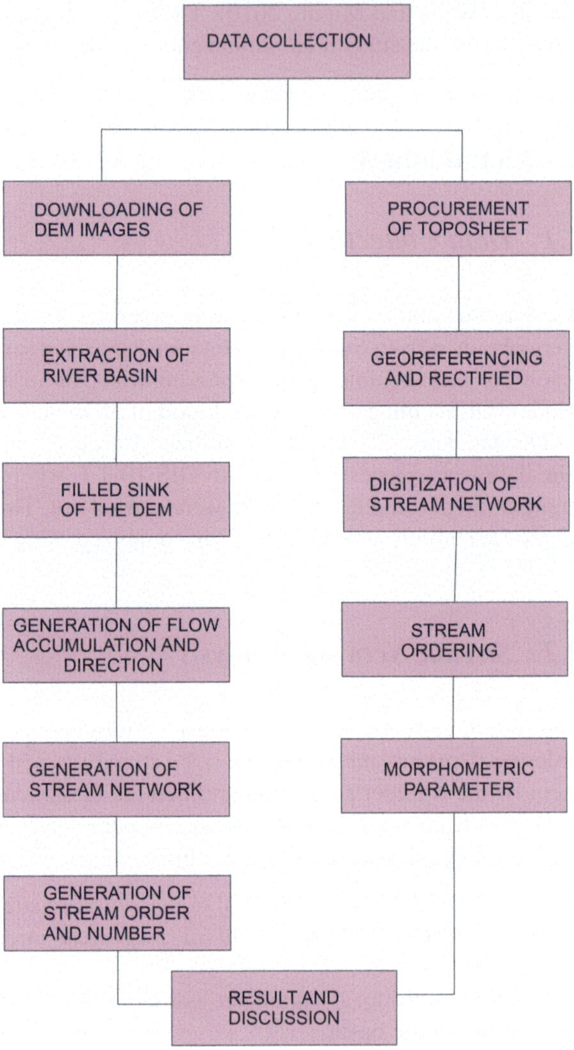

Fig. 4 Flowchart of methodology

2.3 *Relief and Aerial Aspects*

Relief aspects such as basin relief, ruggedness number, and relative relief were integral to understanding the basin's topographic strength and variations. Basin relief was calculated to measure the vertical dimension of the basin, while ruggedness number highlighted the terrain roughness. Relative relief provided insight into the elevation differences across the basin. Similarly, aerial aspects like basin area, drainage density, and circulatory ratio were computed to evaluate the spatial distribution of streams and the overall shape of the basin. GIS-based techniques facilitated these computations, allowing precise measurements and mapping.

2.4 Morphotectonic Analysis

Morphotectonic indices were used to infer tectonic activity in the Kopili River Basin:

- Sinuosity Index (S): Calculated as S = SL/VL, where SL is the length of the stream channel and VL is the straight-line distance between the stream's endpoints. This index helps understand meandering and tectonic influences. (Source: Muller, 1968).
- Transverse Topographic Symmetry Factor (T): Defined as T = Da/Dd, where Da is the distance from the midline of the drainage basin to the midline of the active meander belts, and Dd is the distance from the midline of the drainage basin to the basin split. This parameter assesses the symmetry of the basin, highlighting tectonic shifts. (Source: Cox, 1994).
- Asymmetry Factor (AF): Computed as AF = (Ar/At) * 100, where Ar represents the area of the basin on the right side of the stream trunk, and at is the total area of the basin. This factor indicates the tilt or deviation caused by tectonic forces. (Source: Gardener, 1987).

The integration of GIS and remote sensing tools was pivotal. These technologies enabled precise extraction, mapping, and analysis of geomorphic and tectonic parameters, shedding light on the hydrological and geological dynamics of the basin. Together, these methodologies provided a comprehensive understanding of the Kopili River Basin's morphometric and morphotectonic characteristics.

3 Results and Discussion

3.1 Morphometric Parameters

3.1.1 Linear Aspect

The linear aspects of the Kopili River Basin provide insights into its lithological, hydrological, and neotectonic dynamics through the study of stream order, stream number, stream length, mean stream length, stream length ratio, bifurcation ratios, and Rho coefficients. The basin's hierarchical stream order classification, following Strahler's method, extends up to the eighth order.

- Total stream number: 23,742, with first order streams constituting the majority. This dominance suggests a dendritic drainage pattern, indicative of homogenous geological formations and lesser structural disturbances.

Table 1 Linear aspects of basin

Stream order (U)	Stream number (N_u)	Stream length (L_u) (in kms)	Mean stream length (L_{sm}) (in kms)	Stream length ratio (R_L)	Bifurcation ratio (Rb)	Mean bifurcation ratio (R_{bm})	Rho coefficient
1st	18,500	7753.32	0.42	0.44	4.55	4.11	0.10
2nd	4064	3411.46	0.84	0.50	4.51		0.11
3rd	901	1712.84	1.90	0.50	4.17		0.12
4th	216	852.805	3.95	0.51	4.8		0.11
5th	45	439.87	9.77	0.58	3.75		0.15
6th	12	256.52	21.38	0.51	4		0.13
7th	3	130.23	43.41	0.98	3		0.33
8th	1	127.43	127.43				
Total	237,42	14,684.47					

- Total stream length: 14,684.47 km, where the shorter streams correspond to steep slopes, while longer streams indicate regions with lower gradients, reflecting varied terrain influences across the basin.
- Stream length ratio: Ranges from 0.44 to 0.98, demonstrating the influence of slope and basin topography on stream dynamics.
- Bifurcation ratios: Values between 3 and 4.55 highlight the minimal impact of geological controls on the basin's drainage network.
- Rho coefficient: Ranging from 0.10 to 0.33, this underscores the basin's low water retention capacity during flood events, a crucial factor in managing hydrological hazards (Table 1) and (Figs.5, 6 and 7).

3.1.2 Aerial Aspect

Aerial parameters, encompassing basin shape, drainage density, stream frequency, and texture, reflect the two-dimensional geomorphic attributes of the Kopili Basin.

- Basin area: 4868.26 km^2, delineating the spatial extent of the basin.
- Basin perimeter: 527.61 km, providing an understanding of the basin's boundary intricacies.
- Drainage density: 3.02 km/km^2, indicative of impermeable subsurface materials and moderately high runoff during precipitation events.
- Stream frequency: 4.88 km/km^2, reflecting the combined effects of rainfall, lithology, and topography on stream distribution.
- Drainage texture: 44.99 km^{-1}, classified as extremely fine, suggesting the presence of dense vegetation or impermeable soils in parts of the basin.

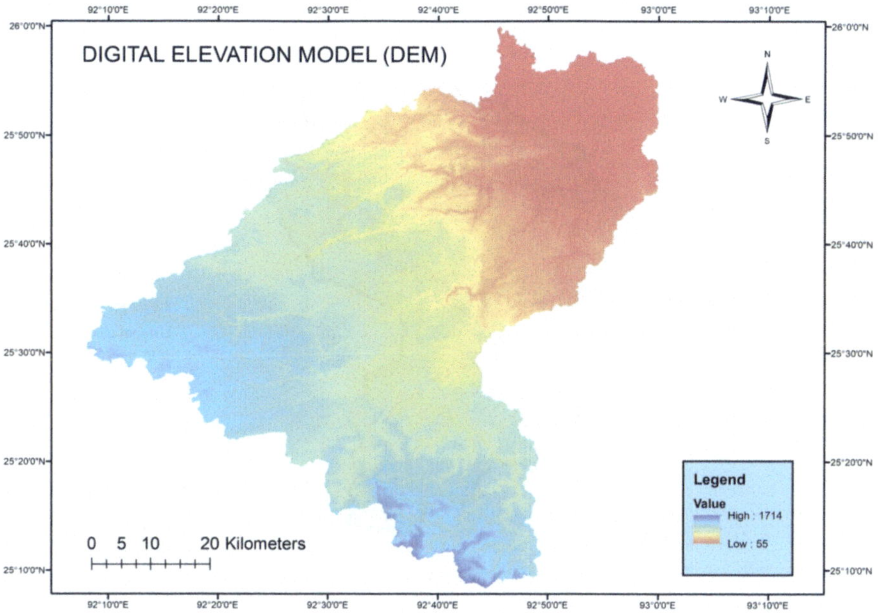

Fig. 5 DEM of Kopili River Basin

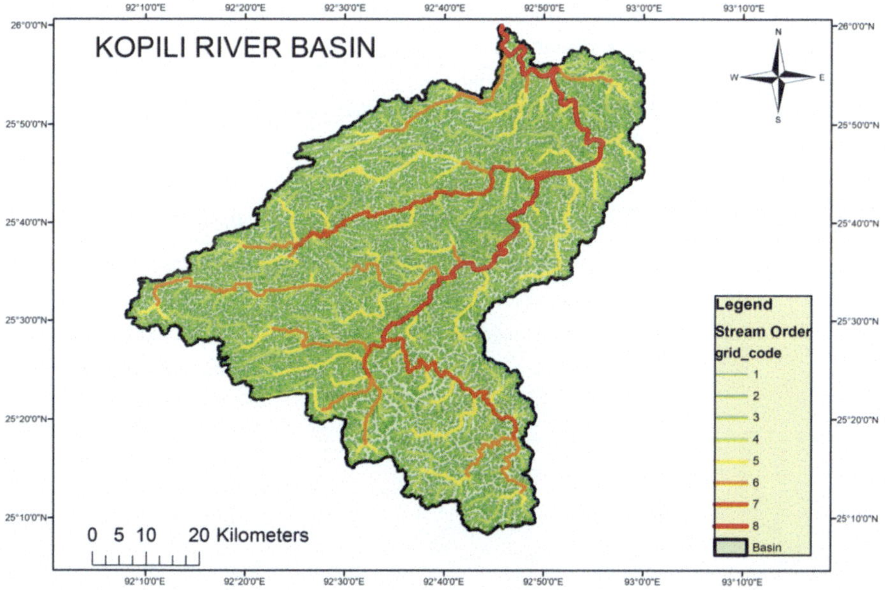

Fig. 6 Drainage map of Kopili River Basin

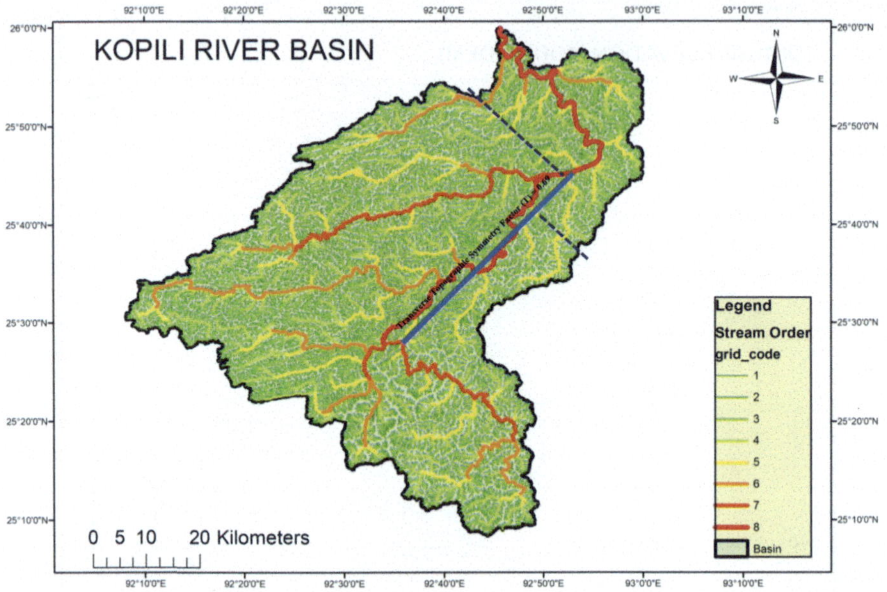

Fig. 7 Drainage map with transverse topographic symmetry factor

- Form factor: 0.56, which denotes a near-circular basin geometry, affecting flood potential and water flow dynamics.
- Circulatory ratio: 0.22, showing that the Kopili Basin is in the youth stage of drainage development, with evolving hydrological systems (Table 2).

Table 2 Aerial aspects of the basin

Morphometric parameter	Result
Basin area (A)	4868.26 (km^2)
Basin perimeter (P)	527.61 (km)
Basin length (L_b)	93.1902 (km)
Drainage density (D_d)	3.02 (km/km^2)
Stream frequency (F_s)	4.88 (km/km^2)
Drainage texture (R)	44.99 (km^{-1})
Circulatory ratio (R_c)	0.22
Form factor (F_f)	0.56

3.1.3 Relief Aspect

Relief parameters provide crucial insights into the three-dimensional morphological characteristics of the basin. These attributes are significant in understanding erosion patterns, sediment transport mechanisms, and terrain ruggedness.

- Basin relief: 1659 m, indicating substantial elevation differences and high erosional energy within the basin.
- Relief ratio: 0.017, pointing to steep slopes and structurally resistant rock formations contributing to sediment stability.
- Ruggedness number: 5.01 km^2, highlighting the basin's steep slopes and extreme surface morphology, which influence erosion rates and runoff behavior.

3.2 Morphotectonic Parameters

3.2.1 Sinuosity Index

The sinuosity index, measured at 2.07, reveals the meandering nature of the Kopili River. This high value is attributed to both tectonic activity and terrain fluctuations that have shaped the River's course over time. The meanders suggest the River's adaptation to changing geological and hydrological conditions.

3.2.2 Transverse Topographic Symmetry Factor

With a transverse topographic symmetry factor (T) value of 0.69, the basin demonstrates notable asymmetry. This asymmetry likely results from ongoing tectonic activities, which have caused lateral shifts in the basin's geomorphology.

3.2.3 Asymmetry Factor

The asymmetry factor (AF) of 23.21 indicates a pronounced tilt within the Kopili Basin. This tilt has led to the main channel's alignment toward the downstream right side of the drainage basin. Such shifts are often linked to tectonic forces or long-term geomorphic adjustments.

By integrating morphometric and morphotectonic analyses, supported by advanced GIS and remote sensing tools, the study effectively decodes the Kopili River Basin's geomorphic processes, tectonic influences, and hydrological behavior. The findings offer a well-rounded perspective on the basin's structural dynamics and contribute to informed water resource management and developmental planning.

4 Conclusions

The comprehensive analysis of the Kopili River Basin underscores the significance of employing morphometric parameters alongside advanced remote sensing and GIS techniques. These methodologies are pivotal for understanding the drainage characteristics and geomorphic processes of the region. Such insights are invaluable for addressing flood hazards, implementing effective control measures, and advancing water resource management and developmental planning.

An in-depth examination of the drainage basin reveals that the Kopili River Basin exhibits a stream order extending up to the eighth order. The first order streams have the highest frequency, whereas the eighth order streams are the least numerous, with the total stream count amounting to 23,742. This distribution clearly indicates a decrease in stream frequency as the stream order rises.

The assessment of stream length ratios demonstrates variations in slope and topographic conditions throughout the basin, while the bifurcation ratio indicates minimal influence from geological structures on the drainage network. The Rho coefficient further reveals the basin's low water storage capacity, highlighting a heightened risk of flooding in the region.

Evaluating aerial morphometric parameters such as drainage density, stream frequency, drainage texture, and form factor suggests that the Kopili River Basin is ideally circular, indicative of the youth stage of drainage development. It's extremely fine texture results from the region's abundant streams and predominantly impermeable subsurface material. Coupled with high relief and dense vegetation, these characteristics contribute to the basin's dynamic hydrological behavior.

The relief aspects, including basin relief and ruggedness number, reflect high flow velocity and significant runoff potential. The basin's steep slopes and extreme surface morphology further amplify these hydrological tendencies.

Morphotectonic parameters, such as the sinuosity index and transverse topographic symmetry factor, provide compelling evidence of the basin's asymmetry and the meandering nature of the Kopili River. The asymmetry factor indicates a pronounced tilt in the basin, with the main channel shifting toward the downstream right side. These findings strongly suggest tectonic activity, potentially linked to the Kopili Fault present in the study area. However, further investigations into neotectonic activity are warranted to confirm the precise causes of these tectonic influences.

In Conclusions, the Kopili River Basin emerges as a geomorphologically complex and hydrologically dynamic region shaped by intricate interactions between morphometric and morphotectonic processes. The integration of advanced analytical tools has facilitated a deeper understanding of its structural dynamics, paving the way for informed decision-making in resource management and sustainable development. Future studies focusing on neotectonic analyses will be crucial for unraveling the ongoing tectonic activity and its implications for the basin's geomorphology and hydrology.

References

Avinash, K., Deepika, B., & Jayappa, K. S. (2014). Basin geomorphology and drainage morphometry parameters used as indicators for groundwater prospect: insight from geographical information system (GIS) technique. Journal of Earth Science, 25, 1018–1032.

Bhat, F. A., Bhat, I. M., Sana, H., Iqbal, M., & Mir, A. R. (2013). Identification of geomorphic signatures of active tectonics in the West Lidder Watershed, Kashmir Himalayas: Using Remote Sensing and GIS. International Journal of Geomatics and Geosciences, 4(1), 164–176.

Bhatt, C. M., Chopra, R., & Sharma, P. K. (2007). Morphotectonic analysis in Anandpur Sahib area, Punjab (India) using remote sensing and GIS approach. Journal of the Indian Society of Remote Sensing, 35, 129–139.

Doornkamp JC, Cuchlaine AMK (1971). Numerical analysis in Geomorphology—An Introduction. Edward Arnold, London, pp. 3–20.

Flint, J. J. (1974). Stream gradient as a function of order, magnitude, and discharge. Water Resources Research, 10(5), 969–973.

Horton, R. E. (1932). Drainage-basin characteristics. Transactions, American geophysical union, 13(1), 350–361.

Horton, R. E. (1945). Erosional development of streams and their drainage basins; hydrophysical approach to quantitative morphology. Geological society of America bulletin, 56(3), 275–370.

Kale, V. S., & Shejwalkar, N. (2008). Uplift along the western margin of the Deccan Basalt Province: Is there any geomorphometric evidence? Journal of Earth System Science, 117, 959–971.

Leeder, M. R., & Jackson, J. A. (1993). The interaction between normal faulting and drainage in active extensional basins, with examples from the western United States and central Greece. Basin research, 5(2), 79–102.

Leopold, L. B., & Maddock, T. (1953). The hydraulic geometry of stream channels and some physiographic implications (Vol. 252). US Government Printing Office.

Pophare, A. M., & Balpande, U. S. (2014). Morphometric analysis of Suketi River basin, Himachal Himalaya, India. Journal of earth system science, 123, 1501–1515.

Prakash, K., Mohanty, T., Singh, S., Chaubey, K., & Prakash, P. (2016). Drainage morphometry of the Dhasan River basin, Bundelkhand craton, central India using remote sensing and GIS techniques. Journal of Geomatics, 10(2), 122–132.

Prakash, K., Mohanty, T., Pati, J. K., Singh, S., & Chaubey, K. (2017). Morphotectonics of the Jamini River basin, Bundelkhand Craton, Central India; using remote sensing and GIS technique. Applied Water Science, 7, 3767–3782.

Rai, P. K., Mohan, K., Mishra, S., Ahmad, A., & Mishra, V. N. (2017). A GIS-based approach in drainage morphometric analysis of Kanhar River Basin, India. Applied Water Science, 7, 217–232.

Schumm, S. A. (1956). Evolution of drainage systems and slopes in badlands at Perth Amboy, New Jersey. Geological society of America bulletin, 67(5), 597–646.

Singh, N., Jha, M., Tignath, S., & Singh, B. N. (2020). Morphometric analysis of a badland affected portion of the Mandakini River sub-watershed, central India. Arabian Journal of Geosciences, 13, 1–14.

Strahler, A. N. (1957). Quantitative analysis of watershed geomorphology. Eos, Transactions American Geophysical Union, 38(6), 913–920.

Walia, D., & Mipun, B. S. (2010). Hydrological behavior of Umshing River, East Khasi Hills, Meghalaya. Acta Geophysica, 58, 908–921.

Yunus, A. P., Oguchi, T., & Hayakawa, Y. S. (2014). Morphometric analysis of drainage basins in the Western Arabian Peninsula using multivariate statistics. International Journal of Geosciences

Chapter 6
Climate Changes in the Himalayan Region Based on Tree Ring Proxies—A Review

Amalava Bhattacharyya, Rupesh Dhyani, Mayank Shekhar, and Jagdish Chandra Kuniyal

1 Introduction

The Himalayan climate is dynamic and ever changing as per the influence of solar and other forcings like deforestation, natural disasters, etc. Moreover, warming Himalayan Mountain system plays a great role in regulating climate dynamics not only India but as a whole of South East Asia and even entire globe. This mountain has been warming at a higher rate to global average and the rate is more in higher ranges and in the western part along the 3000 km long Himalayan Arc (Wester et al. 2019). It is evident by 1.5^0 C increases in temperature from 1982 to 2006 (Shrestha et al. 2012). Major variations in climate are strongly linked to the emergence and fall of local cultures and civilizations (Cook et al. 2010). Understanding of past climate is vital for a long climatic record from diversified regions of the globe. Such long climate reconstructions can be used to put current climate changes in context over time, assess the performance of global climate models, and separate anthropogenic imprints from natural climate phenomena (Ljungqvist et al. 2016). Dendroclimatology, a special sub-branch of science, deals with the chronological sequence of tree rings (dendrochronology) and its applications on several aspects of climatic analysis and other applications (Fritts 1976). Tree-ring chronologies in many parts of the globe have yielded hundreds to thousands of years of climate or

A. Bhattacharyya (✉) · M. Shekhar
Birbal Sahni Institute of Palaeosciences, Lucknow, India
e-mail: amalava@yahoo.com

R. Dhyani · J. C. Kuniyal
Department of Geography, Justus-Liebig-University, Gießen, Germany

J. C. Kuniyal
College of Forestry, Veer Chandra Singh Garhwali Uttarakhand University of Horticulture and Forestry, Ranichauri 249 199, Tehri Garhwal, Uttarakhand, India

N. Khare (ed.), *Changing Climate over India*,
https://doi.org/10.1007/978-981-92-1098-5_6

environmental information (Esper et al. 2002, 2018; Cook et al. 2010; Gennaretti et al. 2014; Ljungqvist et al. 2016).

The Himalayan region recorded a great promise for the development of this branch of science (Bhattacharyya et al. 1988; Bhattacharyya and Yadav 1989; Shekhar et al. 2022) because of its great forest diversity along with environment aided by the seasonality of the climate which makes suitable for the development of growth rings in a large number of trees both conifers and broad-leaved trees (Bhattacharyya et al. 1988; Bhattacharyya and Yadav 1999; Shekhar 2014) and even in shrubs. Since long back in India, growth rings are used in diverse forestry aspects, i.e., evaluation of growth rates of trees, wood productivity, quality, rotation cycles (Gamble 1902) or even in exploratory analysis of tree growth /climate relationship (Chowdhury 1939, 1940a, b). However, systematic study with accurate dating of tree rings in terms of their calendar year formation through cross dating began only in 1980s (Bhattacharyya et. al. 1988).

Most regions of the Himalaya are devoid of long-term observational climate data, and the data that is accessible is also often riddled with discrepancies and in homogeneity. The urbanization effect as well as the wind effect on precipitation observations in meteorological observations, both contributes to systematic bias. Thus, the proxy records, especially, tree rings can be used to address such discrepancies since they record a broad variety of climatic variables that can be quantified in a variety of ways and dated with annual precision and accuracy (Bhattacharyya et al. 1992; Bhattacharyya and Yadav 1999; Bhattacharyya and Shah 2009). Climatic parameters, mostly precipitation and temperature reconstructed from the central and Western Himalayan region dated back to centuries (Bhattacharyya and Yadav 1999; Singh et al. 2006; Singh et al. 2009; Yadav 2011a; Yadav 2011b). There are also studies on other parameters linked influence of temperature and precipitation. These are standardized Precipitation Evapotranspiration Index (SPEI) (Shekhar et al. 2018; Dhyani et al. 2022), Palmer Drought Severity Index (PDSI) (Yadav 2013), relative humidity (Dhyani et al. 2021b) and glacier mass balance (Shekhar et al. 2017) to past centuries. The climate of the Himalaya is unique due to the effect of seasonal wind patterns linked to south-west monsoon, north-east and western disturbances and their corresponding fluctuations in precipitation, temperature, and other. Recent observations reveal, several climate drivers of tropical and extra-tropical origin, such as the El-Nino Southern Oscillation (ENSO), North Atlantic Oscillation (NAO), Indian Ocean Dipole (IOD), Madden–Julian Oscillation (MJO), and Arctic Oscillation, influence the region's weather patterns on various spatio-temporal scales (Krishnan et al. 2019 and others). Western disturbances are becoming more variable, with a larger likelihood of snowfall in the Karakoram and Western Himalaya, changes that are anticipated to lead to increases in glacier mass in those places (Krishnan et al. 2019). The available meteorological records of the Himalayan region are still short for both spatial and temporal scale for the reliable estimation of future climate change from the diversified geographical regions of the Himalaya.

Attempts are going on tree ring proxy based climatic reconstruction from sites under a grip of particular climatic zones to understand details of long-term climatic behavior. Several evidences of climate change have been recorded in tree ring-based climate reconstructions. For example, the Indian Western Himalayan region had an unprecedented surge in spring season precipitation over the latter few decades of the twentieth century (Singh and Yadav 2005; Singh et al. 2006). Similarly, winter temperature reconstructions using *Pinus wallichiana* tree-ring data from the Kashmir Himalaya revealed a substantial warming trend that began in the late twentieth century and has continued to the present (Shah et al. 2018). Summer drought (June–September) reconstruction from high altitude region of Western Himalaya revealed recent droughts since later part of twentieth century (Dhyani et al. 2022). Spring relative humidity reconstruction from low elevation site of the similar zone also revealed decline trend of relative humidity since 1980 CE representing regional phenomena (Dhyani et al. 2021b). Furthermore, mass balance reconstruction from the Western Himalaya showed later phase of Little Ice Age was substantially briefer and weaker in the Himalaya (Shekhar et al. 2017). These reconstructions from the Western Himalaya have been found to have a strong link with ENSO characteristics. Despite this, the availability of extensive tree-ring chronologies of various conifer and broad-leaf species from the Western Himalaya suggests the possibility of building long-term climate reconstructions (Singh et al. 2004; Yadav et al. 2006; Dhyani et al. 2021a) and its possible linkage to global climate system.

This chapter provides a summary of notable research that have documented considerable information based on tree ring proxy on climatic changes of the Himalayan trees and their utility in gaining a more thorough understanding of climate and its significance in climate modeling inputs.

2 Climatic Signals in Trees of the Himalaya

The Himalaya, due to a wide range of geographical diversity exhibit distinct pattern of climate can be divided into climatic sub-zones that are closely tied to the region's topographical distribution from its foothills to higher altitudes, close to permanent ice and snow-covered peaks. These zones are characterized by diversified forest types ranging, respectively, from low to high altitude wise, subtropical, temperate, and alpine (Bhattacharyya et al. 1997). The annual cycle of temperature and precipitation differs significantly from one zone to other. Along the southern front of the range, the amount of annual rainfall increases from west to east, from western to eastern part. Climate varied from High monsoon to low monsoon (Fig. 1). The eastern part of the Himalayas is warmer than the Western Himalaya since they are located at lower latitude. Furthermore, responses of trees to climate are specific in this forest types.

In dendroclimatology, trees sensitivity to climate are generally assessed by Mean Sensitivity (MS) value of their tree ring sequence which is defined as difference of widths of consecutive rings for year t to $t + 1$(Fritts 1976). In general trees growing at the dry arid sites are more climatically sensitive than to moister sites. The monsoon arrives early in June, far eastern India, Bhutan, and Nepal, where it stays for the longest. In the Western Himalayas, the monsoon arrives later in the year, so it is significantly drier than the Eastern Himalayas. This is evident by the general

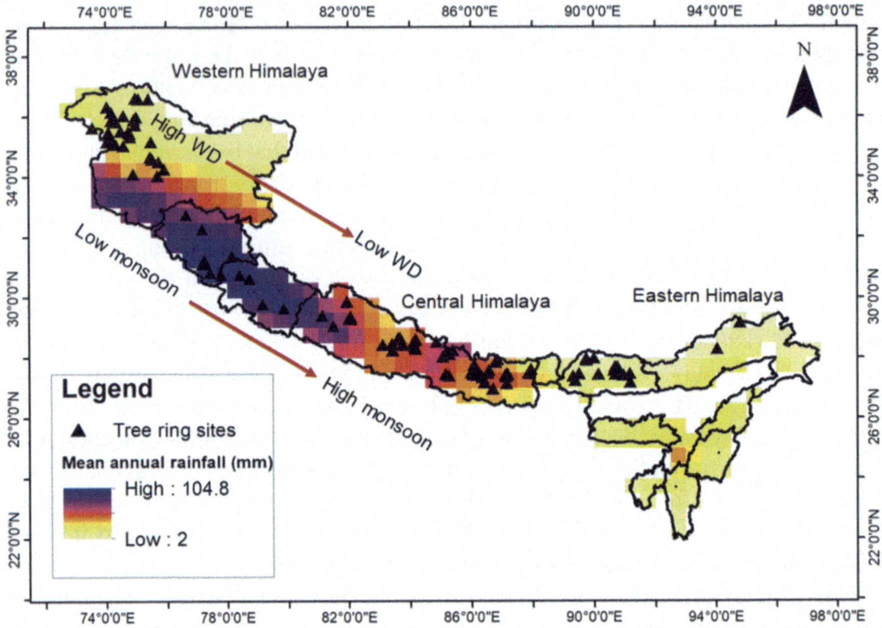

Fig. 1 Broad outline of climate and network of tree ring studies from the Himalayan region (western to eastern)

sensitivity of tree ring species decline from western to Eastern Himalaya, higher to lower longitude in the Himalayan region (Fig. 2).

The basic idea to derive climate signals from trees are based on uniformitarian principle which states that physical and biological processes that link present environmental processes to current tree growth patterns must have existed in the past (Fritts 1976). The responses of trees largely depend on climatic conditions of the site they grow, and several trees/shrubs responded to climate in similar ways across the Himalayan region, though dendroclimatic potentiality shrubs yet to be established. Trees growing at the western and central Himalayan region are most responsive to spring temperatures and precipitation that limit their tree growths (Bhattacharyya and Yadav 1999; Yadav et al. 1999; Cook et al. 2003; Singh et al. 2006; Singh et al. 2009; Yadav 2009; Yadav 2011a; Yadav 2011b; Thapa et al. 2015; Panthi et al. 2017). On the other hand, in few sites tree growth are recoded responsive to the summer (Yadav et al. 2011; Sano et al. 2012, 2017; Dhyani et al. 2022) and winter (Shekhar et al. 2018; Shah et al. 2019) season temperature. In the monsoon dominated Eastern Himalaya region the number of studies are few and the role of late summer temperature is more significant (Chaudhary and Bhattacharyya 2000; Yadava et al. 2015; Borgaonkar et al. 2018) rather than precipitation in limiting tree growth. In some sites, tree-ring data are not much sensitive to climatic changes but indirectly linked with the combined influence of precipitation and temperatures are there as in the reconstruction of River discharge (Shah et al. 2014; Shekhar and Bhattacharyya

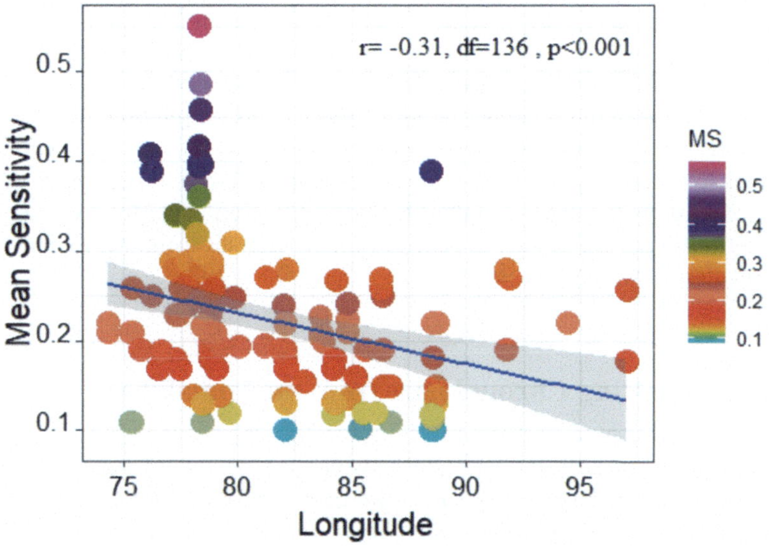

Fig. 2 Mean sensitivity values of tree ring species from the Western Himalaya to the Eastern Himalaya

2015) or good indicator of glacial mass balance records (Bhattacharyya et al. 2006; Shekhar et al. 2017). The dramatic retreat of the Gangotri glacier, Western Himalaya, has coincided with enhanced tree growth of Himalayan birch in recent years (Bhattacharyya et al. 2006).In general, trees are mostly controlled by tem perature at higher elevations and precipitation at lower elevations. There instances at many sites precipitation signals alone are not significant butPDSI or SPEI combined index of temperature and precipitation are found suitable for the reconstruction of droughts (Dhyani et al. 2022). Other tree ring parameters, such as tree-ring δ18O records, were also consistently found to contain a robust summer moisture signal from the western and central Himalayan region (Sano et al. 2012, 2017; Singh et al. 2019).

3 Tree Ring Parameters

The dating of tree rings and the analysis of the ring characteristics (width, density, and isotopic composition) are common practices for the dendroclimatic analysis. Although samples are generally taken using increment borers, or sometimes cross-sections if available from fallen or logged trees. The synthesis of studies revealed that > 90% studies in the Himalayan region used the ring width parameter. Tree-ring width is a simple measure of the radial increment of a tree, and this parameter has contributed to the largest proportion of dendroclimatological research in the

Himalayan region, and more generally, worldwide. In addition, intra-annual analyses of the early- or late-wood width have been conducted (Huges and Davis 1986; Huges 1992; Borgaonkar et al. 2011, 2018). The analyses of stable isotope ratios of hydrogen, oxygen, and carbon in annual rings of *Abies pindrow* from Western Himalaya (Ramesh et al. 1985, 1986) have demonstrated potential applicability in developing long-term records of climatic and drought variables affecting tree growth. These studies show that the wood density and isotope variables along with conventional ring width parameter should provide reliable climatic estimates to understand long-term climate variability.

4 Long Tree Ring Chronologies from the Himalaya

The length of tree-ring chronologies depends on the longevity of trees and samples taken from isolated and undisturbed areas. Further extension of these chronologies is possible with cross dating with the availability of samples from dead woods or timbers used in the construction of houses. In the Himalayan region old living trees that provide long tree ring sequence are found in trees growing both stressed and favorable environment conditions. In stress condition, old trees are confined in rocky steep slopes with thin soil cover. These trees are usually characterized by stripped bark, crown die back, and heavy upper branches. On the contrary, at favorable conditions, trees attain large size and huge girth but such trees are rarely available because of large scale logging (Bhattacharyya et al. 2023). Until now, the wood used in constructing tree-ring chronologies has mostly come from living trees in climate-stressed areas (Bhattacharyya et al. 1992). Tree species in the Western Himalaya are recorded long lifespan, lasting over a thousand years (Singh and Yadav 2005; Singh et al. 2006). More than two millennia old *Juniperus polycarpos* trees (AD 940–2006) have been grown in the Western Himalayan region (Yadav et al. 2006). *Cedrus deodara* and *Pinus gerardiana*, which thrive in moisture-stressed areas are also reached millennia in age (Yadav et al. 2006; Singh et al. 2009). However, tree species preferring relatively wetter sites such as *Abiespindrow*, *Abies. spectabilis*, *Pinus wallichiana,* and *Picea smithiana* are usually younger, occasionally extending as long as 500–600 years (Fig. 3).

From the Central Himalayan region, the longest tree-ring width record, which dates back to AD 856, comes from *Tsuga dumosa* growing in a temperate forest (Cook et al. 2003). Such long-term tree ring records should be utilized in climate models because they provide a high-resolution annual time series and provide more reliable data by eliminating natural disruptions. Most of the long records of tree ring chronologies of the Himalayan region are from conifer trees. Recently, the longest record of broadleaf tree *Betula utilis* spanning AD 1331–2017 (Dhyani et al. 2021a) has been developed from subalpine zone of Western Himalayan region (Fig. 4) highlighted tree growth suppression and release during three major phases of the recent era, viz. Medieval Warm Period (MWP), Little Ice Age (LIA), and warming during the last several decades.

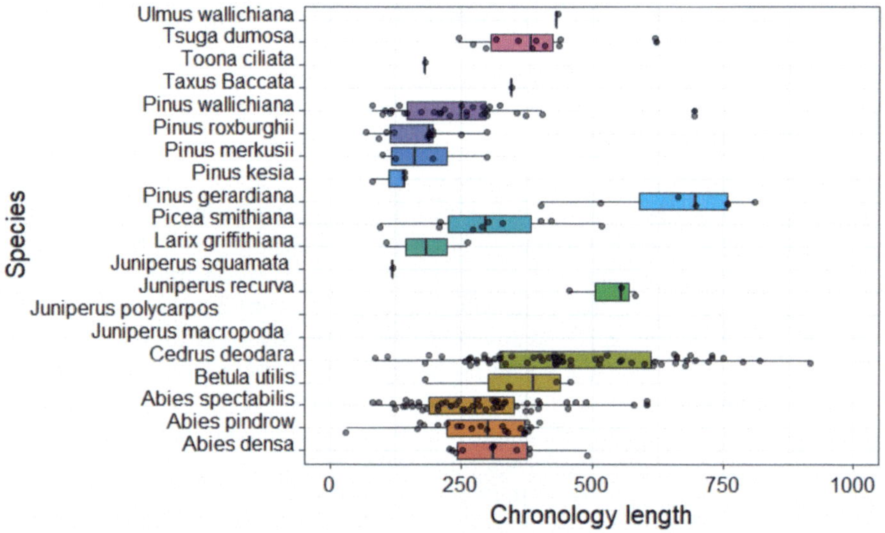

Fig. 3 Showing chronology length of tree ring species across the Himalayan region (Chronology of Junipers > 1500 years is not shown in the diagram)

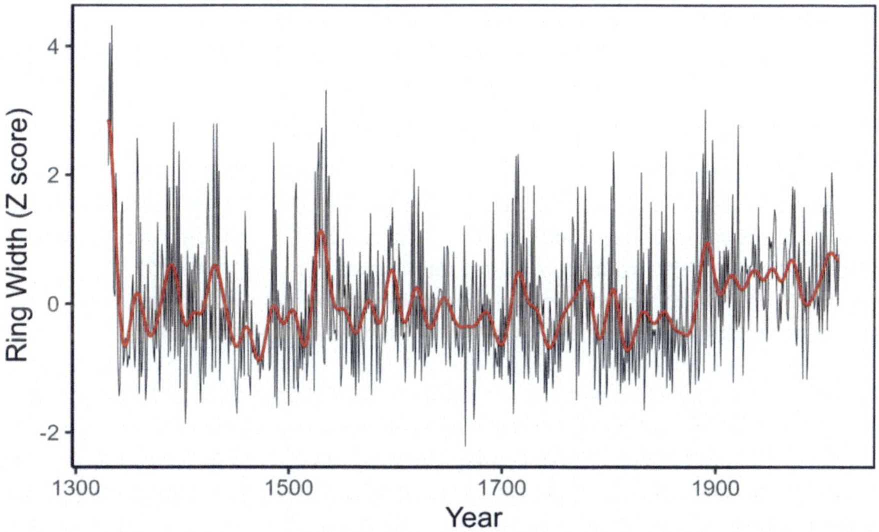

Fig. 4 Tree ring width chronology of the oldest birch tree spanning AD 1331–2017 (smoothed by 30 year low pass filter) from the subalpine zone of the Western Himalaya (modified from Dhyani et al. (2021a, b))

5 Long-Term Climate Variability Extracted from Tree Rings

5.1 Temperature Records

There are several studies, mostly on the reconstruction mean values of temperature. Initial work on based on low sample replication. (Yadav et al. 1997, 1999). Subsequently, with more replication of samples, temperatures reconstructed for multi-months or seasons using tree ring width parameters. Spring temperature reconstructions using *Cedrus deodara* covering time from AD 1600 to 1985 revealed a long-term cooling trend that began in the late seventeenth century and lasted in the early twentieth century, with the warmest period 1945–1974 (Yadav and Singh 2002). February–June mean temperature reconstruction utilizing ring-width of same species of *Cedrus deodara* from Tons valley, Western Himalaya, India revealed the coolest period during AD 1911–1940 and the warmest period during AD 1941–1970 (Singh and Yadav 2014) in the twentieth century. During LIA temperature was much cooler in the Northern hemisphere. But it was not much conspicuous in the Himalayan region. The LIA cooling in the Western Himalaya is represented by the cool events also reported in spring temperature reconstruction spanning AD 1226–2000 using tree ring *Cedrus deodara* around 1573–1622, 1731–1780, and 1817–1846 (Yadav et al. 2004). Tree ring width data of *Betula utilis* from high-elevation cold arid region of Lahaul-Spiti, Himachal Pradesh reported potential for summer (June–July) temperature reconstruction. The reconstructed summer temperature extending back to AD 1752 (263 years) showed annual to decadal variability.

From the central Himalaya, a temperature reconstruction from February to June indicated an exceptionally cooling trend from 1815 to 1822 CE, which corresponds to the eruption of Tambora in Indonesia (Cook et al. 2003). Another reconstruction of spring temperature from the Central Himalayan region (1600–2017 CE) using the species *Tsuga dumosa* shows that the spring season has not shown any warming trend over the last four centuries (Aryal et al. 2020). Spring temperature reconstructions from AD 1591 to 2012 using *Picea smithiana* from the far Western Central Himalaya did not reveal a substantial cooling pattern during the Little Ice Age, but there were a few cold years (Thapa et al. 2015). A reconstruction of March–September precipitation from 1752 to 2000 CE using tree-ring data from *Abies spectabilis* from Nepal's timberline zone shows a warming trend from the 1750s until around 1790, then cooling until 1810, then a gradual warming trend until 1950, and a notable cold period lasting until the present (Sano et al. 2005). However, these reconstructions did not show much coherency and did not show a warming trend in the twentieth century due to low frequency variations associated with any recent warming. This inconsistency could be explained by variance caused by various climatic and disturbance factors.

Tree-ring data from some sites are also found suitable for the winter temperature reconstruction. Tree rings of *Pinus wallichiana* spanning 1840–2014 CE from North-west Himalaya in the winter temperature reconstruction using exhibited remarkable warm period during 1982–2012 CE the latter part of twentieth century to till now (Shah et al. 2019). Similarly, a winter season (previous October–current February) minimum temperature reconstruction using *Abies spectabilis* tree rings for Central Himalaya covering the period 1780–2008 CE revealed that minimum temperature of winter has been steadily increasing since the early twentieth century, with unprecedented rapid warming in the latter half of the century, and the cold episodes in the reconstruction coincided with major volcanic eruptions in the Northern Hemisphere and tropical regions (Gaire et al. 2020).

Trees in the North-East Himalaya, in general, remain climatically less sensitive in dendroclimatic view. Chaudhary et al. (1999), evaluated the dendroclimatic potentiality of seven conifer species, viz., *Abies densa, Juniperus indica, Larix griffithiana, Pinus roxburghii, P. wallichiana, Taxus baccata,* and *Tsuga dumosa growing* in diverse ecological sites of Eastern Himalaya and reported all these taxa are datable and suitable for chronologies building. A sub-alpine deciduous conifer *Larix griffthiana*, growing in Sange, Arunachal Pradesh, has been recorded suitable for May-temperature reconstruction (Chaudhary and Bhattacharyya 2000).

Pinus kesiya growing on and around the Shillong plateau, Meghalaya has also been analyzed to evaluate its suitability for tree-ring analysis. Five well-replicated, 80 to 142 years long site chronologies were developed for *Pinus kesiya* and correlated with climate records of Shillong plateau but there was no common response of the growth to climate was found. Few temperature reconstruction has also been possible from this part of the Himalaya and mostly are found suitable for the late summer temperature (Bhattacharyya and Chaudhary 2003; Yadava et al. 2015; Borgaonkar et al. 2018). Tree-ring chronology of *Abies densa* growing near the tree-line in the Eastern Himalayan region from 1757 to 1994 reveals yearly to multi-year oscillations, with the warmest and coldest 10-year intervals occurring in 1978–1987 and 1801–1810, respectively (Bhattacharyya and Chaudhary 2003). Similarly, tree-ring data of *Larix griffithiana* growing in North Sikkim, Eastern Himalaya, spanning AD 1852–2005, was used for late mean summer temperature reconstruction and showed warming since the 1930s, with 1996–2005 being the driest in the preceding 150 years (Yadava et al. 2015).

In a pioneering attempt tree-ring data of the Himalayan region are also reported to be correlated with sea-surface temperature and also have tele-connections with different climate phenomena. Buckley et al. compared *Pinus merkusii* ring-width chronologies from Mishmi Hills, Arunachal Pradesh along with Thai and Bhutan pine chronologies with the global surface temperatures for the past 150 years, shifting the correlations through three seasonal averages: two seasons preceding the monsoon (Dec–Feb and Mar–May), and the monsoon season itself (Jun–Sep). They showed a strong correlation of *Pinus merkusii* with tropical Indian and Pacifc Ocean bands in seasons preceding the summer monsoon.

5.2 Hydroclimatic Records

5.2.1 Precipitation Records

Majority of studies are on pre-monsoon precipitation reconstruction using tree-ring chronologies of several tree species. Both wet and dry spring years were noted during the Little Ice Age (Yadav and Park 2000; Dhyani et al. 2023; Singh and Yadav 2005; Singh et al. 2006). In Western Himalayan region, a spring precipitation reconstruction using tree-ring data of *Cedrus deodara* spanning AD 1731–1986 showed wettest period in 1977–1986 and driest in 1932–1941 (Singh and Yadav 2005). A similar reconstruction using tree-ring data of *Cedrus deodara* caught inter decadal patterns, the increase in precipitation during the last few decades of the twentieth century is the most remarkable (Singh et al. 2006). A multi-months March–July precipitation reconstruction spanning AD 1310–2004 from Kinnaur, Himachal Pradesh captured wettest period in the twentieth century during AD 1963–1992 and the driest in the eighteenth century AD 1773–1802 (Singh et al. 2009). An extended annual precipitation reconstruction (August–July) from the Western Himalaya's monsoon shadow zone from AD 1330 to 2008 demonstrates multi-decadal droughts of high magnitude in the fourteenth and fifteenth centuries, as well as increased wet conditions in the twentieth century (Yadav 2011b). It has been recorded that the long-term wet (dry) phases in the reconstructed precipitation series are linked to El-Nino (La-Nina) events. Using tree-ring width data time span of AD 1730 to 2012 from deodar trees February–May precipitation was recorded. It showed twentieth-century decadal droughts in the 1920s and 1960s to early 1970s, as well as pluvial in the 1910s and 1980s (Yadav et al. 2014). April–June precipitation reconstruction for the Lidder Valley, Kashmir Himalaya, using tree rings of *Cedrus deodara* found a lengthy decadal dry period between 1822 and 1887 C.E., the driest era in the reconstruction (Shah et al. 2018). Rainfall reconstruction using an isotope tree-ring chronology spanning 273-year (1743–2015 CE) for June–July for the Dingad valley in Uttarakhand revealed a decreasing rainfall trend since 1743 CE, with the driest period in the late twentieth and early twenty-first centuries (Singh et al. 2019).

Reconstruction of precipitation in the non-monsoon months (previous October to current May) back to AD 1171 revealed that the wettest and the driest non-monsoon months occurred in the fourteenth and the thirteenth century, respectively. Tree ring-based precipitation in the Central Himalayan region is mostly shown common trend to that of the Western Himalayan region. A March–June precipitation reconstruction from Central Himalaya found longest wet periods occurred during 1850–1862, 1878–1886, 1909–1917, 1971–1984, and 2000–2008 while dry periods were usually shorter and occurred during 1873–1877, 1921–1923, 1925–1929, 1951–1956, 1958–1962, and 1994–1996 (Gaire et al. 2017) which are consistent with Feb–May precipitation reconstruction from the Western Himalaya (Yadav et al. 2014). A tree ring $\delta^{18}O$ based May–September precipitation reconstruction from Bhutan Himalaya spanning 1743–2011 C.E. shows normal conditions during the twentieth century, whereas records from western Nepal and the southern/eastern Tibetan Plateau show weakening trends in monsoon intensity during the last 100–200 years (Sano et al. 2013).

5.2.2 Humidity Records

Recently, a tree ring-based reconstruction of February–May relative humidity from the subtropical zone of the Western Himalaya spanning 1800–2015 C.E. revealed a decline in relative humidity over the past three decades (Fig. 6), indicating that global warming is not evident as predicted by climate models (Dhyani et al. 2021b).

5.2.3 SPEI Records

In some trees growing in sites where signal of either temperature or precipitation alone are not sufficient for their reconstruction statistically. However, SPEI index in some sites is found suitable for the reconstruction because of its muti-scalar nature which represents combined influence of both precipitation and temperature. In Western Himalayan region, using tree-ring width data from *Pinus roxburghii* (Chir pine) from subtropical belt, a winter November–March SPEI reconstruction for the time period 1767–2013 CE revealed several intervals of high and low intensity drought since the seventeenth century and its strong linkage with the Western Disturbance, indicating that the Western Disturbance may play a role in modulating droughts in this region (Shekhar et al. 2018). In an another study February–May SPEI reconstruction covering AD 1773–2016 using *Cedrus deodara* revealing multi-year droughts occurred in 1818–1822, 1798–1802, 1813–1817, 1793–1797, and 1958–1962 and pluvial in 1783–1787, 1838–1842, 1788–1792, 1933–1937, and 1808–1812, respectively, in which some severe droughts were also associated with famines and crop failures in the Himalayan region (Yadava et al. 2021). Later an extensive study based on network of ring-width chronologies deodar from seven moisture stressed high-elevation sites in Kishtwar, Jammu and Kashmir were used to develop the annual Standardized Precipitation Index (SPI from previous year June to current year May (SPI12-May)) extending back to 1613 CE. The SPI reconstruction revealed more frequent and of larger magnitude during 1760s–2017 CE in comparison with the earlier part of the reconstruction (1613–1750s CE). In general, it shows strong consistency with other tree-ring-based hydrological records of the Western Himalaya and strong linkage of the SPI with Tropical Pacific Ocean surface temperature. A long tree ring-based summer (June–September) SPEI record spanning 1696–2017 CE from Darma Valley (Fig. 5), Western Himalaya captured two broad phases on the basis of trend in SPEI time series, wet phase around 1696–1903 CE with punctuated, relatively intensified sub-phase around 1806–1815 CE.; and dryer phase 1927–2017 C with more severity around 1990–2017 CE (Dhyani et al. 2022). This study also showed a notable drying trend since 1990 indicated warming induced drought stress for the region (Fig. 5).

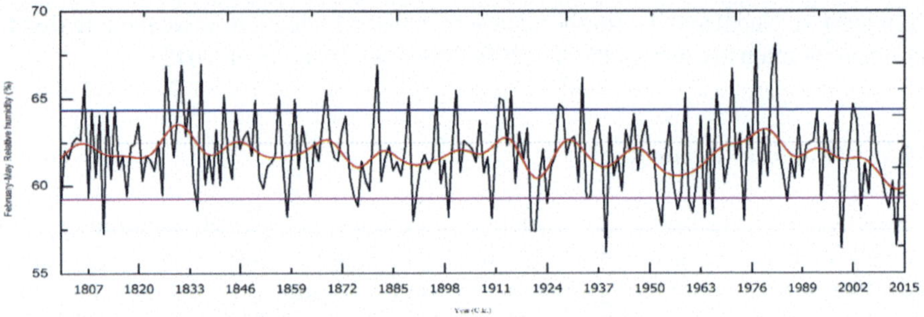

Fig. 5 Reconstruction of the February–May Relative Humidity from 1800 to 2015 CE (thin curve) (Dhyani et al. 2021b). The bold red curve represents the 10 yr FFT (Fast Fourier Transform) smoothing curve of the reconstruction and blue upper and pink lower horizontal lines, respectively, represent as high RH years (mean + SD), and low RH years (mean –SD). Reproduced with permission from Elsevier

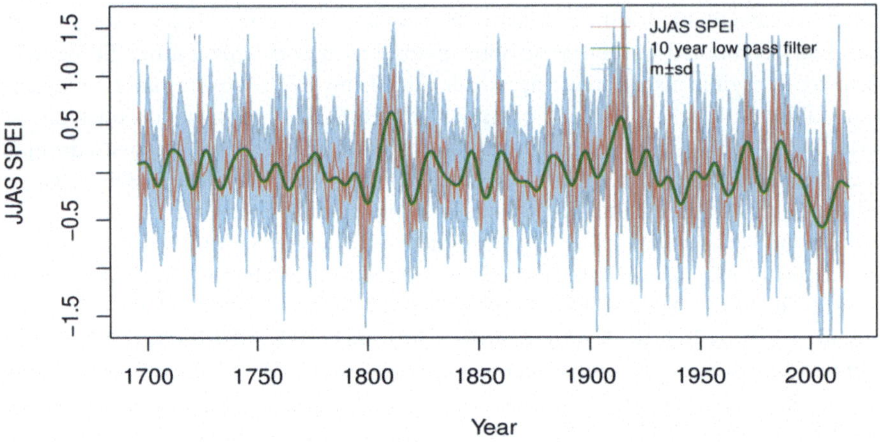

Fig. 6 Reconstructed JJAS SPEI from Darma Valley, Western Himalaya during the period 1696–2017 CE (Dhyani et al. 2022). (*light pink curve*). The *bold* green *curve* represents the 10 year low pass filter of the reconstruction and the shaded sky blue density plot represents the (mean ± SD). Reproduced with permission from Elsevier

In Central Himalaya, a March–May reconstruction from AD 1707 to 2013 using tree rings of *Tsuga dumosa* documented a protracted dry period from 1921 to 1960, while the longest wet period occurred from 1975 to 1993 (Bhandari et al. 2019).

5.2.4 PDSI Records

Another parameter, Palmer Drought Severity Index (PDSI) is used readily available temperature and precipitation data to estimate relative dryness. Using ring width chronologies of *Cedrus deodara* and *Pinus gerardiana* together from a network of moisture-stressed sites, a October–May PDSI 1310–2005 AD was reconstructed which captured dominant decadal-scale pluvial phases (1981–1995, 1952–1968, and 1918–1934) as compared to the severe droughts in the early seventeenth century (1617–1640) as well as late fifteenth to early sixteenth (1491–1526) centuries and these drought anomalies are positively (negatively) associated with central Pacific (Indo-Pacific Warm Pool) sea surface temperature anomalies (Lal et al. 2025; Yadav 2013).

In central Himalaya, using a tree-ring $\delta^{18}O$ chronology of *Abies spectabilis* from the Nepal Himalaya summer monsoon season PDSI spanning AD 1778–2000 was reconstructed which demonstrates weakening trend of monsoon severity over the past two centuries (Sano et al. 2012). A similar summer monsoon season PDSI reconstruction from the Indian Western Himalaya, spanning AD 1767–2008, revealed similar patterns related to moisture supply from the Bay of Bengal (Sano et al. 2017). The Central Himalaya experienced prolonged drought during the nineteenth and twentieth centuries, as evidenced by a February–August PDSI reconstruction using tree rings of *Pinus wallichiana* for the trans-Himalaya region, which captured historic mega-drought events in Asia such as the Strange Parallels drought (1756–1768), the East India drought (1790–1796), and the late Victorian Great Drought (1876–1878) (Gaire et al. 2019).

A similar spring (March–May) PDSI reconstruction from the Central Himalaya using tree rings of *Picea smithiana* from AD 1725 to 2013 reveals strong inter-annual to multi-decadal drought variations, with the most prominent spring wet episode occurring in 1883–1891, which coincided with high precipitation periods in the central (Gaire et al. 2017) and Western Himalayas (Yadav et al. 2014). The reconstructed spring PDSI variations also revealed a continuously shift toward drier conditions after the 1980s, a tendency also noted in several other tree-ring studies from nearby regions (Sano et al. 2012; Yadav et al. 2014; Gaire et al. 2017). Moreover, Lal et al. 2025) highlight the importance of long-term drought assessment for sustainable water management in the Western Himalaya. Their reconstruction identifies major droughts (1794–1825, 1888–1980, 2000–2010) and wetter phases (1825–1875, 1982–1998), with increasing drought frequency in recent decades. Spectral analysis reveals periodicities at ~ 80, 33, 6, 2.3, and 2 years, linked to ENSO, PDO, and AMO teleconnections. The study underscores rising pre-monsoon drought risk and the need for proactive adaptation strategies across the region.

5.2.5 Stream Flow Records

Climate projections show that potential change in precipitation and temperature brought about by climate change could affect runoff. A 711 years AD 1295–2005 long previous year December to current year July Satluj discharge reconstruction from Western Himalaya revealed low and high River discharge in the eighteenth (1771–1820) and nineteenth century (1819–1868), respectively, followed by the decreasing tendency in River discharge since the 1990s which is consistent with the decreasing trend in winter precipitation in the region (Singh and Yadav 2013).

A March–April stream flow was reconstructed using tree-ring data from *Cedrus deodara* in the Kullu valley, Himachal Pradesh, in the Western Himalaya, covering the years AD 1834–1984, with the highest streamflow during AD 1840–1843, 1893–1897, 1922–1926, 1935–1940, and 1942–1945, and the lowest during AD 1847–1851, 1863–66, 1927–1929, 1960–1962, and 1972–1974 (Shah et al. 2013). A tree ring-based stream flow reconstruction *Larix griffithiana* from Lachen, North Sikkim, Eastern Himalaya captured monsoon failure (or East India Drought) of AD 1792–1796 and past great droughts of AD 1876–1878 as well as the Tambora volcanic eruption of Indonesia in AD 1815 in the Eastern Himalayan region (Shah et al. 2014). A tree ring reconstruction of *Abies densa* of the Zema Sikkim Himalaya from AD 1775 to 2007 revealed a decreasing trend in River flow since the 1990s, which is consistent with previous discharge data and has major consequences for the region's socio-economics (Shekhar and Bhattacharyya 2015). Further, a seven-century (1309–2004 CE) tree-ring–based reconstruction of monsoon-season Brahmaputra discharge shows that the early instrumental period (1956–1986 CE) was among the driest intervals of the past 700 years. This study demonstrates that flood hazard, inferred from the frequency of high-discharge years, is underestimated by about 24–38% when relying only on recent observations. Although models project stronger monsoon rainfall and higher flood risk under warming, the short instrumental record does not capture the full range of natural variability (Rao et al. 2020).

6 Climate Change and Glacial History

A comprehensive investigation is currently underway to evaluate the influence of regional climatic variability and glacier mass balance on the growth patterns of *Betula* (birch) across its distribution in the Himalaya and the Tibetan Plateau. This work aims to address existing spatial and temporal gaps in glacier-related chronologies, thereby improving our understanding of the relationship between climate variability and glacier dynamics and their implications for water resources and other societal concerns. Birch is particularly sensitive to environmental changes because its growth is strongly affected by cold air masses descending from nearby glaciers. Preliminary examination of the tree-ring chronology and its relationship with temperature records indicates that this species preserves clear signals of major climatic phases over the past millennium, including the warming associated with the Medieval Warm Period (MWP), the cooling during the Little Ice Age (LIA), and the recent warming phase (Dhyani et al. 2021a; Fig. 4). The chronology further suggests that the LIA was not a uniformly cold interval but was punctuated by short episodes of relatively warmer conditions, reflected by brief periods of enhanced growth. The higher frequency of such growth surges during the seventeenth century indicates that the later stage of the LIA in the Himalayan region was comparatively weaker, consistent with earlier observations (Shekhar et al. 2017); Moreover, tree-ring $\delta^{18}O$ serves as a sensitive indicator of regional climate, while $\delta^{13}C$ captures local ecohydrology. Using ice-core and tree-ring $\delta^{18}O$ data from the central Himalaya (CH). The results indicate a growing east–west contrast in regional climate conditions since the 1960s. In addition, $\delta^{13}C$ records obtained from tree rings in transitional Western Himalayan valleys provide the basis for reconstructing glacier mass balance (GMB) variability over approximately the past three centuries (Singh et al. 2021). The annually resolved GMB (1743 CE–present) reveals three distinct phases: positive mass balance until the mid-nineteenth century, a stable but slightly negative phase between 1870 and 1960, and rapid ice loss thereafter. The recent acceleration in glacier mass balance variability is largely linked to ongoing anthropogenic warming, along with a reduction in the strength of the mid-latitude westerlies and the Arabian Sea branch of the Indian summer monsoon. Isotopic–climate relationships also suggest a progressive eastward decrease in the influence of the westerlies across the region. These findings collectively provide an important long-term perspective for interpreting recent glacier mass balance changes (Singh et al. 2021).

7 Future Outlook

An expanding network of tree-ring records from the Himalayan region particularly from upper temperate and subalpine zones has significantly enhanced the spatial coverage and site density available for dendroclimatic studies. This broader dataset now enables more robust assessments of past climate variability and its linkages with major circulation systems such as ENSO, the East Asian monsoon, and the Atlantic Multidecadal Oscillation. Despite these advances, a major methodological challenge remains: many Himalayan species encode complex and not yet fully resolved climatic or environmental signals. Moreover, most existing reconstructions still rely on single-site datasets and use ring width as the primary proxy. Looking forward, greater emphasis should be placed on combining multiple tree-ring parameters such as stable isotopes, wood anatomical features, and other high-resolution metrics to explore their usefulness for reconstructing seasonal climate patterns. Improvements in chronology development are also becoming feasible through larger sample depths and more refined detrending approaches. At the same time, inconsistencies observed in carbon-isotope-based climate signals warrant further investigation. Finally, integrating tree-ring–derived climate information into vegetation and climate models will enhance future risk projections and strengthen their societal relevance.

Acknowledgements We acknowledge the Director, Birbal Sahni Institute of Palaeosciences, Lucknow for providing facilities and encouraging inter-institutional R&D collaborations. M.S., A.B., and R.D. acknowledge the National Mission on Himalayan Studies (NMHS), Ministry of Environment, Forest and Climate Change, Government of India, under the grant NMHS 2024–25/SC-XIII/MG/SL-16. J.D.C. and R.D. also express their sincere thanks to the Department of Science and Technology (DST), Government of India, for financial support provided under Task Force 3, Second Phase of the National Mission for Sustaining the Himalayan Ecosystem (NMSHE), titled "Forest Resources and Plant Biodiversity."

References

Bhandari S, Gaire NP, Shah SK, Speer JH, Bhuju DR, Thapa UK (2019) A 307-Year Tree-Ring SPEI Reconstruction Indicates Modern Drought in Western Nepal Himalayas. Tree-Ring Research 75:73–85. https://doi.org/10.3959/1536-1098-75.2.73

Bhattacharyya A, Chaudhary V (2003) Late-Summer Temperature Reconstruction of the Eastern Himalayan Region Based on Tree-Ring Data of Abiesdensa. Arctic, Antarctic, and Alpine Research 35:196–202. https://doi.org/10.1657/1523-0430(2003)035[0196:LTROTE]2.0.CO;2

Bhattacharyya A, Shah SK (2009) Tree-ring studies in India past appraisal, present status and future prospects. IAWA journal 30:361–370

Bhattacharyya A, Yadav RR (1999) Climatic reconstructions using tree-ring data from tropical and temperate regions of India-a review. IAWA journal 20:311–316

Bhattacharyya A, Shah SK, Chaudhary V (2006) Would tree ring data of Betulautilis be potential for the analysis of Himalayan glacial fluctuations? Current Science 91:754–761

Bhattacharyya, A., Dhyani, R., Joshi, R., Shekhar, M., Kuniyal, J.C., Ranhotra, P.S., & Singh, S.P. (2023). Is survival of Himalayan Cedar (*Cedrus deodara*) threatened? An evaluation based on predicted scenarios of its growth trend under future climate change. Science of The Total Environment, 882, 163630.

Bhattacharyya A, LaMarche Jr VC, Hughes MK (1992) Tree-ring chronologies from Nepal

Borgaonkar HP, Sikder AB, Ram S (2011) High altitude forest sensitivity to the recent warming: A tree-ring analysis of conifers from Western Himalaya, India. Quaternary International 236:158–166. https://doi.org/10.1016/j.quaint.2010.01.016

Borgaonkar HP, Gandhi N, Ram S, Krishnan R (2018) Tree-ring reconstruction of late summer temperatures in northern Sikkim (eastern Himalayas). Palaeogeography, Palaeoclimatology, Palaeoecology 504:125–135

Chaudhary V, Bhattacharyya A (2000) Tree ring analysis of Larixgriffithiana from the Eastern Himalayas in the reconstruction of past temperature. Current Science. 79:1712–1715

Cook ER, Krusic PJ, Jones PD (2003) Dendroclimatic signals in long tree-ring chronologies from the Himalayas of Nepal. International Journal of Climatology 23:707–732. https://doi.org/10.1002/joc.911

Cook ER, Anchukaitis KJ, Buckley BM, D'Arrigo RD, Jacoby GC, Wright WE (2010) Asian monsoon failure and megadrought during the last millennium. Science 328:486–489

Dhyani, R., Bhattacharyya, A., Joshi, R., Shekhar, M., Kuniyal, JC., &Ranhotra, PS. (2023). Tree rings of *Rhododendron arboreum* portray signal of monsoon precipitation in the Himalayan region. Frontiers in Forests and Global Change, 5, 1044182.

Dhyani R, Bhattacharyya A, Rawal RS (2021a) The oldest Birch tree in the Western Himalaya depicting signature of major climatic events of the recent era in its tree rings. Current Science 121:350–353. https://www.currentscience.ac.in/Volumes/121/03/0350.pdf

Dhyani R, Shekhar M, Joshi R, Bhattacharyya A, Ranhotra PS, Pal AK, Thakur S, Nandi SK (2021b) Reconstruction of pre-monsoon relative humidity since 1800 C.E. based on tree-ring data of PinusroxburghiiSarg. (chir–pine) from Pithoragarh, Western Himalaya. Quaternary International. https://doi.org/10.1016/j.quaint.2021.04.026

Dhyani R, Bhattacharyya A, Singh Rawal R, Joshi R, Shekhar M, Singh Ranhotra P (2022) Is tree ring chronology of blue pine (*Pinuswallichiana A. B. Jackson*) prospective for summer drought reconstruction in the Western Himalaya? Journal of Asian Earth Sciences 229:105142. https://doi.org/10.1016/j.jseaes.2022.105142

Esper J, Schweingruber FH, Winiger M (2002) 1300 years of climatic history for Western Central Asia inferred from tree-rings. The Holocene 12:267–277

Esper J, George SSt, Anchukaitis K, D'Arrigo R, Ljungqvist FC, Luterbacher J, Schneider L, Stoffel M, Wilson R, Büntgen U (2018) Large-scale, millennial-length temperature reconstructions from tree-rings. Dendrochronologia 50:81–90. https://doi.org/10.1016/j.dendro.2018.06.001

Gaire NP, Bhuju DR, Koirala M, Shah SK, Carrer M, Timilsena R (2017) Tree-ring based spring precipitation reconstruction in western Nepal Himalaya since AD 1840. Dendrochronologia 42:21–30

Gaire NP, Dhakal YR, Shah SK, Fan Z-X, Bräuning A, Thapa UK, Bhandari S, Aryal S, Bhuju DR (2019) Drought (scPDSI) reconstruction of trans-Himalayan region of central Himalaya using Pinuswallichiana tree-rings. Palaeogeography, Palaeoclimatology, Palaeoecology 514:251–264. https://doi.org/10.1016/j.palaeo.2018.10.026

Gaire NP, Fan Z-X, Shah SK, Thapa UK, Rokaya MB (2020) Tree-ring record of winter temperature from Humla, Karnali, in central Himalaya: a 229 years-long perspective for recent warming trend. GeografiskaAnnaler: Series A, Physical Geography 102:297–316. https://doi.org/10.1080/04353676.2020.1751446

Gennaretti F, Arseneault D, Nicault A, Perreault L, Bégin Y (2014) Volcano-induced regime shifts in millennial tree-ring chronologies from northeastern North America. PNAS 111:10077–10082. https://doi.org/10.1073/pnas.1324220111

Krishnan R, Shrestha AB, Ren G, Rajbhandari R, Saeed S, Sanjay J, Syed MdA, Vellore R, Xu Y, You Q, Ren Y (2019) Unravelling Climate Change in the Hindu Kush Himalaya: Rapid Warming in the Mountains and Increasing Extremes. In: Wester P, Mishra A, Mukherji A, Shrestha AB (eds) The Hindu Kush Himalaya Assessment: Mountains, Climate Change, Sustainability and People. Springer International Publishing, Cham, pp 57–97

Lal, D., Pandey, P., Shekhar, M., Chand, P., Dhyani, R., Ahmed, I., Kumar, R., Bhattacharyya, A. and Ranhotra, P.S., (2025). Tree-Ring-Based Drought Reconstruction Reveals Increased Pre-Monsoon Droughts Over the Past Two Centuries in the Lug Valley Kullu, Northwest Himalayas. *International Journal of Climatology*, 70104. https://doi.org/10.1002/joc.70104

Ljungqvist FC, Krusic PJ, Sundqvist HS, Zorita E, Brattström G, Frank D (2016) Northern Hemisphere hydroclimate variability over the past twelve centuries. Nature 532:94–98

Panthi S, Bräuning A, Zhou Z-K, Fan Z-X (2017) Tree rings reveal recent intensified spring drought in the central Himalaya, Nepal. Global and Planetary Change 157:26–34. https://doi.org/10.1016/j.gloplacha.2017.08.012

Sano M, Furuta F, Kobayashi O, Sweda T (2005) Temperature variations since the mid-18th century for western Nepal, as reconstructed from tree-ring width and density of Abiesspectabilis. Dendrochronologia 23:83–92. https://doi.org/10.1016/j.dendro.2005.08.003

Sano M, Ramesh R, Sheshshayee M, Sukumar R (2012) Increasing aridity over the past 223 years in the Nepal Himalaya inferred from a tree-ring δ18O chronology. The Holocene 22:809–817. https://doi.org/10.1177/0959683611430338

Sano M, Tshering P, Komori J, Fujita K, Xu C, Nakatsuka T (2013) May–September precipitation in the Bhutan Himalaya since 1743 as reconstructed from tree ring cellulose δ18O. Journal of Geophysical Research: Atmospheres 118:8399–8410. https://doi.org/10.1002/jgrd.50664

Sano M, Dimri AP, Ramesh R, Xu C, Li Z, Nakatsuka T (2017) Moisture source signals preserved in a 242-year tree-ring δ18O chronology in the western Himalaya. Global and Planetary Change 157:73–82. https://doi.org/10.1016/j.gloplacha.2017.08.009

Shah SK, Bhattacharyya A, Shekhar M (2013) Reconstructing discharge of Beas River basin, Kullu valley, western Himalaya, based on tree-ring data. Quaternary International 286:138–147. https://doi.org/10.1016/j.quaint.2012.09.029

Shah SK, Bhattacharyya A, Chaudhary V (2014) Streamflow reconstruction of Eastern Himalaya River, Lachen 'Chhu', North Sikkim, based on tree-ring data of Larixgriffithiana from Zemu Glacier basin. Dendrochronologia 32:97–106

Shah SK, Pandey U, Mehrotra N (2018) Precipitation reconstruction for the Lidder Valley, Kashmir Himalaya using tree-rings of Cedrusdeodara. International Journal of Climatology 38:e758–e773. https://doi.org/10.1002/joc.5405

Shah SK, Pandey U, Mehrotra N, Wiles GC, Chandra R (2019) A winter temperature reconstruction for the Lidder Valley, Kashmir, Northwest Himalaya based on tree-rings of Pinuswallichiana. Climate Dynamics 53:4059–4075

Shekhar M, Bhattacharyya A (2015) Reconstruction of January–April discharge of ZemuChuu–A first stage of Teesta River, North Sikkim, Eastern Himalaya, based on tree-ring data of fir. Journal of Hydrology: Regional Studies 4:776–786

Shekhar M, Bhardwaj A, Singh S, Ranhotra PS, Bhattacharyya A, Pal AK, Roy I, Martín-Torres FJ, Zorzano M-P (2017) Himalayan glaciers experienced significant mass loss during later phases of little ice age. Sci Rep 7:10305. https://doi.org/10.1038/s41598-017-09212-2

Shekhar M, Pal AK, Bhattacharyya A, Ranhotra PS, Roy I (2018) Tree-ring based reconstruction of winter drought since 1767 CE from Uttarkashi, Western Himalaya. Quaternary International 479:58 69. https://doi.org/10.1016/j.quaint.2017.08.029

Shekhar M., Ranhotra P.S., Bhattacharyya A., Singh A., Dhyani R., Singh S. (2022) Tree-Ring-Based Hydrological Records Reconstructions of the Himalayan Rivers: Challenges and Opportunities. In: Rani S., Kumar R. (eds) *Climate Change. Springer Climate. Springer, Cham.* https://doi.org/10.1007/978-3-030-92782-0_3

Shekhar, M. (2014). Application of Multi-proxy tree-ring parameters in the reconstruction of climatevis-à-vis glacial fluctuation from the Eastern Himalaya (Doctoral dissertation, Ph. D thesis. Department of Botnay, University of Lucknow, Lucknow

Shrestha, U.B., Gautam, S., Bawa, K.S., 2012. Widespread climate change in the Himalayas and associated changes in local ecosystems. PloS One 7, 5

Singh J, Yadav RR (2013) Tree-ring-based seven century long flow records of Satluj River, western Himalaya, India. Quaternary International 304:156–162. https://doi.org/10.1016/j.quaint.2013.03.024

Singh J, Yadav RR (2014) February–June temperature variability in western Himalaya, India, since AD 1455. Quaternary International 349:98–104. https://doi.org/10.1016/j.quaint.2014.06.066

Singh J, Park W-K, Yadav RR (2006) Tree-ring-based hydrological records for western Himalaya, India, since AD 1560. Climate Dynamics 26:295–303

Singh J, Yadav RR, Wilmking M (2009a) A 694-year tree-ring based rainfall reconstruction from Himachal Pradesh, India. Climate Dynamics 33:1149–1158

Singh J, Singh N, Chauhan P, Yadav RR, Bräuning A, Mayr C, Rastogi T (2019) Tree-ring δ18O records of abating June–July monsoon rainfall over the Himalayan region in the last 273 years. Quaternary International 532:48–56. https://doi.org/10.1016/j.quaint.2019.09.030

Singh, N., Shekhar, M., Singh, J., Gupta, A. K., Bräuning, A., Mayr, C., & Singhal, M. (2021). Central Himalayan tree-ring isotopes reveal increasing regional heterogeneity and enhancement in ice mass loss since the 1960s. *The Cryosphere*, *15*(1), 95-112.

Singh J, Yadav RR (2005) Spring precipitation variations over the western Himalaya, India, since AD 1731 as deduced from tree rings. Journal of Geophysical Research: Atmospheres 110

Singh J, Yadav RR, Dubey B, Chaturvedi R (2004) Millennium-long ring-width chronology of Himalayan cedar from Garhwal Himalaya and its potential in climate change studies. Current Science 590–593

Thapa UK, Shah SK, Gaire NP, Bhuju DR (2015) Spring temperatures in the far-western Nepal Himalaya since AD 1640 reconstructed from Piceasmithiana tree-ring widths. Climate dynamics 45:2069–2081

Wester, Philippus, Arabinda Mishra, Aditi Mukherji, and Arun Bhakta Shrestha. The Hindu Kush Himalaya assessment: mountains, climate change, sustainability and people. Springer Nature, 2019.

Yadav RR (2009) Tree ring imprints of long-term changes in climate in western Himalaya, India. J Biosci 34:699–707. https://doi.org/10.1007/s12038-009-0058-7

Yadav RR (2011a) Long-term hydroclimatic variability in monsoon shadow zone of western Himalaya, India. Climate Dynamics 36:1453–1462

Yadav RR (2013) Tree ring–based seven-century drought records for the Western Himalaya, India. Journal of Geophysical Research: Atmospheres 118:4318–4325. https://doi.org/10.1002/jgrd.50265

Yadav RR, Singh J (2002) Tree-ring-based spring temperature patterns over the past four centuries in western Himalaya. Quaternary Research 57:299–305

Yadav RR, Park W-K, Bhattacharyya A (1997) Dendroclimatic Reconstruction of April–May Temperature Fluctuations in the Western Himalaya of India Since A.D. 1698. Quaternary Research 48:187–191. https://doi.org/10.1006/qres.1997.1919

Yadav RR, Park W-K, Bhattacharyya A (1999) Spring-temperature variations in western Himalaya, India, as reconstructed from tree-rings: AD 1390-1987. The Holocene 9:85–90

Yadav RR, Braeuning A, Singh J (2011) Tree ring inferred summer temperature variations over the last millennium in western Himalaya, India. Climate Dynamics 36:1545–1554

Yadav RR, Misra KG, Kotlia BS, Upreti N (2014) Premonsoon precipitation variability in Kumaon Himalaya, India over a perspective of 300 years. Quaternary International 325:213–219

Yadav RR, Park W-K, Singh J, Dubey B (2004) Do the western Himalayas defy global warming? Geophysical Research Letters 31. https://doi.org/10.1029/2004GL020201

Yadav RR, Singh J, Dubey B, Misra KG (2006) A 1584-year ring width chronology of juniper from Lahul, Himachal Pradesh: Prospects of developing millennia-long climate records. Current Science 1122–1126

Yadav RR (2011b) Tree ring evidence of a 20th century precipitation surge in the monsoon shadow zone of the western Himalaya, India. Journal of Geophysical Research: Atmospheres 116

Yadava AK, Yadav RR, Misra KG, Singh J, Singh D (2015) Tree ring evidence of late summer warming in Sikkim, northeast India. Quaternary International 371:175–180

Yadava AK, Misra KG, Singh V, Misra S, Sharma YK, Kotlia BS (2021) 244-year long tree-ring based drought records from Uttarakhand, western Himalaya, India. Quaternary International 599–600:128–137. https://doi.org/10.1016/j.quaint.2020.12.038

Chapter 7
Multiproxy Sediment Core Record of Precipitation/Drought Cycles and Hydrodynamic Changes in Gangapur Reservoir, Maharashtra, India

Sumedh K. Humane, Rajeshree R. Ukey, Samaya S. Humane, and Pranit B. Gajbhiye

Abstract A multi-proxy study of 70cm long sediment core has consistently provided the signatures of watershed changes and envisaged further the transformations caused by anthropogenic activities in the Gangapur Reservoir between ~1931CE and ~2012CE. A variety of temporal dataset such as sediment geochemisty, diatoms, and particle size variations have been used and correlated with the precipitation trends of study area to discern the impact of natural phenomena and anthropogenic activities on the reservoir. The heavy precipitation events of the region were compared with the cycles of water discharge from the Gangapur Reservoir and diatom diversity and its flux rate. Geochemical ratios normalized with erosional indicators (i.e., Ti and Al) show the atmospheric transport of pollutants into the reservoir from time to time. The grain size analysis and the fluctuations in diatom species revealed changing cycles of precipitation/droughts in the catchment of the studied reservoir, indicating the feasibility of reservoir to record the climatic variability and land use changes in the watershed.

1 Introduction

Reservoirs are artificial water bodies and experience various anthropogenic stresses from their initiation (Schroeder et al. 2016; Costa-Böddeker et al. 2012; Bennion et al. 2011). Many reservoirs preserve these records of environmental and hydrodynamic changes of recent times quite well covering the span of a few decades to a

S. K. Humane · S. S. Humane (✉) · P. B. Gajbhiye
Department of Geology, Rashtrasant Tukadoji Maharaj Nagpur University, Nagpur, India
e-mail: samaya.humane@gmail.com

R. R. Ukey
Geological Division, Central Design Organization, Maharashtra Engineering Research Institute (MERI), Nashik, India

N. Khare (ed.), *Changing Climate over India*,
https://doi.org/10.1007/978-981-92-1098-5_7

century (Cardoso-Silva et al. 2022; Nadiga 2021; Liu et al. 2012; Shotbolt et al. 2006; Yeloff et al. 2005). Recent studies have also shown the significance of diatoms recovered from sediments of tropical reservoirs to understand the environmental changes (Amorim and Moura 2022; Bartozek et al. 2019; Santos et al. 2017).

The reservoir management plans need the information about the rivers of various watershed changes affecting water quality along with the external stressors like nutrients and pollutants deposited through air (Schroeder et al. 2016; Woodridge et al. 2014). The responses of diatom community to climate change have been investigated by several workers (Jones and Birk 2004; Solovieva et al. 2008; Anderson et al. 2012; Chen et al. 2014; Lee et al. 2018). The study of diatom assemblages and geochemical investigations of the reservoir sediments could establish the linkages between changes in the catchment and water quality status along with the historical data (Hansika and Yatigammana 2019; Liu et al. 2012; Woodridge et al. 2014; Xu et al. 2007).

The present work reports about environmental and hydrodynamic changes in the watershed of the Gangapur Reservoir (Nashik District, India) based on stratigraphic variations in diatom assemblages, geochemistry, grain size analysis, and historical water discharge records. The aim of this study was also to discern the impact of the anthropogenic activities and the extreme weather events on the watershed of the Gangapur Reservoir since its inception.

1.1 Study Area

The Gangapur Reservoir falls in the Nashik district on the northwest part of the Maharashtra state (Fig. 1a-b). The catchment of the Gangapur Reservoir is mainly comprised of the Deccan Trap Basalt of the Deccan Volcanic Province (DRM 2001). Physiographically, the western part of the district is represented by highly rugged and dissected terrain of the Western Ghats with the peaks like Kalsubai and Trimbak and the low-lying plains mark the eastern and southeastern parts of the district (Avhad et al. 2019). The easterly flowing Godavari and Girna Rivers with their tributaries constitute the main drainage system in the district (Avhad et al. 2019).

1.2 Land Use Pattern

Land use and land management practices have a significant impact on natural resources including water, soil, nutrients, plants, and animals (Dale et al. 2000). Land use maps provide information about types of activities including farming, mining, residential use, water storage, soil, nutrients, plants, and animals. The major portion of the watershed of the Gangapur Reservoir has been covered by the agricultural land followed by the fallow land and the human settlements (Fig. 1c).

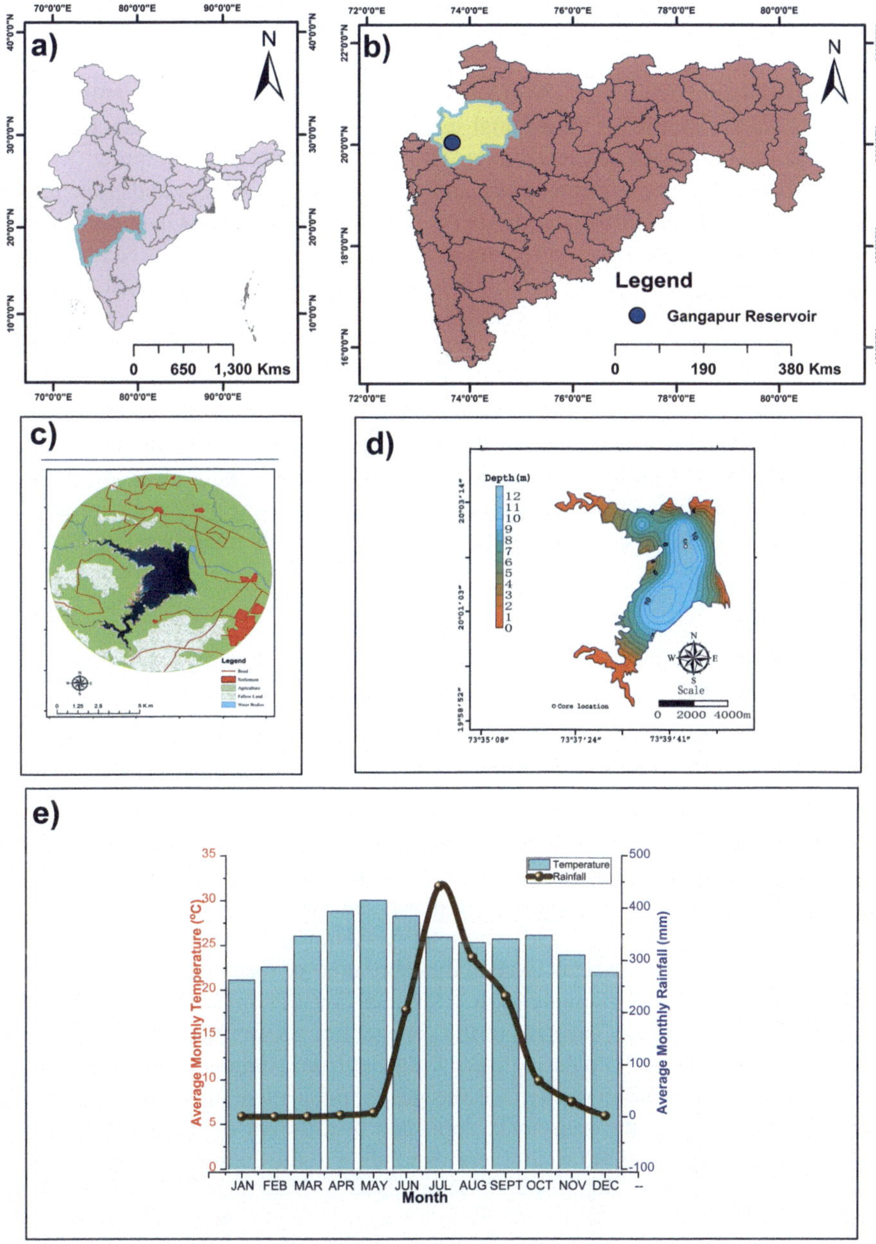

Fig. 1 (a) Map of India showing Maharashtra state, **b** location of study area in Nashik district, **c** land use around Gangapur Reservoir, **d** sediment core location (**c**) in the Gangapur Reservoir, and **e** average monthly temperature and rainfall variation in Nashik district during different months

1.3 Rainfall and Climate

The Nashik District experiences warm and dry climatic condition for the entire year, except during the southwest monsoonal season. The climate of this region has been broadly divided into four seasons, winter (December to February), summer (March to May), monsoon (June to September), and post-monsoon (October and November) (Dhonde 2014). The location of sediment core recovered from the Gangapur Reservoir is shown in Fig. 1d. The average monthly rainfall of the district has been highest during the month of July (Fig. 1e). Within the narrow strip of the district, at the close proximity of the Western Ghats, the rainfall is extremely heavier than in the remainder of the district. However, the rainfall within the plateau region of the Western Ghats generally declines from the west to east due to some local variations in the topography (Avhad et al. 2019). In the region of the Western Ghats the temperature is also much less than that in the rest of the Nasik district. Depending upon the elevations, temperatures begin to extend rapidly from the latter half of February and May which is the hottest month with a mean monthly maximum temperature of ~30 °C. December is the coldest month with a mean monthly minimum temperature of ~20 °C (Fig. 1e).

2 Methodology

2.1 Coring and Sample Processing

A sediment core (Lat: N 20° 02′28″ N; Long: E 73° 4o' 10.08″ E) of 72 cm length was retrieved from the maximum depth of the Gangapur Reservoir with the help of a gravity corer (Fig. 1d). The core section was cut vertically into two halves and further sub-sectioned at 0.5 cm and 1 cm intervals (Humane et al. 2023). One half of the core was preserved as a repository in the laboratory of the department, while the other half was used for XRF analysis, ICPMS analysis, diatom studies, particle size analysis, and ^{210}Pb dating. The ages of the core sediments samples of the Gangapur Reservoir derived from ^{210}Pb dating were followed from Humane et al. (2023). The bulk sediment accumulation rates of various geochemical elements (g/m^2/year) have been determined using the sediments deposited at different depths (based on the constant rate supply, i.e., CRS model) corresponding to the quantity (mg/g) of every constituent in the bulk sediment (Fig. 2; Humane et al. 2023; Garrison and Laliberte 2010).

Geochemical Analysis

The core samples collected from the Gangapur Reservoir were dried first in the oven at about 60 °C (Humane et al. 2023). These samples were then crushed/ powdered to about -170 mesh size to make it homogenized and used for the geochemical analysis (Humane et al. 2023).

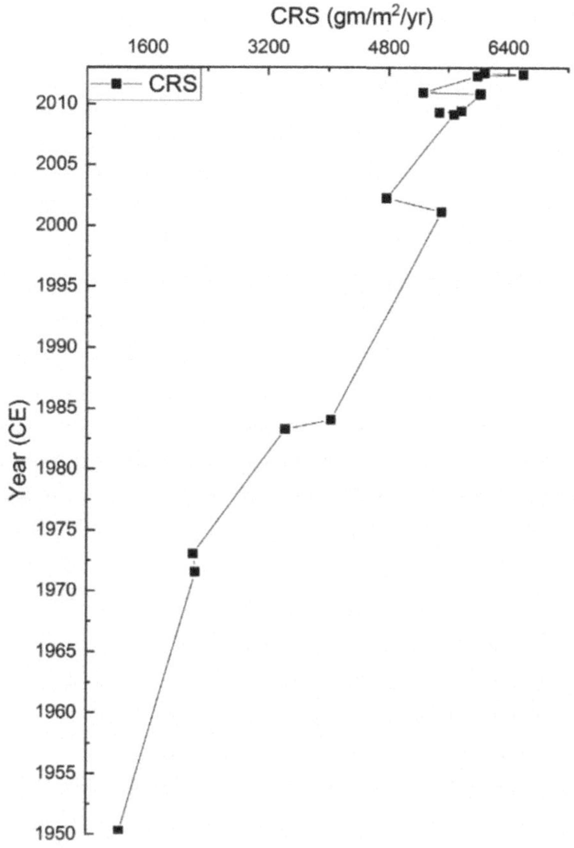

Fig. 2 Sediment accumulation rate (g/m^2/year) of Gangapur Reservoir since 1950 CE

Sample Preparation and Analysis for Major and Trace Element Using XRF

For XRF analysis, the finely powdered sediment samples were transformed to pellets using hand pellet press (Humane et al. 2023). These pellets were then directly used for XRF analysis. A total of 24 samples from the Gangapur Reservoir were analyzed for major oxides such as Na_2O, MgO, Al_2O_3, SiO_2, P_2O_5, K_2O, CaO, TiO_2, MnO, and Fe_2O_3 and minor elements such as Rb, V, Ba, Sc, Ni, Cr, Cu, Zn, Ga, U, Nb, Zr, Sr, Y, Pb, Th, and Co using XRF (Model: Brucker 58 Tiger) at Wadia Institute of Himalayan Geology, Dehradun (WIHG) (Humane et al. 2023).

Chemical Weathering Intensity (CWI) and CaO/MgO

The changes in the chemical weathering intensity (CWI) show the fluctuations in weathering under the varying climatic conditions (Sun et al. 2009). The CWI is calculated using the following formula, *i.e.*, $\{(CaO + MgO + Na_2O)/Al_2O_3\}$. CaO/MgO ratio has been used to understand the changing climatic condition over the study area.

2.2 *Grain Size Analysis*

A total of 11 samples from the core sediments of the Gangapur Reservoir were analyzed to know the concentrations of different fractions of the particle size from Wadia Institute of Himalayan Geology, Dehradun, India, using the Particle Size Analyzer (Model: Master Sizer 2000 APA 2000).

2.3 *Diatom Study*

About 1gm of dried core sediments obtained through conning and quartering were taken in a beaker for isolating siliceous material especially diatoms and chrysophycean cysts. The standard maceration procedure was followed to process the core sediment samples for the investigation of diatoms (Battarbee et al. 2001). The diatom identification was done using Leica DM 350 Biological Research Microscope at the magnification of 40x, 63x, and 100x (oil immersion) along with the microphotography (John 2012; Karthick et al. 2013; https://www.algaebase.org/search/species; Metzeltin et al. 2005; Metzeltin and Lange-Bertalot 2007).

3 Results

3.1 *Sediment Core Lithology and Texture*

The entire core was light brown in color. Overall, the core consists of silty clay, with intermediate variation from slight sandy, silty, to clayey fractions. Intercalations of a few dark bands with light bands are also seen throughout the core section. The end member triangular classification proposes the distribution of sand, silt, and clay (Trefethen 1950). The distributions of sand-silt–clay of a sedimentary core of the Gangapur Reservoir show the dominance of silty clay as per the triangular classification (Fig. 3).

3.2 *Geochemical Study*

The geochemical content (major and trace elements) from the sediment core of the Gangapur Reservoir has been used to understand hydrodynamic changes in the watershed along with the varying strength of the rainfall. The accrual of the geochemical content has provided the precise information of the changing climatic conditions over the study area. The rate of accumulation of the geochemical elements has provided

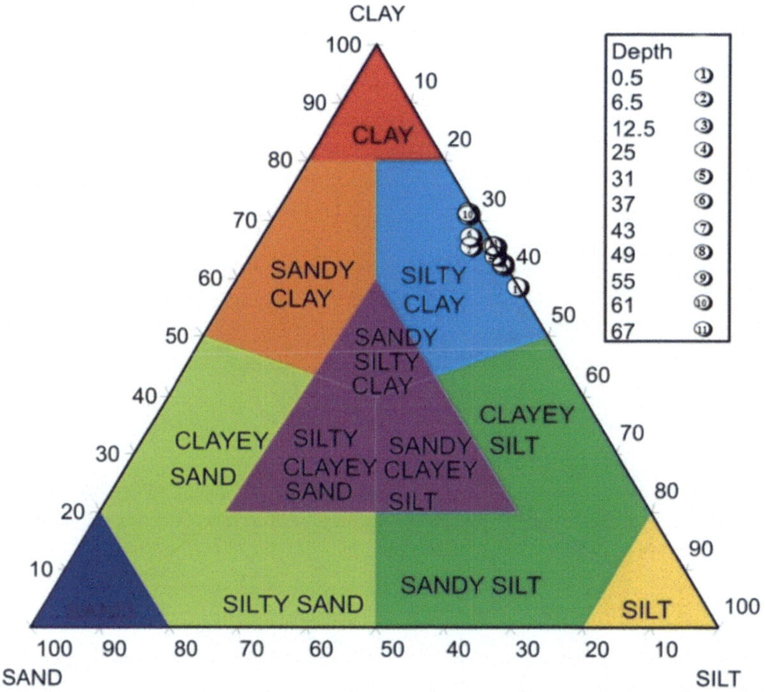

Fig. 3 Distribution of Sand, Silt, and Clay in the sediment core of Gangapur Reservoir (Trefethen 1950)

the information of historical changes in the strength of soil erosion in the catchment of the reservoir coupled with the anthropogenic activities (Fig. 4).

3.3 Stratigraphic Distribution of Diatoms

The core sediments of Gangapur Reservoir illustrate the presence of mainly 32 diatom species representing 21 genera. The common diatom taxa of the Gangapur Reservoir in the form of stratigraphic data are demonstrated (Fig. 5).

12 diatom taxa (out of 32 taxa) have attained a maximum abundance of >0.2% in at least one sample in the core sediment of the Gangapur Reservoir and have been used for the diatom stratigraphic study. Six diatom zones and seven sub-zones were identified for major species abundance using constrained cluster analysis using computer program Tilia (Grim 1991; Fig. 5). Diatom assemblages were categorized as 1) Planktonic diatoms, such as *Aulacoseira granulata* (Ehr.) Simonsen (12.5–100%), *Aulacoseira ambigua* (Grunow) Simonsen (1.42–100%), *Cyclotella meneghiniana* Kutzing (0.76–66.66%), *Pantocsekiella ocellata* Kiss & Ács (0.76–20%), and *Cyclotella comensis* Grunow in Van Heurck (0.35–5.35%); and 2) the benthic diatoms

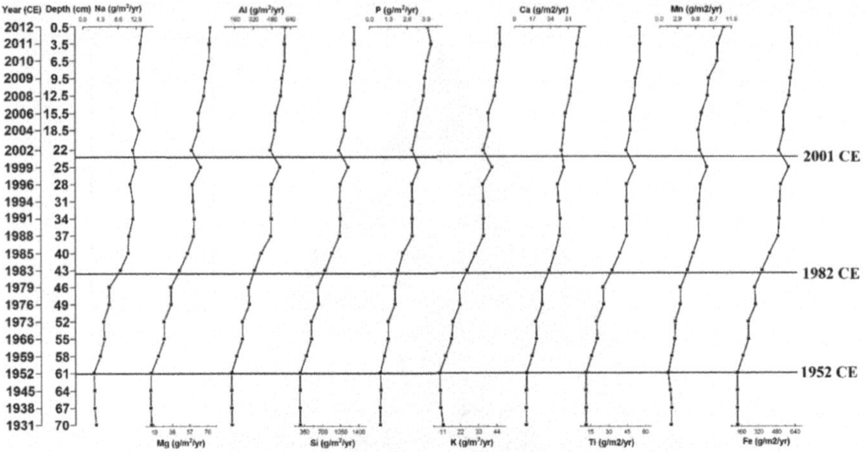

Fig. 4 Vertical Profile of accumulation rate of geochemical elements in the sediment core of Gangapur Reservoir (DY: Drought Year-Severe, De et al. 2005; reproduced using supplementary data from Humane et al. 2023)

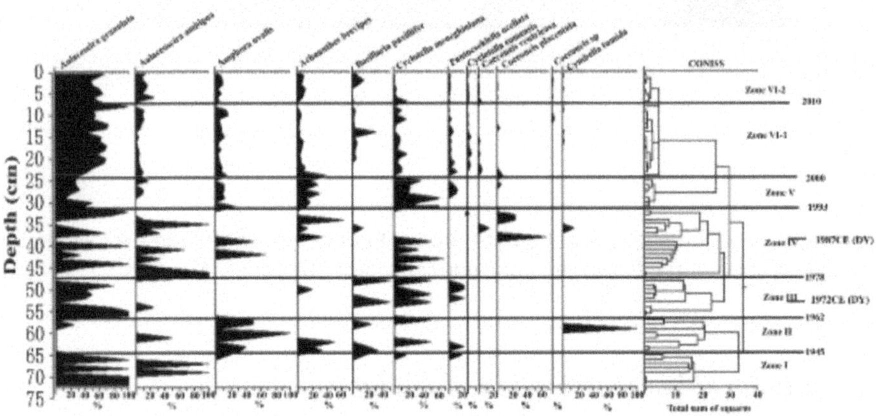

Fig. 5 Diatom Succession in Gangapur Reservoir since ~1931 CE–2012 CE. Only the major taxa (species with ≥1% in at least one sample) were shown. The right side shows the constrained incremental sum of square (CONISS)

such as *Amphora ovalis* Kützing (3.50–100%), *Achnanthes brevipes* Bory (2.35–62.50%), *Bacillaria paxillifer* (Muller) Marsson (1.75–50%), *Caloneis ventricoa* (Ehr.) Meister (0.35–14.28%), *Cocconeis placentula* Ehrenberg (0.54–66.66%), and *Cymbella tumida* Kützing (0.38–100%). The planktonic diatoms were dominated by *A. granulata* (Fig. 6a, f). The benthic diatoms were dominated by *Amphora ovalis* (Fig. 6b) and *Achnanthes brevipes* (Fig. 6e). Zone I (65–70 cm: ~1931–1945 CE) was characterized by predominance of planktonic diatom species *Aulacoseira granulata* (~66.66% on average), *Aulacoseira ambigua* (~100% on average), and *C. meneghiniana* (~16.66% on average). Zone II (57–64 cm: ~1946–1962 CE) was distinguished by dominance of the benthic diatom species *A. ovalis* (~16.66%, on average) and *A. brevipes* (~25%, on average). Other abundant planktonic species were *C. meneghiniana* (~50% on average) and *A. ambigua* (~50% on average). The average ratio of planktonic and benthic taxa (P/B) was 2.4 in this zone. Zone III (47–56 cm: ~1963–1978 CE) shows prevalence of the planktonic diatom species *A. granulata* (~70% on average) and *C. meneghiniana* (~35% on average). The benthic species *A. ovalis* was absent in this zone but other benthic species was *B. paxillifer* (~35% on average). The P/B ratio was 1.5. Zone IV (32–46 cm: ~1979–1993 CE) was distinguished by the abundance of the planktonic species *A. granulata* (~64.285% on average), *A. ambigua* (~56.25% on average), *C. meneghiniana* (~49.99% on average; Fig. 6c–d). The benthic diatom species dominated by *A. ovalis* (~58.33% on average), *A. brevipes* (~47.91% on average), and *C. placentula* (~45.83% on average). The average of P/B ratio was 1.12. Zone V (24–31 cm; ~1994–2000 CE) was represented by the decline in the *A. granulata* (~31.25% on average) with the increase in *C. meneghiniana* (~41.25% on average). The other planktonic species found in this zone were *A. ambigua* (~14.09% on average) and *P. ocellata* (~8.33% on average; Fig. 6g). The benthic diatom species abundant in this zone were *A. brevipes* (~24.78% on average) and *A. ovalis* (~17.5% on average). The average P/B ratio was 1.12 in this zone. Zone VI (0–23 cm; ~2001–2012 CE) was classified into two sub-zones. This zone is characterized by dominance of planktonic diatoms. Zone VI-1 (07–23 cm; ~2001–2010 CE) shows dominance of benthic diatom species *A. granulata* (~74.91% on average), *C. meneghiniana* (~10.21% on average), *A. ambigua* (~7.195% on average), and *P. ocellata* (~6.80% on average). The benthic diatom species dominating in this zone were *A. ovalis* (~10.61% on average), *A. brevipes* (~5.55% on average), and *B. paxillifer* (~16.94% on average). The ratio of planktonic and benthic taxa (P/B) had an average of 0.74. Zone VI-2 (0–6 cm; ~2011–2012 CE) was represented again by abundance of planktonic diatoms *A. granulata* (~58.75% on average) and *A. ambigua* (~17.88% on average). The concentration of *C. meneghiniana* (~2.88% on average) was declined in this zone. The abundant benthic diatom species were *A. ovalis* (~6.7% on average), *A. brevipes* (~7.87% on average), and *B. paxillifer* (~10.66% on average). The ratio of planktonic and benthic taxa (P/B) had an average of 3.12.

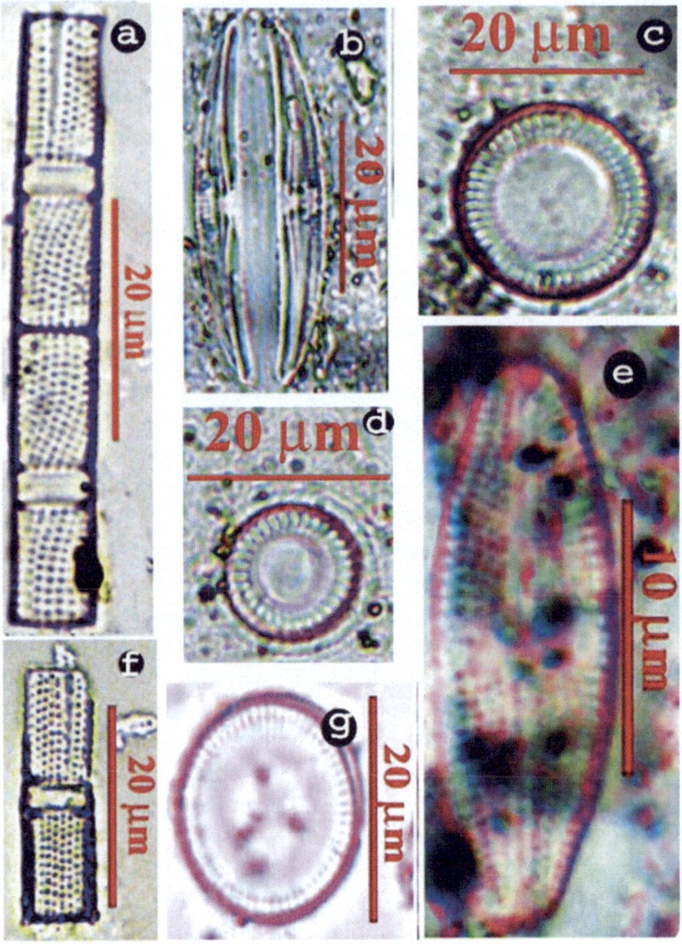

Fig. 6 Light microscopic images of diatom species recovered from sediment core of the Gangapur Reservoir a, **f** *Aulacoseira granulata* (Ehr.) Simonsen, **b** *Amphora ovalis* Kützing, **c–d** *Cyclotella meneghiniana* Kützing, **e** *Achnanthes brevipes* Bory, and **g** *Pantocsekiella ocellata* Kiss and Ács

4 Discussion

4.1 Geochemical Variability and Seasonal Processes

The reservoir sediments are occasionally used as environmental archives considering the sediment disturbance by fluctuating water levels (Shotbolt et al. 2006). However, with rapid rates of sedimentation, closeness to urban areas and presence of management data set, the reservoirs become the potentially important resources for reconstructing recent, anthropogenic derived environmental change (Shotbolt et al. 2006).

The geochemical variables from the core sediments of the lakes/reservoirs are used to understand their impact in the drainage basin (Garrison and Laliberte 2007, 2010). The fluctuation in the profile of chemical elements titanium (Ti) and aluminum (Al) reflects changes in the rate of soil erosion as they are formed from the detrital aluminosilicates (Garrison and Laliberte 2010). The growth of aquatic plants and algae are mainly governed by the nutrients phosphorous and nitrogen. The organic matter profiles will signify the lake productivity. Potassium (K) is the important component of the synthetic fertilizers and soils. Thus, the changes in the profile of potassium (K) reflect the soil erosion and the input of commercial fertilizers in the drainage basin. Zinc (Zn) is the important component of galvanized roofs and tires and represents the urban runoff (Garrison and Laliberte 2007). The zinc (Zn) profile of the cores also reflects smelting of lead–zinc-ores in the region (Dean 2002). The calcium (Ca) profiles reflect the use of soil modifications for the development of facilities adjacent to water bodies (Garrison and Laliberte 2007). The manganese (Mn) profiles reflect changes in the oxygen levels in the bottom waters (Garrison 2008). The rate of accumulation of the geochemical elements in the Gangapur Reservoir was calculated by adding the geochemical concentration with the rate of sedimentation (using ^{210}Pb dating and CRS model). The deposition of elements in the reservoir with time was established using the accumulation rate. The accumulation rate gives information about the impact of processes in the drainage basin (Garrison and Laliberte 2007, 2010).

The Gangapur Reservoir shows progressively increasing rate of soil erosion from ~1931 till ~2000 CE. The little decline in the elemental concentrations was noticed at ~2002 CE and ~2006 CE with further continuous rise in the soil erosion till ~2012 CE (Fig. 4). The major increase in the soil erosion in the drainage basin of the Gangapur Reservoir was noticed after ~1982 CE. The peaks of phosphorous (P) are similar to the other erosional indicators in the Gangapur Reservoir (Fig. 4). The gradual rise in the P content of the Gangapur Reservoir was noticed since ~1952 CE (Fig. 4) which indicates that soils/sediments have released P as the major source of nutrients in the drainage basin of the Gangapur Reservoir. The verticle profile of phosporous and nitrogen may not reflect the lakes eutrophication history due to diagenesis (Anderson and Rippey 1994, Schelske et al. 1972 and Fitzpatrick et al. 2003) and the breakdown of phosphorous and nitrogen into the inorganic components and some of the material then recycled in to the water column and removed out of the sediments (Garrison and Laliberte 2010). The calcium (Ca) profile of the Gangapur Reservoir is similar to erosional indicators, i.e., Titanium and Aluminum (Fig. 4). Thus, it can be concluded that the major part of the Ca of the Gangapur Reservoir was derived from soil particles of the drainage basin (Fig. 4). The geochemical elements like silica, manganese, and sodium are showing similar increased trend like titanium and aluminum, which reflect that the major concentration of these elements are derived from the soils and sediments in the watershed of the Gangapur Reservoir (Fig. 4).

The anthropogenic inputs other than mineral sediments can be discerned using the chemical weathering intensity and normalize geochemical ratios with the erosional indicators such as Al and Ti (Engstrom and Wright 1984; Garrison and Laliberte 2010; Veena et al. 2014). Therefore, the selected geochemical elements from the

Gangapur Reservoir were normalized to aluminum (Al) to differentiate the anthropogenic inputs other than mineral sediments to know the impact of the elements on the reservoirs other than the soil particles. The elemental profiles normalized to Al show the anthropogenic impact both from the concerned watershed and atmospheric transport through time on the sediments of the Gangapur Reservoirs is very much evident. The ratio of these elements would be the same throughout the core if soil particles were the only source of potassium, as both aluminum (Al) and potassium (K) are found in sediments/soil particles in the watershed (Garrison and Laliberte 2007). In the Gangapur Reservoir, there is a gradual depletion in the K:Al ratio since ~1931–2012 CE (Fig. 7). The gradual decline in the K:Al ratio in contrast to the higher concentrations of erosional indicators (Ti and Al) suggests that potassium is not mainly derived through the erosional processes in the watershed but the other anthropogenic sources could be the reason for its rise (Fig. 7). The earthen dam was built on the Gangapur Reservoir between ~1946 and ~1957 CE. Thus, the first disturbance in the drainage basin has been attributed to the initial work for dam construction after ~1947 CE and represented by the decline in K:Al ratio (Fig. 7). The gradually increasing trend of the P:Al ratio (like erosional indicators Ti and Al) in the Gangapur Reservoir clearly indicates elevated trend of soil erosion with the minor decline after the construction of the dam ~1957 CE (Fig. 7). The Zn:Al ratio in the Gangapur Reservoir was progressively increased from ~1931–2012 CE indicating increased erosional processes in the watershed (Fig. 7). The algal growth in the aquatic system is mainly controlled by the nutrients like phosphorous and nitrogen (Garrison and Laliberte 2010). The K:P ratio of the Gangapur Reservoir also shows increased peak during ~1931 CE, ~1966–1973 CE, and ~1982–1984 CE indicating the enhanced soil erosion in the drainage basin. But, during ~1988–2012 CE, the gradual declining trend of K:P ratio was noted pointing its other important source besides soil erosion and fertilizers (Fig. 7). The possible reason behind the increased P may be the internal loading of phosphorus.

Once the bottom water loses the oxygen, manganese (Mn) preferentially migrates with respect to iron (Engstrom et al. 1985). This manganese generally moves from the sediments into the deepest water causing the environment of manganese in the deep-water sediments (Garrison 2008). With the preferential movement of the iron and manganese, they remain in suspension for a long period together (Jones and Bowser 1978). Thus, the ratio of iron to manganese (Fe:Mn) declines (Mn increases) with the loss of oxygen (Garrison 2008). The decline in Fe:Mn ratio after the dam construction, i.e., ~1967–2000 CE and ~2002–2012 CE in the Gangapur Reservoir, clearly indicates that the hypolimnetic oxygen of the Gangapur Reservoir has been declined during the above two periods linked with the increased eutrophication in the reservoir (Fig. 7).

In general, the continental weathering processes are regulated and controlled by several factors including climate (Liu et al. 2012; Oliva et al. 2003). A stronger chemical weathering rate correlates to more humid and warm climates. Meanwhile, weaker chemical weathering relates to drier climates (Colin et al. 2006; Liu et al. 2005; Jiwarungrueangkul et al. 2019). Consequently, the interaction of chemical

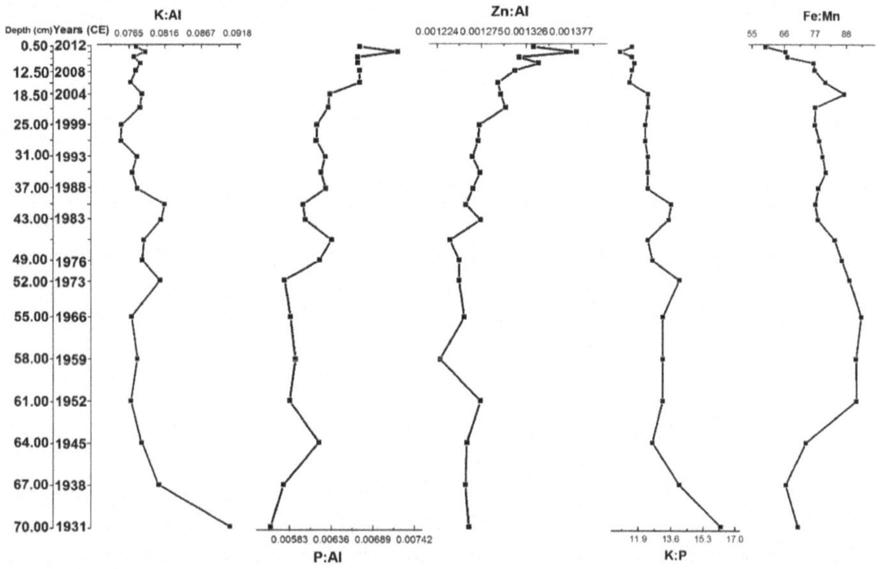

Fig. 7 Vertical profile of sediment core of Gangapur Reservoir showing K:Al, P:Al, Zn:Al, K:P, and Fe:Mn; calculated using data from Humane et al. (2023)

weathering with atmospheric condition remains elusive and desires more investigation (Jiwarungrueangkul et al. 2019). The oxide such as CaO, MgO, and Na_2O are more soluble and movable, while Al_2O_3, TiO_2, and SiO_2 are used as the most resistant and insoluble elements in the environment during weathering and erosion (Engstrom and Wright 1984; Veena et al. 2014; Garrison and Laliberte 2007). The more weathering is represented by higher values of CWI (Fig. 8). This can be correlated to wet conditions and the heavy monsoonal rainfall in the study area. The higher CaO/MgO ratios in sediments indicate more strong growth of anthropogenic carbonate in the soils of the drainage basin, which indicates warm climate with alternating wet and dry periods (Veena et al. 2014; Wang et al. 1990). The different proxies have been used by the various workers to investigate past monsoonal fluctuations such as grain size analysis, i.e., sand, silt, and clay ratios (Veena et al. 2014; Kashiwaya et al. 1988). The relationship of the diatoms and other proxies of the sediment cores of the Gangapur Reservoir were compared with the actual rainfall data (https://www.tropmet.res.in; Harris et al. 2020; https://rdcu.be/b3nUI), i.e., the Mean Annual Rainfall (MAR) over the Godavari Sub-basin, Maharashtra, was compared in the stratigraphic profiles (Fig. 8). The proxy data of the sediment core of the Gangapur Reservoir was also compared with the water discharge record of the Gangapur Reservoir (WRD 2022).

The CWI and CaO/MgO ratio and their relationship with the depth (cm) in the Gangapur Reservoir core were evaluated (Fig. 8). This core clearly shows fluctuating trends of major dry and wet periods in the warm tropical region of the study area since the last 81 years. (I) ~1931–1959 CE: The lower values of CWI with a little

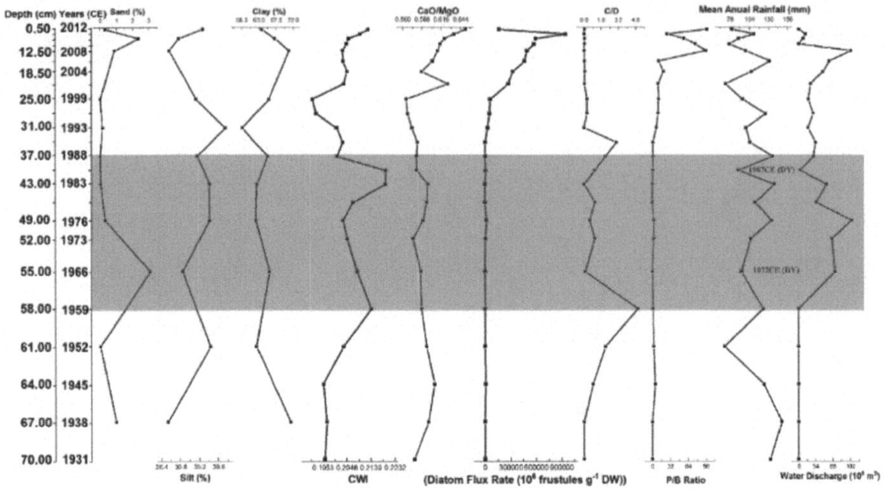

Fig. 8 Sand, silt, and clay ratio, CWI, CaO/MgO ratio, diatom flux rate, cyst/diatom (C/D) ratio, planktonic/benthic (P/B) ratio of diatoms, mean annual rainfall (mm) record (https://www.tropmet.res.in), and water discharge data of Gangapur Reservoir (WRD 2022) (DY: Drought Year-Severe, De et al. 2005; CIW and CaO/MgO ratio: calculated using data from Humane et al. 2023)

rise in CaO/MgO ratio indicates the presence of dry/low rainfall period (Fig. 8). It is also corroborated by the low mean annual rainfall (MAR) and low sand and clay content along with the moderate silt (Fig. 8). (II) ~1960–1988 CE: During this period, the CWI initially declined and reached maximum during ~1983–1985 CE. This initial decline in the CWI points to the existence of low rainfall/dry period in the catchment of the Gangapur Reservoir (Fig. 8). The record of extreme weather event, i.e., presence of drought year (~1972 CE; De et al. 2005) and low Cyst/Diatom (C/D) ratio also corroborate this finding. The decline in the CWI corresponds to rise in CaO/MgO ratio (Fig. 8). However, the presence of maximum peak of the CWI at the top of this unit and decline in CaO/MgO ratio (~1983–1985 CE) indicate increased rainfall in the region. This rise in the rainfall has been subsequently changed to drought year (~1987 CE; De et al. 2005) as shown by the sudden fall in the CWI (Fig. 8). The moderate level of discharge of water from the Gangapur Reservoir during this severe drought could be attributed to its use for the domestic purpose. However, there was no discharge of water from this reservoir in another severe drought (~1987 CE) pointing to the possibility of the non-availability of water in the reservoir for domestic use. (III) ~1989–2012 CE: The continuous decrease in the CWI in this zone indicates low precipitation events since ~1989 CE– ~2004 CE except ~1995 CE as also supported by the decline in the sand and silt ratio (with increase in clay content) and the low water discharge from the reservoir (Fig. 8). The topmost part of this unit (~2005– ~2012 CE) shows the moderate rise in the CWI with increased strength of rainfall supported by higher water discharge from the reservoir and increased P/B ratio (~2008 CE, ~2012 CE) of diatoms (Fig. 8). The diatom flux rate has been progressively increased during ~2000–2011 CE indicating increasing strength of rainfall in the catchment of the Gangapur Reservoir (Fig. 8).

4.2 Diatoms Response to Hydrodynamic Change

The maximum flux rate of diatoms in the Gangapur Reservoir was recorded at the top of the core (~2011 CE). The temperature variation in the study area was not significant as the Gangapur Reservoir falls under tropical condition. However, the diatom species may respond to variation in the precipitation (rainfall) and dry periods (droughts) of the region. The planktonic to benthic (P/B) ratio of the Gangapur Reservoir shows dominance of the planktonic diatom assemblages at 0–6 cm depth (~2010–2012 CE), whereas the diatom flux ratio has been increased from ~1999 CE–2012 CE corresponding to the increased precipitation in the region (Fig. 8). The substantial change in the species composition of diatoms was distinguished by program CONISS. Diatoms of the Gangapur Reservoir show considerable change of the centric diatom A. granulata by the benthic species *A. ovalis* during ~1945–1962 CE. *A. granulata* declined in concentration during ~1993–2000 CE with the rise in the content of *C. meneghiniana* and benthic species *A. brevipes* (Fig. 5).

The climate warming may be represented by the significant shift of diatoms from benthic to planktonic diatom assemblages (Chen et al. 2014). The longer and stronger thermal stratification is indicated by the small size and fast-growing planktonic diatoms (Chen et al. 2014; Rühland et al. 2010). The diatom shift in the Gangapur Reservoir could be due to the nutrient supply. *A. granulata* was dominant during summer in the Great Lake (Stoermer and Ladewski 1976). *A. granulata* is most plentiful in shallow and more eutrophic waters, which are less stratified during summer (Stoermer et al. 1995). *A. granulata* is filament-forming taxa with strongly silicified frustules (Gabito et al. 2013 and Kilham et al. 1986) and indicate eutrophic environments (Bennion and Simpson 2010; Gong et al. 2009). *A. granulata* also prevails in high nitrogen environments (Kilham and Kilham 1978). The different species of *Aulocoseira* are abundant in the less transparent eutrophic waters due to their lower light saturation intensity (Talling 1957). The increased concentration of *A. ambigua* indicates low to moderate phosphorus levels (Garrison and Laliberte 2007).

The Gangapur Reservoir also shows distinct changing patterns of the planktonic and benthic (P/B) diatom species (Figs. 5 and 8). *A. granulata* was predominated over a maximum period in the sediment core with the intermittent rise of *A. ambigua, A. ovalis, A. brevipes, C. meneghiniana,* etc. The major periods of abundance of *A. granulata* in the Gangapur Reservoir core were during ~1931–1945 CE, ~1962–1965 CE, ~1990–1993 CE, and with maximum abundance during ~2000–2012 CE. Thus, domination of *A. granulata* in the Gangapur Reservoir during the above periods suggests the less stratification in the core with highly eutrophic waters with elevated nitrogen content. During ~1945–1962 CE, *A. granulata* disappeared completely. The possible reason behind the disappearance of *A. granulata* may be due to the less supply of nutrients like phosphorus and nitrogen in addition to less rainfall and existence of continuous dry period in this duration (Fig. 5). *A. granulata* was replaced by benthic diatom species *A. ovalis, C. tumida,* and *A. brevipes*. Thus, the dominance of *A. ovalis* indicates alkaliphilous water and optimum range of filterable P between 0.35 and 1 mg^{-1} (www. environement_agency.gov.uk). *A. granulata* was also mainly declined during ~1967–1974 CE and ~1993–2000 CE with the dominance of *C. meneghiniana*. The profusion of *C. meneghiniana* points to the presence of eutrophic, electrolyte-rich waters (www.ces.iisc.ernet.in) during the above periods. The decline in *A. granulata* also corresponds to increase in *A. ambigua* during ~1978–1979 CE and ~1987–1990 CE representing low to moderate phosphorus levels in contrast with high phosphorous levels of *A. granulata* (Garrison and Laliberte 2007; Fig. 5). The Gangapur Reservoir clearly exhibits changing diatom species diversity since the last 81 years, with maximum diversity during ~1996–2002 CE. Thus, it can be surmised that the changing diatom species diversity in the Gangapur Reservoir could be in response to changes in wet and dry periods existed in the drainage basin (Fig. 5).

5 Conclusions

The present study of the proxy records revealed the three major events of dry period (drought) and two major events of high precipitation since the last 81 years over the catchment of the Gangapur Reservoir which can be fairly well corroborated by the instrumental records of precipitation the Godavari Sub-basin of Nashik district, Maharashtra. The first disturbance in the drainage basin was noticed in the form of decline in K:Al ratio since the initial work for dam construction after ~1947 CE. The gradually increasing trend of the P:Al ratio (like erosional indicators Ti and Al) in the Gangapur Reservoir clearly indicates increasing trend of soil erosion with minor decline noticed after the construction of the dam ~1957 CE. The major episode of the soil erosion in the drainage basin of the Gangapur Reservoir was noticed after ~1982 CE. The changing diatom species diversity in the Gangapur Reservoir was mainly in response to the changing strength of the wet and dry climatic cycles over its catchment area coupled with the variability in the nutrient concentrations. The water discharge record of the Gangapur Reservoir also corroborates quite well in understanding the impact of the hydrodynamic changes on diatom diversity.

Acknowledgements We (SKH and SSH) acknowledge with thanks the University Grants Commission (UGC), New Delhi, for their financial support to the department under UGC-SAP DRS-II (No.F.550/2/DRS-II/2016; SAP-I; 03/05/2016). The authors thank the Sub-divisional Engineer, Left Bank Canal, Water Resource Department, Government of Maharashtra, Nashik, for providing the water discharge data of the Gangapur Reservoir, Nashik District (MS). The authors gratefully acknowledge the consent received from the Geological Society of India to use some of the data of our paper (https://doi.org/10.1007/s12594-023-2489-4) published in Journal of Geological Society (2023).

References

Amorim, C.A. and Moura, A.D.N. (2022). Habitat templates of phytoplankton functional groups in tropical reservoirs as a tool to understand environmental changes. Hydrobiologia, v. 849(5), pp.1095-1113.

Anderson, N.J. and Rippey, B. (1994). Monitoring lake recovery from point source eutrophication: the use diatom-inferred epilimentic total phosphorus and sediment chemistry freshwater Biology. Hydrobiologia, v. 32, pp. 625-639.

Anderson, N.J., Foy, R.H., Engstrom, D.R., Rippey, B. and Alamgir, F. (2012). Climate forcing of diatom productivity in a low land, eutrophic lake: White Lough revisited. Freshwat. Biol., v. 57, pp. 2030 – 2043.

Avhad, Mahesh, Awasarkar Abhishek, Lilake Hrishikesh, Pawar, Sumedh and Ghumare, P.N. (2019). Morphomertic Analysis of Darna Catchment Area using QGIS. Inter. Jour. Res. Engineer. Sci. Managt., v. 2(5), pp. 456-460.

Bartozek, E.C., da Silva-Lehmkuhl, A.M., Gregory-Eaves, I. and Bicudo, D.C. (2019). Environmental and spatial dRivers of diatom assemblages in the water column and surface sediment of tropical reservoirs. Jour. Paleolimnol., v. 62(3), pp. 245-257.

Battarbee, R.W., Jones, V.J., Flower, R.J., Cameron, N.G., Bennion, H., Carvalho, L. and Juggins, S. (2001). Diatoms. *In*: Smol JP, Birks HJB and Last WM (*Eds.*), *Tracking Environmental Change Using Lake Sediments: Terrestrial, Algal, and Siliceous Indicators.* Kluwer Academic Publishers, Dordrecht, The Netherlands, 3, pp. 155–202.

Bennion, H., Battarbee, R.W., Sayer, C.D., Simpson, G.L. and Davidson, T.A. (2011). Defining reference conditions and restoration targets for lake ecosystems using paleolimnology: a synthesis. Jour. Paleolimnol., v. 45, pp. 533–544.

Bennion, H., and Simpson, G. L. (2010). The use of diatom records to establish reference conditions for UK lakes subject to eutrophication. Jour. Paleolimnol., 10.1007/ s10933-010-9422-8.

Cardoso-Silva, S., Mizael, J.O.S.S., Frascareli, D., Figueira, R.C.L., Pompêo, M., Vicente and Moschini-Carlos, V. (2022). Geochemistry and sedimentary photopigments E as proxies to reconstruct past environmental changes in a subtropical reservoir. Environment. Sci. Poll. Res., v. 29(19), pp. 28495–28509.

Chen C., Zhao L., Zhu C., Wang J., Jiang J and Yang S. (2014). Response of diatom community in Lugu lake (Yunnan- Guizhou Plateau, China) to climate change over the past century. Jour. Paleolimnol., v. 51(3), pp. 357-373.

Colin, C., Turpin, L., Blamart, D., Frank, N., Kissel, C. and Duchamp, S. (2006). Evolution of weathering patterns in the Indo-Burman Ranges over the last 280 kyr: Effects of sediment provenance on $^{87}Sr/^{86}Sr$ ratios tracer. Geochem., Geophys., Geosyst., v. 7, Q03007. https://doi.org/10.1029/2005GC000962

Costa-Bödeker S, Bennion H, Jesus TA, Albuquerque ALS, Figueira RCL, Bicudo DC (2012) Paleolimnologically iferred eutrophication of a shallow, tropical, urban reservoir in southeast Brazil. Jour. Paleolimnol., v.48, pp. 751-766.

Dale, V.H., Brown, S., Haeuber, R.A., Hobbs, N.T., Huntly, N., Naiman, R.J., Riebsame, W.E., Turner, M.G. and Valone, T.J. (2000). Ecological principles and guidelines for managing the use of land. Ecolog. Applicat., v. 10(3), pp. 639–670.

De, U.S., Dube, R.K. and Prakasa Rao, G.S. (2005). Extreme weather events over India in the last 100 years. Indian Geophy. Union, v. 9, pp. 173-187

Dean W. (2002). A 1500-year record of climatic and environmental change in Elk Lake, Clearwater County, Minnesota II: geochemistry, mineralogy, and stable isotopes. Jour. Paleolimnol., v. 27, pp. 301-319.

Dhonde UV (2014) Ground water information of Nashik distirct, Maharashtra. A report submitted to Govt. of India, Ministry of Water Resources, Central Ground Water Board, Central Region, Nagpur, 1–16.

DRM (2001) District Resource Map of Nasik District. Geological Survey of India.

Engstrom, D.R., Swain, E.B. and Kingston, J.C. (1985). A paleolimnological record of human disturbance from Harvey's Lake, Vermont: geochemistry, pigments, and diatoms. Freshwat. Biol., v.15, pp. 261-288.

Engstrom, D.R. and Wright, H.E. (1984) Chemical stratigraphy of lake sediments as a record of environmental change. In: Hayworth Lund, Jm WC (Eds.), Lake Sediments- Environmental History, Leicester University Press, pp. 11–67.

Fitzpatrick, F.A., Garrison, P.J., Fitzgerald, S.A. and Elder, J.F. (2003). Nutrient, trace-element, and ecological history of Musky Bay, Lac Courte Oreilles, Wisconsin, as inferred from sediment cores. US Geolog. Surv. Water-Res. Investigat. Rep., 02-4225: 141p.

Gabito, L., Bonilla S. and Antoniades, D. (2013). Paleolimnological reconstruction of change in a subtropical lake: a comparison of the subfossil record to limnological data. Limnetica, v. 32(2), pp. 175-188.

Garrison, P.J. and Laliberte, G.D. (2007). Paleoecological study of Big Round Lake, Polk County.Wisconsin Depart Natl Resour Bureau Sci Serv, pp 1–17.

Garrison, P.J. and Laliberte, G.D. (2010). Paleoecological Study of lake Chetac, Sawyer Country.Wisconsin Depart. Natl. Resour. Bureau Integrat. Sci., pp 1–17.

Garrison, P.J. (2008). Paleoecological study of Grindstone Lake, Sawyerl County. Wisconsin Depart Natl Resour Bureau Sci Serv, pp 1–16.

Gong, Z.J., Li Y.I., Shen, J. and Xie, P. (2009). Diatom community succession in the recent history of a eutrophic Yunnan Plateau Lake, Lake Dianchi, in subtropical China. Limnology, v. 10, pp. 247 -253.

Grimm, E. (1991) Tilia, Version 2.0. Illinois State Museum, Illinois.

Hansika, R.V.H. and Yatigammana, S.K. (2019). Distribution of diatom assemblages in the surface sediments in Sri Lankan reservoirs located in the main climatic regions and potential of using them as environmental predictors. Trop. Ecol., v. 60(3), pp. 415-425.

Harris, I., Osborn, T.J., Jones, P. and Lister, D.H. (2020). Version 4 of the CRU TS monthly high resolution gridded multivariate climate dataset Sci data 7, 109.

Humane, S.K., Ukey, R.R., Humane, S.S., Gajbhiye, P.B. and Aparajit, N. (2023). ^{210}Pb Geochronology and Heavy Metal Fluxes in Darna and Gangapur Reservoirs, Maharashtra, India. Jour. Geol. Soc. India (2023) 99:1427-1437. https://doi.org/10.1007/s12594-023-2489-4

Jiwarungrueangkul, T., Liu, Z., Stattegger, K. and Sang, P.N. (2019). Reconstructing chemical weathering intensity in the Mekong River Basin since the last glacial maximum. Paleoceonogra. Paleoclim., v. 34, pp. 1710–1725. doi. https://doi.org/10.1029/2019PA003608

John, J. 2012. Diatoms in the Swan River Estuary, Western Australia: Taxonomy and Ecology. Koeltz Scientific Books, 456p.

Jones, B.F. and Bowser, C.J. (1978). The Mineralogy and Related Chemistry of Lake Sediments. *In*: Lerman A., (Ed.), *Lakes. Chemistry, Geology and Physics*. Springer Verlag, New York, pp. 179–235.

Jones, V.J., and Birks, H.J.B. (2004). Lake-sediment records of recent environmental change on Svalbard: results of diatom analysis. Jour. Paleolimnol., v. 31, pp. 445-466,

Karthick, B., Hamilton, P.B. and Kociolek, J.P. (2013). An illustrated guide to common diatoms of Peninsular India. Gubbi labs, 206p.

Kashiwaya, K., Yamamoto, A., Fukuyama, K. (1988). Statistical analysis of grain size distribution in Pleistocene Sediments from lake Biwa, Japan. Quatern. Res., v. 30, pp. 12-18.

Kilham, S.S. and Kilham, P. (1978). Natural community bioassays: predictions of results based on nutrient physiology and competition. Verh. Internatl. Ver. Limnol., v. 20, pp. 68–74.

Kilham P., Kilham S. S., and Hecky R. E. (1986). Hypothesized resource relationships among African planktonic diatoms. Limnol. Oceanogra., v. 31, pp. 1169–1181.

Lee, H., Yun, S.M., Lee, J.Y., Lee, S.D., Lim, J. and Cho, P.Y. (2018). Late Holocene climate changes from diatom records in the historical Reservoir Gonggeomji, Korea. Jour. Appl. Phycol., v. 30(6), pp. 3205-3219.

Liu, Z., Colin, C., Trentesaux, A., Siani, G., Frank, N., Blamart, D. and Farid, S. (2005). Late Quaternary climatic control on erosion and weathering in the eastern Tibetan Plateau and the

Mekong River Basin. Quarter. Res., v. 63, pp. 316-328. https://doi.org/10.1016/j.yqres.2005.02.005
Liu J, Zhangwen L, Huanjun Z, Han BP (2012) Hydrodynamic change recorded by diatoms in sediments of Liuxihe Reservoir, southern China. Jour. Paleolimnol., v.47, pp.17-27.
Metzeltin, D., Lange-Bertalot, H. and Garcia-Rodriguez, F. (2005). Diatoms of Uruguay. Iconographia Diatomologica, v. 15, pp. 1-736.
Metzeltin, D. and Lange-Bertalot, H. (2007). Tropical diatoms of South America, 2. *In: Iconographia diatomologica, annotated diatom micrographs*, (*Ed.*) H. Lange-Bertalot, 15: 736p. Koeltz Scientific Books, Sttutgart.
Nadiga, B.T. (2021). Reservoir computing as a tool for climate predictability studies. Jour. Advanc. Model. Earth Syst., v. 13(4), e2020MS002290.
Oliva, P., Viers, J. and Dupre, B. (2003). Chemical weathering in granitic environments. Chem. Geol., v. 202, pp. 225-256.
Rühland K. M., Paterson A. M., Hargan K, Jenkin A, Clark B. J., Smol J. P. (2010) Reorganization of algal communities in the Lake of the Woods (Ontario, Canada) in response to turnof- the-century damming and recent warming. Limnol. Oceanograph., v. 55, pp. 2433–2451.
Santos, J.A., Marins, R.V., Aguiar, J.E., Chalar, G., Silva, F.A.T.F. and Lacerda, L.D. (2017) Hydrochemistry and trophic state change in a large reservoir in the Brazilian northeast region under intense drought conditions. Jour. Limnol., v. 76(1), pp. 41-51
Schelske, C.L., Feldt, L.E., Santiago, M.A., Stoermer, E.F. (1972). Nutrient enrichment and its effect on phytoplankton production and species composition in Lake Superior. In: Proceedings 15th Conference of Great Lakes research. International Association of Great Lakes Resources, Ann Arbor, MI, pp 149–163.
Schroeder, L.A., Martin, S.C., Kerns, G.J. and McLean, C.E. (2016). Diatom assemblages in a reservoir sediment core track land-use changes in the watershed. Jour. Paleolimnol., v. 55, pp. 17-33.
Shotbolt, L., Hutchinson, S.M. and Thomas, A.D. (2006). Sediment stratigraphy and heavy metal fluxes to reservoirs in the southern Pennine uplands, UK. Jour. Paleolimnol., v. 35, pp. 305–322.
Solovieva, N., Jones, V., Birks, J.H.B., Appleby, P. and Nazarova, L. (2008). Diatom responses to 20th century climate warming in lakes from the northern Urals, Russia. Palaeogeogra., Palaeoclimatol., Palaeoecol., v. 259, pp. 96-106.
Stoermer, E.F., Emmert, G., Julius, M.L. and Schelske, C.L. (1995). Paleolimnologic evidence of rapid recent chance in Lake Erie's trophic status. Department of Fisheries and Aquatic Sciences, 7922NW 71st stress, university of Florida, Gainesvile, FL 32653, USA. JL 3070, pp. 45–53.
Stoermer, E.F. and Ladewski, T.B. (1976). Apparent optimal temperatures for the occurrence of some common phytoplankton species in southern Lake Michigan. Publ. Great Lakes Res. Div. No. 18.
Sun, Q., Wang, S., Zhou, J., Chen, Z., Shen, J., Xie, X., Wu, F. and Chen, P. (2009). Sediment geochemistry of lake Daihai, north- central China: implications for catchment weathering and climate change during Holocene. Jour. Paliolimnol., v. 43, pp. 75-87.
Talling J.F. (1957). Photosynthetic characteristics of some freshwater plankton diatoms in relation to underwater radiation. New Phytologist, v. 56, pp. 38-50.
Trefethen, J.M. (1950) Classification of sediments. Amer. Jour. Sci., v. 248, pp. 52-62.
Veena MP, Achyuthan H, Eastoe C, Farooqui A (2014) A multi-proxy reconstruction of monsoon variability in the late Holocene, South India. Quarter. Internatl., v. 325, pp. 63-73.
Wang, S.M., Tu, Y.S., Wu, R.J. and Freng, M. (1990). The Daihai Lake: Enviromnent Evolution and Climate Change. *University of Science and Technology of China Press* Hefei, pp. 1–191 (in Chinese).
Woodridge J, Davies H, Blake W, Fyfe R (2014) Recent environmental change in an upland reservoir catchment: a paleolimnological perspective. Jour. Paleolimnol., v. 52, pp. 229–244.
WRD (2022). Water Resource Division, Left Bank Canal, Nashik District, Government of Maharashtra.

Xu, H., Hou, Z.H.., Ai, L. and Tan, L.C. (2007). Precipitation at Lake Qinghai, NE Qinghai-Tibet Plateau, and its relation to Asian summer monsoons on decadal/ interdecadal scales during the past 500 years. Palaeogeogra. Palaeoclimatol. Palaeoecol., v. 254, pp.541-549.

Yeloff, D.E., Labadz, J.C. and Hunt, C.O. (2005). The suitability of a south Pennine (UK) reservoir as an archive of recent environmental change. Jour. Paleolimnol., v.34: 339–348.

Part IV
Climate Changes in Archaeological Sites

Chapter 8
Geological Tools Deciphering Tsunami as the Guiding Factor for Shaping the Destiny of the Ancient Harappan City, Dholavira, Gujarat, India

Rajiv Nigam, Rupal Dubey, Victor J. Loveson, Rajeev Saraswat, Vasant S. Shinde, P. D. Sabale, and Ajit

Abstract Dholavira, located in the Rann of Kutch region of Gujarat, India, is an ancient archaeological site that provides a fascinating glimpse into the Indus Valley Civilization (IVC). It is one of the five largest Harappan sites and gained further attention after it was declared by UNESCO as the 40th site of world cultural heritage in India. Here, we report the oldest known evidence of a possible impact and protective measures employed to thwart tsunamis, from the Harappan port town of Dholavira, in the state of Gujarat, India. Dholavira had a citadel with fortification, with a maximum wall thickness of up to 18 m, which is extraordinary. Based on (i) the location of Dholavira along a tsunami/storm-prone coast, (ii) the climatic scenario during the Mid Holocene with special reference to a higher sea level and (iii) the lack of proto-historic military evidence, we surmised that the walls may have acted as a defence against tsunamis. The presence of extremely thick protective walls thus implies that the Dholavirans were probably aware of the magnitude of destruction caused by tsunami waves. Further, in order to assess whether a tsunami had occurred at this site, we did surveys in the middle town area. A homogenous, thick sediment layer was mapped in a detailed ground-penetrating radar survey. The layer contained reworked benthic foraminiferal shells deposited. The layer also contained the bones of domestic animals. AMS dates of these bones suggested an age of ~1851±26 years BP. The homogenous nature of the layer and the presence of marine microfossils in it clearly suggest one-time transport of sediments from the marginal marine region, against the gravity, under the influence of a tsunami. In view of the foregoing, we propose that the unusually thick walls surrounding the Dholavira settlement were a protective measure against extreme oceanic events. A tsunami did occur in this region

R. Nigam (✉) · R. Dubey · V. J. Loveson · R. Saraswat
National Institute of Oceanography, Goa, India
e-mail: rajivnigam1954@gmail.com

V. S. Shinde · P. D. Sabale · Ajit
Deccan College, Pune, India

N. Khare (ed.), *Changing Climate over India*,
https://doi.org/10.1007/978-981-92-1098-5_8

around 1850 years before the present, affecting the outside citadel. Whether an unusually thick wall could protect the citadel, detailed surveys by GPR and analyses of sediments through coring/trenching inside the citadel area are highly recommended.

1 Introduction

The Holocene period is marked by the rise and demise of several well-developed human civilizations. The climatic changes, especially the monsoon intensity, extent and seasonality, were crucial for the survival and proliferation of these ancient civilizations (Giosan et al. 2012; Saraswat et al. 2016; Sarkar et al. 2016; Khonde et al. 2017). One such civilization in the Indus Valley, the Harappan Civilization, is known for its magnificent growth and prosperity, evident in their efficiency and accuracy in architecture, agricultural practices, trade supremacy and urbanization (Corbett and Winebrake, 2008; Possehl 2002). An important Harappan settlement known for being the centre of maritime trade is Dholavira, located in the north-western state of Gujarat, India (Bisht 2015). It is the second largest archaeological site in India and the largest port-town of Harappan times, found in India. The Harappan settlement at Dholavira lasted for about 1500 years (from 5000 to 3450 years BP). The archaeological excavations unearthed that the township comprised three parts, namely the Castle, the Middle Town, and the Lower Town. The three architectural units were built during different stages of development (Bisht 2015). Although the hierarchical habitation is a part of the Harappan architecture (Petersen 2012), that at Dholavira is different than other Harappan sites, in being protected by city walls (Bisht 1984; Ghosh 1989; Nath 2015; Rao 1979). These walls are extraordinarily thick (12–18 m), apparently built as a protective measure. Intriguingly, walls of such thickness are not reported even during the historical times when the conflicts were more frequent, and the weapons became increasingly destructive. Therefore, the real purpose of the massive walls in Dholavira has been a topic of considerable debate. Based upon the proximity of the town to the palaeo-shoreline, the status of sea level at the time of Dholavira and reports of tsunami in the north-eastern Arabian Sea, it was hypothesized that the thick walls were built to protect the town from extreme oceanic events such as tsunami (Nigam et al. 2016).

A tsunami transports a huge amount of marine sediments onto land, against the gravity. The unique nature of marine sediments implies its easy characterization. Only the nature of sediments, however, is not sufficient to infer transport from the deeper region. The presence of marine microfossils, especially foraminifera, is a definite indicator of the gradual or sudden marine ingress. The anomalous presence of benthic foraminifera in inverted sediments of the shallower region has been used to infer past storm/tsunami events (Nigam and Chaturvedi 2006; Mamo et. al., 2009). Additionally, a gradual transgression will result in definite layering in buried sediments. Contrary to that, a sudden sporadic event like the tsunami will dump the entire material in a short duration resulting in the lack of definite layering.

The objective of this paper is to use a multi-proxy approach to study whether any tsunami happened at the site of Dholavira. If yes, whether the fortification by construction of thick wall around the city could provide any protection.

2 Study Area

Dholavira is located in the north-western corner of the Khadir Bet, an island in the Great Rann of Kuchchh, Gujarat, India (Fig. 1). The Khadir Bet Island lies almost in the centre of the Rann and is uplifted in its northern part, as a result of a fault (Chowksey et al. 2011; Merh 1995). This upliftment exposed the oldest Mesozoic strata representing the beginning of Mesozoic sedimentation on the Precambrian basement. The southern half of Khadir Bet Island slopes down into the Rann of Kuchchh. Kuchchh lies just below the Tropic of Cancer and therefore experiences sub-tropical climate. The study site is characterized by a highly undulatory terrain sloping towards south-west direction. The settlements are located just 1.7 km away from the present coastline. Two dried rivulets surround the Dholavira archaeological remains. The rivulets were rain-fed and used for navigation, in ancient times. The western rivulet was used for transport and the minor eastern river was utilized for domestic water usages. Several rectangular tanks and check dams to store water for domestic use were excavated along the eastern rivulet (Bisht 2015).

3 Methodology

3.1 Sub-Surface Survey Using Ground Penetrating Radar (GPR)

To ascertain the hypothesis of thick walls being a protective measure against storm surges and/or tsunami, a detailed field survey was carried out in the year 2016. Only a part of the Dholavira settlement has been left exposed after the excavation (from the year 1989–90 to 2004–05) under the aegis of the Archaeological Survey of India (ASI). Therefore, a ground penetrating radar (GPR) survey was done in the Middle Town and the Lower Town area, to understand the sub-surface structures and outline a potential area for sample collection (Fig. 1). The SIR 20 GPR system developed by the GSSI, USA, was used for the survey. Before the survey, the system was initialized and tuned to the field conditions by setting the relevant parameters (gain, dielectric constant, scan scale, filters setting, etc.) and validating the measuring wheel for accurate horizontal readings. The GPR sub-surface profiles were generated along north–south direction by using 200 MHz antennae with 6 m (100 ns) depth penetration (Fig. 2a). The sub-surface profiles were processed with RADAN software (v7). The low-pass filtering was applied to remove high-frequency noises and subsequently,

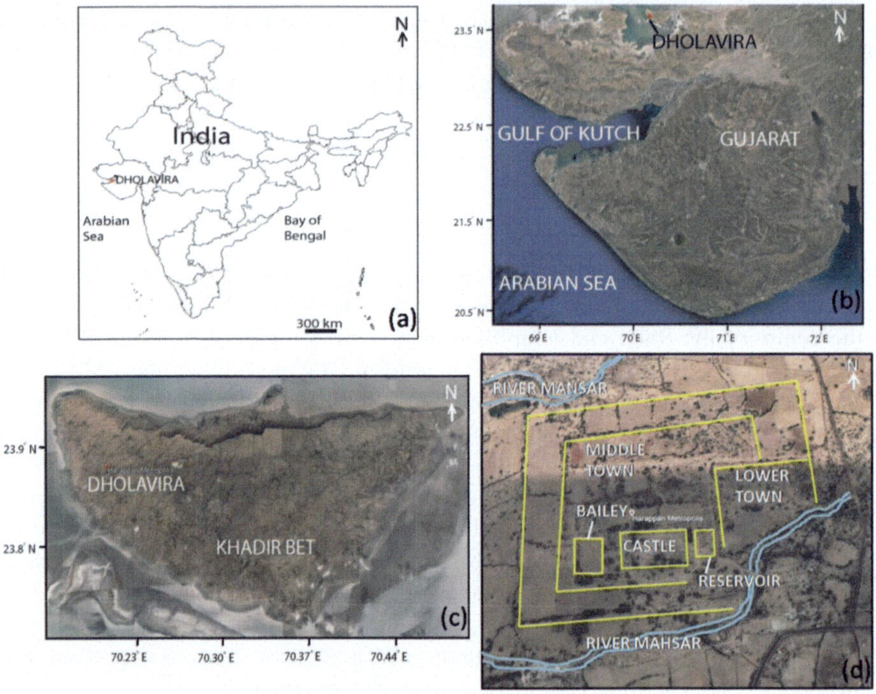

Fig. 1 Location and study area map of Dholavira, Gujarat, India. **a** India; **b** Gujarat state of India; **c** Khadir Bet in Gujarat State of India; **d** Dholavira

the flat-lying ringing system noise was eliminated by a high-pass filtering technique. The deconvolution and migration techniques were used to get the ringing multiples and the diffraction, respectively. The normalized time domain data was enhanced with Hilbert transformation algorithms. By using the algorithm, the time domain radar data were transformed into magnitude and frequency domain data to enhance some of the subtle subsurface features. The converted information typically exhibits significant dielectric changes and variation of some of the features based on the respective magnitude reflection characteristics.

3.2 *AMS Radio Carbon Dating of Bone Sample*

For verification of the GPR depth profile, a trial pit of 2.50 m × 2.50 m × 3.25 m was dug (Fig. 3). From this pit some bone samples were collected. As per the GPR profiles and visual observations, the sediments in this pit looks to be one time deposit, establishing chronological setting was not required.

The age of one bone sample was established by Accelerator Mass Spectrometer radiocarbon analysis at the Direct AMS Radiocarbon Dating Service, USA. The

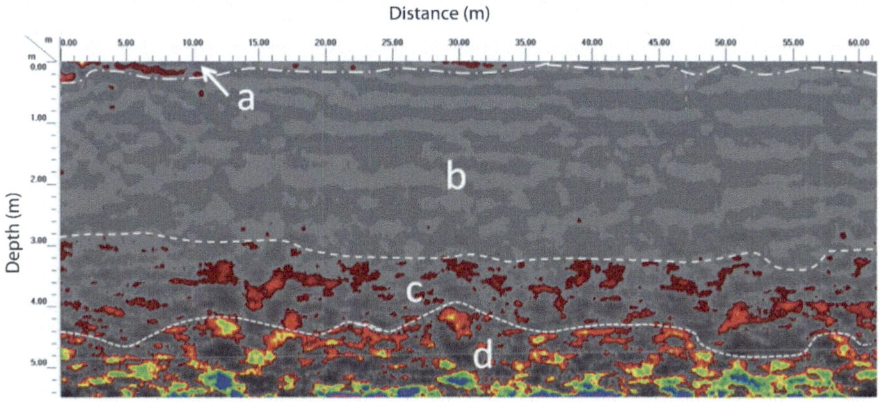

Fig. 2 Time domain sub-surface profile of GPR at middle town at trench site, Dholavira. **a** Top soil; **b** Thick homogeneous sedimentary layer; **c** Middle sedimentary layer indicating scattered fragments of boulders; and **d** Bottom sedimentary layer showing demolished structural remains

Fig. 3 The trench dug up to collect sediment samples

Table 1 The age of one bone sample was established by Accelerator Mass Spectrometer radiocarbon analysis at the Direct AMS Radiocarbon Dating Service, USA. The radiocarbon age (1851 ± 26 yr) was calibrated by using IntCal20 calibration curve in Calib8.2 program (Murray and Wintle, 2003). The calibrated age was 1761 yr BP (2 σ age range 1709–1825 yr BP)

DirectAMS code	Submitter ID	Sample type	Fraction of modern		Radiocarbon age	
			pMC	1σ error	pMC	1σ error
D-AMS 031,928	DHL/ 6–8-2016	Animal bone (calcined)	79.42	0.26	1851	26

radiocarbon age (1851 ± 26 yr) was calibrated by using the IntCal20 calibration curve in the Calib8.2 Program. The calibrated age (Table 1) was 1761 yr BP (2 σ age range 1709–1825 yr BP).

3.3 *Study of Foraminifera*

The sediment samples were processed by following the standard procedure for foraminiferal studies. An appropriate aliquot (50 g) of the sediment was freeze-dried. The dried sediment was weighed and then soaked in distilled water overnight. Subsequently, the sediments were wet-sieved using a 63 μm sieve. The coarse fraction retained on the sieve was finally oven-dried, weighed and stored in plastic vials. Foraminifera were picked from ~ 25 g coarse fraction, obtained by coning and quartering.

3.4 *Archaeological Artefacts (Broken Pottery and Bones)*

Broken pottery and bones were also recovered from the pit (Fig. 4). The pottery pieces and bones, however, did not occur continuously throughout the pit. The depth intervals of pottery pieces and bones were recorded. In the laboratory, the pottery and bones were washed carefully under gentle shower using a fine brush to thoroughly clean the material. The maximum and minimum dimension of all the pottery pieces were measured, to designate a size range for each piece of pottery. In case of bone, the maximum length was considered as the index for bone size. The bone shape and colour were also recorded. Identification of the bone pieces was attempted to generate archaeo-zoological records. However, very small bone fragments that were beyond the recognizable size limit could not be identified. An aggregate of cut-stone boulders was also found in the pit, between 0.9 and 1.3 m depth. The dimensions of cut-sandstone boulders (stone slabs) were recorded.

Fig. 4 Various potteries (**a**) and mammal bones (**b**) collected at the excavation trench at Dholavira

4 Results

4.1 Sub-Surface Information from GPR Profiles

Four sedimentary horizons were delineated based on the reflection amplitude characteristics of the GPR profiles. The horizons include (i) top soil (~20–40 cm thick) (Layer-a); (ii) thick sedimentary layer (~2.8 m thick) (Layer-b); (iii) middle layer (~1.2 m thick) with medium-bright reflections (Layer-c); and (iv) bottom layer (~1.4 m thick) (Layer-d) (Fig. 2). The Layer-b has a uniform and comparatively less amplitude reflection, suggesting a homogenous composition throughout the entire layer. But the Layers-c and d have more reflection amplitude in patches throughout. The patchy reflection mainly indicates scattered fragments of buried structures. Layer-d is full of demolished ancient structures as indicated by continuous strong and bright reflection at 4.0 m depth. To validate the reflection characteristics in various layers, the litholog of the trial pit was correlated with the GPR profile. The lithology of the freshly exposed sediment section matches well with the GPR profile.

4.2 Sub-Surface Information from Trenching

The broken pottery pieces were found from 2.25 m downwards. The broken pottery pieces (size ranges from 1.0 to 15.5 cm) were recovered between 2.25 and 2.60 m pit depth (Fig. 4a). The bone pieces were recovered between 2.25 and 2.65 m pit depth.

The very small bone fragments could not be identified, whereas those identified were found to be broken parts of mammalian bones (Fig. 4b). The age of bones shows an age of 1709–1825 years BP.

A careful examination of the coarse fraction (>63 μm) revealed the presence of calcareous benthic foraminifera in the sediments (Fig. 5). The benthic foraminiferal shells were reworked, with dull and abraded surface without ornamentation. The reworked nature of the benthic foraminifera suggests their allochthonous origin. All the foraminiferal shells are rounded symmetrical.

5 Discussion

In the GPR subsurface time domain profile, the remarkable feature is the homogenous sediment layer (zone b) with a lesser amplitude response up to 3.0 m depth. The layer indicates a saline loamy soil. This sediment layer is also seen in other nearby areas, especially in the Middle Town located closer (1.5 km) to the sea. The homogeneous nature of the layer suggests that the sediments were deposited by a single event. This could probably imply deposition by an extreme event like tsunami. At 1.3 m, a few cut-blocks (stone slabs) of sandstones were found which do not seem to define any structure (Fig. 3). There are two possibilities of its occurrences, viz. (i) either the dislocated/detached part of any nearby unearthed structure, or (ii) transported to the site of burial. The former possibility could be due to haphazardness associated with the natural burial processes. But the latter is possible only in case of mass sediment transport under high energy conditions.

None of the pottery pieces is intact length-wise from the rim to the base. The pieces of pottery represent a part of the rim, curved body or base. A complete rim or base of the pot (as a whole piece) has not been found. The pottery pieces represent < 25% of intact pottery preservation. It was observed that pottery pieces have both the smooth and sharp edges. The sharp edges are attributed to shear and tear during transport under high energy condition. Thus, the recovery of very small pottery pieces also gives a clue of some kind of destruction that the region must have witnessed. The bone fragments were also found from the pit. The fragmentation of the bones much like the pottery must have occurred under severe destructive forces. This can be said with confidence because (a) Harappans had mastered the practice of domestication and would in no case leave their dead animals unburied (Bisht 2015) and (b) if animals died their natural death due to abandonment of the site or any other cause, then also preservation of whole animal skeleton would occur at one place. In either case, the intact animal skeleton is expected. The presence of scattered boulders (zone c) only from ~ 3.0 to 4.5 m depth further confirms the single time demolition. Heaps of boulders are not observed around the structural remains, rather, uniform thickness of distribution of scattered boulders are noticed. In case of Dholavira, the mass transport of sediments (and boulders) can occur by (i) flash floods or (ii) tsunami (Dawson and Shi, 2000). The possibility of any major flood disaster from the area has been

Fig. 5 Benthic foraminifera, recovered from the sediments collected in the trench, excavated in Dholavira. **a** Lagena sp.; **b** Eponides repandus (Fichtel and Moll 1798); **c** Pararotalia boltovskoyi Jain and Bhatia 1978; **d** and **e** Pararotalia calcar (d'Orbigny 1926); **f** Pararotalia sp 1; **g** Pararotalia sp 2; **h** and **i** Pararotalia sp 3; **j** and **k** Ammonia sp 1; **l** and **m** Ammonia sp 2; **n** Ammonia sp 3; **o** Asterorotalia dentata (Parker and Jones 1865); **p** and **q** Asterorotalia sp.; **r**, **s** and **t** Pseudorotalia sp.; **u** Rotalidium sp.; **v** Elphidium sp 1; **w** Elphidium sp 2; **x** Elphidium sp 3; **y** Unidentified foraminifera. Scale bar is equal to 50 μm

negated (Bisht 2015). Due to its proximity to the sea, tsunami can be a possibility. But to ascertain the idea, marine signatures should be present in the sediments.

The sediments of the homogenous layer contain reworked calcareous benthic foraminiferal shells. Foraminifera are preferentially marine unicellular microorganisms with an exoskeleton, abundantly present in beaches to bathyal zones (Boltovskoy and Wright 1976; Murray 1991). The shell bearing benthic foraminifera being exclusively marine are an excellent indicator of the marine environment in depositional sequences (Saraswat and Nigam 2013). The reworked nature of foraminiferal shells points towards its transport from another region, before deposition (Murray 1991). The presence of foraminiferal shells in the sediment samples collected from the Dholavira can be explained in several ways. The elevated sea level during the past is one such scenario. The palaeo-sea-level records of the region, however, suggest that at the time of Dholavira settlement, the sea level was only ~ 4 m higher as compared to today (Hashimi et al. 1995). The high sea level stand during the Dholavira time, although mean that the sea was closer to the settlement, but was still not close enough to lead to direct deposition of the sediments. The reworked foraminiferal shells suggest that it might have been blown by the winds over a long period. The lack of layering in the sediments, however, rules out the possible deposition of these sediments either during the time of elevated sea level or by winds. Therefore, marine sediments can be deposited in the Middle Town area only by extreme events.

The extreme oceanic events cause re-suspension, mixing, landward transportation and dispersal of marine sediments (Madsen et al. 1993; Wright et al. 1994; Loveson et al. 2014a, b). The vulnerability of Dholavira port-town to oceanic calamities is due to its location and elevation (Nigam et al. 2016). The Makran subduction zone is the foci of the tsunami, in the northern Arabian Sea, the effect of which is seen along the Gujarat coast (Hafeez 2007; Mahmood et al. 2012). The geomorphology of the coastal area in Kuchchh is such that it increases the effect of tsunami waves when coupled with high tides (Maurya et al. 2008; Shukla et al. 2010). Tsunamis are known to have hit the north-western region of India during the geological past (Nigam and Chaturvedi 2006; Rajendran et al. 2008; Prizomwala et al. 2015) as well as the historical period (Hafeez 2007). Even though the epicentre of the tsunami is in the deep sea, the gigantic waves generated by the tsunami cause extensive erosion of sediments from the deep sea, shelf (Nigam and Chaturvedi 2006) and also from tidal–intertidal depths (Hemphill-Haley 1996). The inland extent and thickness of each such sea storm and/or tsunamigenic deposit varies (Nishimura and Miyaji 1995; Nanayama et al. 2003; Tuttle et al. 2004; Loveson et al. 2014a, b). The study of ancient and modern tsunami deposits along the Japanese coast suggests that tsunami can transport marine sediments to a few tens to hundreds of meters inland (Nishimura and Miyaji 1995). The foraminiferal assemblage of sandy tsunamigenic layer identified in Pantano Morghella lagoon, Sicily, also comprised reworked fauna (Barbano et al. 2009). The signatures of palaeo-storms/tsunami have also been reported earlier from the Kori creek, Kachchh, Gujarat (Nigam and Chaturvedi 2006).

Moreover, a recent study (Nigam et al. 2016) postulated that the Dholavirans were aware of storms/tsunami, as they constructed an unusual 18 m thick protection wall around the citadel area. Unlike, at other archaeological places like Mohenjo-Daro and Harappa and ancient sites in Oman, the destruction by river flooding is ruled out in case of Dholavira as no big river drained the study area. Since the study area is located close to the sea, it is highly likely that the tsunami waves reached the Dholavira settlement. Moreover, the additional effect of tides that have always been on the higher side (8–11 m at present) in this region is also important to note. The seismicity prone Makran thrust zone that has generated high magnitude earthquake seismic waves throughout the historical times is located very near to Dholavira. The above inferences strongly advocate that the Dholavira civilization was impacted by a storm/tsunami.

The bone samples belong to the domestic animals and may be transported from the exposed or semi-cover sediments and were carried by tsunami water and deposited at the present site. Therefore, this age is close to the approximate time of the tsunami which had inundated the Middle Town area of the Dholavira. This implies that the thick sedimentary layer at the trench site at Dholavira (Middle Town) was deposited during 1709–1825-year BP by an extreme marine event (Table 1).

We further propose that the tsunami also tilted the western wall of Citadel, as the wall is inclined outwards, from its base. Earlier, the tilt of the wall was attributed to an earthquake that struck Dholavira settlement during 4150–4450 years B.P. (Joshi and Bisht 1994; Kovach et al. 2010; Bisht 2015). The tsunamigenic force as the possible cause of the tilted wall is further supported by the negligible damage to the Dholavira settlements by several subsequent earthquakes (Rajendran et al. 2008). It is highly unlikely that although a single earthquake damaged the Dholavira settlements, the structure remained unaffected by several other subsequent earthquakes of similar magnitude that struck this region. Thus, we conclude that the thick homogenous sedimentary layer containing reworked benthic foraminifera, unearthed at the northwest corner of the Middle Town of Dholavira, is of marine origin. In view of the viability of tsunami in the area, the marine sediments represent tsunamigenic deposits.

6 Conclusions

The deposition of ~ 2.0 m thick homogenous layer in the Middle Town area of Dholavira with reworked foraminiferal shells clearly suggests that at least one massive tsunami struck the Dholavira township around 1709–1825 years BP as per the calibrated AMS radiocarbon ages. The presence of transported sandstone boulders, pottery fragments and animal bones further confirm the destruction Dholavira has witnessed as a result of this extreme event. The field evidence confirms the proposed hypothesis that the thick walls at Dholavira were a protection against the potential threats from the tsunami. As this work is confined to outside citadel, we do not know whether this tsunami had scaled the unusually thick 13–18 m wall of citadel, and a detailed investigation is recommended inside the citadel by the same methodology adopted in this present study. But it is sure that outer city wall was destroyed by a tsunami and Middle Town area was inundated by the tsunami and thus may be one of the potential reasons for decline of settlements in this region. It is also clear that the Dholavirans were the pioneers in coastal disaster management.

Acknowledgements We thank Dr. S.W.A. Naqvi, Former Director, CSIR-National Institute of Oceanography, Goa, for his help and encouragement in carrying out this work. We are grateful to Dr. Rakesh Tiwari, Director General, Archaeological Survey of India, for permitting field work at the protected site of Dholavira. We thank Dr. A.S. Gaur, Mr. Sundaresh, Mr. Gurudas Tirodkar, Mr. Luis Ryan and late Mr. Bandodkar for their help during the field work.

References

Barbano, M.S., De Martini, P.M., Gerardi, F., Pirrotta, C., Smedile, A., Pinzi, S. 2009. Identification of tsunami deposits in south-eastern Sicily: evidence of the 365 A.D. crete earthquake? *Gruppo Nazionale di Geofisica della Terra Solida*, 1.1, 83–86.

Bisht, R.S. 1984. Structural remains and town planning at Banawali. *In*: Lal, B.B.; Gupta, S.P. (eds.), *Frontiers of the Indus Civilization: Sir Mortimer Wheeler Commemoration Volume*. Janakpuri: Books & Books, pp.89–97.

Bisht, R.S. 2015. Excavation at Dholavira (1989–90 to 2004–05). Archaeological Survey of India, 922p.

Boltovskoy, E., Wright, R. 1976. *Recent Foraminifera*. The Hague: Junk, 515p.

Chowksey, V., Maurya, D.M., Joshi, P., Khonde, N., Das, A., Chamyal, L.S., 2011. Lithostratigraphic development and neotectonic significance of the Quaternary sediments along the Kachchh Mainland Fault (KMF) zone, western India. *Journal of Earth System Science*, 120, 979–999.

Corbett, J.J., Winebrake, J. 2008. The impacts of globalization on international maritime transport activity: Past trends and future perspectives. *OECD/ITF Global Forum on Transport and Environment in a Globalising World* (Guadalajara, Mexico), pp. 1–3.

Dawson, A.G., Shi, S. 2000. Tsunami deposits. *Pure and Applied Geophysics*, 157, 875-897.

Ghosh, A. 1989. *An Encyclopaedia of Indian Archaeology (Vol. 2)*. New Delhi: Munshiram Manoharlal, 430p.

Giosan, L., P.D. Clift, M.G. Macklin, D.Q. Fuller, S. Constantinescu, J.A. Durcan, T. Stevens, G.A.T. Duller, A.R. Tabrez, K. Gangal, R. Adhikari, A. Alizai, F. Filipe, S. VanLaningham, J.P.M. Syvitski 2012. Fluvial landscapes of the Harappan civilization. Procds. Nat. Acad. Sc. 109, E1688–E1694.

Hafeez, H. 2007. The potential of tsunami generation along Karachi and the Makran coast of Pakistan. *Pakistan Journal of Meteorology*, 4, 25–40.

Hashimi, N.H., Nigam, R., Nair, R.R., Rajagopalan, G. 1995. Holocene sea level fluctuations on western Indian continental margin: an update. *Journal of the Geological Society of India*, 46, 157–162.

Hemphill-Haley, E. 1996. Diatoms as an aid in identifying Late-Holocene tsunami deposits. *The Holocene*, 6, 439–448.

Joshi, V.P., Bisht, R.S. 1994. India and the Indus Civilization, pp. 22–31, National Museum Institute, New Delhi.

Khonde, N., Singh, S.K., Maurya, D.M., Rai, V.K., Chamyal, L.S., Giosan, L. 2017. Tracing the Vedic Saraswati River in the Great Rann of Kachchh. Scientific Reports, 7, 5476, https://doi.org/10.1038/s41598-017-05745-8.

Kovach, R.L., Grijalva, K., Nur, A. 2010. Earthquakes and civilizations of the Indus Valley: A challenge for archaeoseismology. Special Paper of the Geological Society of America, 471, 119-127.

Loveson, V.J., Gujar, A.R., Barnwal, R., Khare, R., Rajamanickam, G.V. 2014a. GPR studies over the tsunami affected Karaikal beach, Tamil Nadu, south India. *Journal of Earth System Science*, 123, 1375–1385.

Loveson, V.J., Gujar, A.R., Iyer, S.I., Luis, R.A.A. 2014b. Indications of an extreme event deposits along the west coast of India: evidences from GPR investigations. *Environmental Earth Sciences*, 72, 4155–4166.

Madsen, O.S., Wright, L.D., Boon, J.D., Chisholm, T.A. 1993. Wind stress, bed roughness and sediment suspension on the inner shelf during an extreme storm event. *Continental Shelf Research*, 13, 1303-1324.

Mahmood, N., Khan, K., Rafi, Z., Lovholt, F. 2012. *Mapping of tsunami hazard along Makran coast of Pakistan*. Islamabad, Pakistan: Pakistan Meteorological Department, *Report No. PMD-20*, 30p.

Mamo, B., Strotza, L., Dominey-Howes, D. 2009. Tsunami sediments and their foraminiferal assemblages. *Earth-Science Review*, 96, 263-278.

Maurya, D.M., Thakkar, M.G., Patidar, A.K., Bhandari, S., Goyal, B., Chamyal, L.S. 2008. Late Quaternary geomorphic evolution of the coastal zone of Kachchh, Western India. *Journal of Coastal Research*, 24, 746-758.

Merh, S.S. 1995. *Geology of Gujarat*. Bangalore: Geological Society of India, 222p.

Murray, J.W. 1991. *Ecology and Palaeoecology of Benthic Foraminifera*. New York: John Wiley & sons Inc., 397p.

Murray, A.S., Wintle A.G. 2003. The single aliquot regenerative dose protocol: potential for improvements in reliability. *Radiation Measurements*, 37:377–381

Nanayama, F., Satake, K., Furukawa, R., Shimokawa, K., Atwater, B.F., Shigeno, K., Yamaki, S. 2003. Unusually large earthquakes inferred from tsunami deposits along the Kuril trench. *Nature*, 424, 660–663.

Nath, A., 2015. *Excavations at Rakhigarhi: 1997–98 and 1999–2000*. Archaeological Survey of India, 396p.

Nigam, R., Chaturvedi, S.K. 2006. Do inverted depositional sequences and allochthonous foraminifers in sediments along the Coast of Kachchh, NW India, indicate palaeostorm and/or tsunami effects? *Geo-Mar. Lett.*, 26, 42-50.

Nigam, R., Dubey, R., Saraswat, R., Sundaresh, Gaur, A. S., Loveson, V.J. 2016. Ancient Indians (Harappan settlement) were aware of tsunami/storm protection measures: a new interpretation of thick walls at Dholavira, Gujarat, India. *Current Science*, 111, 2040–2043.

Nishimura, Y., Miyaji, N. 1995. Tsunami deposits from the 1993 southwest Hokkaido earthquake and the 1640 Hokkaido Komagatake eruption, Northern Japan. *Pure and Applied Geophysics*, 144, 719–733.

Petersen, M.C. 2012. Aggressive architecture: fortification of the Indus valley in the mature Harappan phase: Leiden University, The Netherlands, Master's thesis, 77p.

Possehl, G.L., 2002. The Indus Civilization: A Contemporary Perspective. Altamira Press Lanham: MD.

Prizomwala, S.P., Gandhi, D., Ukey, V., Bhatt, N., Rastogi, B.K. 2015. Coastal boulders as evidences of high energy marine events from Diu Island, west coast of India. *Natural Hazards*, 75, 1187–1203.

Rajendran, C.P., Rajendran, K., Thakkar, M., Goyal, B. 2008. Assessing the previous activity at the source zone of the 2001 Bhuj earthquake based on the near-source and distant paleoseismological indicators. J. Geophys. Res., 113, B05311, https://doi.org/10.1029/2006JB004845.

Rao, S.R., 1979. *Lothal, a Harappan Port Town (1955–62). Volume 78*. New Delhi Memoirs of Archaeological Survey of India, 23 p.

Saraswat, R., Naik, D.K., Nigam, R., Gaur, A.S. 2016. Timing, cause and consequences of mid-Holocene climate transition in the Arabian Sea. Quaternary Research, 86, 162-169.

Saraswat, R., Nigam, R. 2013. Benthic Foraminifera. In: Elias, S.A. (ed.) The Encyclopaedia of Quaternary Science, Amsterdam: Elsevier. v. 2, pp. 765–774.

Sarkar, A., Mukherjee, A.D., Bera, M.K., Das, B., Juyal, N., Morthekai, P., Deshpande, R. D., Shinde, V.S., Rao, L.S. 2016. Oxygen isotope in archaeological bioapatites from India: Implications to climate change and decline of Bronze Age Harappan civilization. Scientific Reports 6, 26555, https://doi.org/10.1038/srep26555.

Shukla, S.B., Prizomwala, S.P., Ukey, V., Bhatt, N., Chamyal, L.S. 2010. Coastal geomorphology and tsunami hazard scenario along the Kachchh coast, western India. *Indian Journal of Geo-Marine Sciences*, 39, 549-556.

Tuttle, M.P., Ruffman, A., Anderson, T., Jeter, H. 2004. Distinguishing tsunami from storm deposits in eastern North America: The 1929 Grand Banks tsunami versus the 1991 Halloween storm. *Seismological Research Letters*, 75, 117–131.

Wright, L.D., Xu, J.P., Madsen, O.S. 1994. Across shelf benthic transports on the inner shelf of the Middle Atlantic Bight during the "Halloween Storm" of 1991. *Marine Geology*, 118, 61–77.

Chapter 9
Site Catchment Analysis of Math Pimpri Archaeological Site, Deccan: Past and Present Land Use Patterns to Understand the Palaeoenvironment

P. D. Sabale

1 Introduction

Site catchment analysis applies the concept of a catchment originally used for watersheds to archaeology, defining the area from which a site obtained its resources. The method examines the relationship between technology and the natural resources available within a site's economic reach. Two approaches exist: one works inward from the landscape towards the site using estimated travel ranges, while the other works outward from materials found at the site to identify their nearest sources. It provides the important information regarding the people of particular culture, their subsistence practices, and economic and social organization (Bailey 2005; Hen et al. 2005).

The technique highlights how man–land relationships, especially the distribution and seasonality of resources, influence site location (Schiffer 1975). It serves as an alternative to central place theory by focusing more on resource availability than on settlement spacing or population density. Ultimately, site catchment analysis views archaeological sites as centres of economic activity and evaluates the surrounding area to better understand past locational decisions (Ropar 1979).

In the present paper, the site catchment analyses of Math Pimpri—a quaternary site in Sina river basin in south Ahmednagar district—were carried out by using a scientific method.

Study Area

Quaternary excavation at Math Pimpri a multi-cultural site was conducted in two consecutive seasons, i.e. 2013–14 and 2014–15. The site is located on latitude 18°53.334' N and longitude 74° 52.712′ E and situated on the left bank, in between

P. D. Sabale (✉)
Department of Archaeology, Deccan College, Pune, India
e-mail: sabalepd@yahoo.com

N. Khare (ed.), *Changing Climate over India*,
https://doi.org/10.1007/978-981-92-1098-5_9

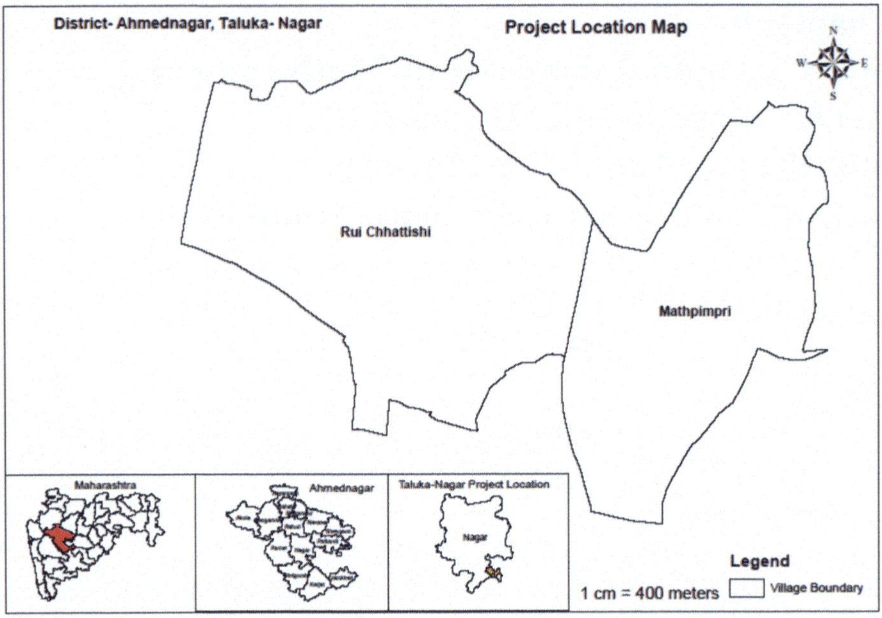

Fig. 1 Location map of the study area

the terrace of an obtuse meander of Sina river (Fig. 1). This locality lies 25 km south of Ahmednagar city headquarters and is well accessible (2 km) from Rui Chattisi, a weekly market nearby the site present on Ahmednagar-Solapur Highway. The river originates in Balaghat hill ranges, north of Ahmednagar city and flows through the highly draught prone regions, like Bid, Osmanabad, and Solapur districts, and, therefore, it is highly seasonal.

The area is seen on Quaternary map 47 J and toposheet map No. 47 J/13 of 1:250,000 and 1:50,000 scales respectively, of Survey of India (SOI 1973 and 1974). Figure 1 shows the study area has been divided into two zones, i. Math Pimpri and ii. Rui Chattisi. A smaller inset map at the bottom center shows detailed village locations within Nagar Taluka. The legend indicates village boundaries with blue lines. A scale bar shows that 1 centimeter equals 400 meters. The map helps identify the specific position and village layout of Nagar Taluka within the larger district area. In addition to Math Pimpri, some areas of Rui Chattisi is taken for the detail site catchment analysis of this site. It is clearly seen in Figs. 1 and 2 that the Rui Chattisi area is bigger than the Math Pimpri area, in all respects. This area is taken only to compare the main study area Math Pimpri, in various ways. Therefore, in this paper only field analyses of Math Pimpri were taken for study.

Previous Work

A detailed geoarchaeological exploration of Sina River from its source to destination was undertaken during field season 2009–10, which includes the parts of Ahmednagar, Bid Osmanabad, and Solapur districts of Maharashtra (Sabale and Kshirsagar 2013). Similarly, a large number of flood-affected sites along the river channels in

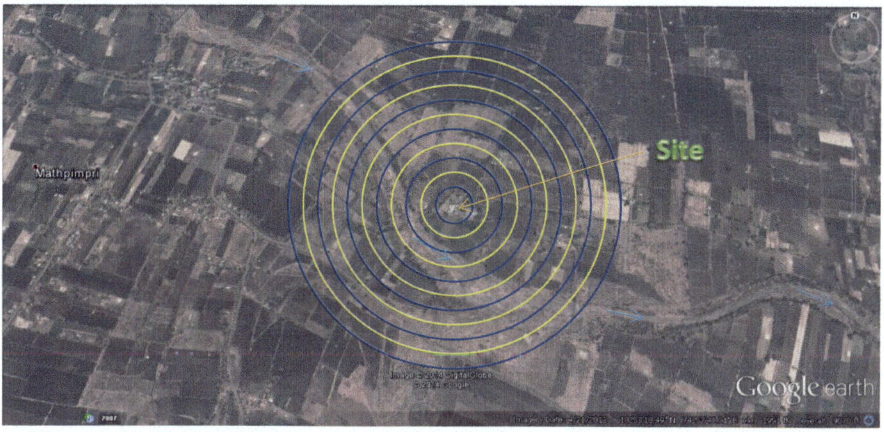

Fig. 2 Local and detailed map of the study area

middle Bhima basin were documented and reported by P. D. Sabale (IAR 2010–11; UGC MRP, New Delhi Report, 2013–14 and ICSSR MRP, New Delhi Report, 2016–17).

In 2012–13 and 2013–14, in two consecutive seasons a detailed quaternary geoarchaeological excavation was carried out at Math Pimpri (Sabale et al. 2014). In this excavation, along with cultural material, animal fossils (Sabale et al. 2021 (in Press)) and human skeletons were also recovered (Tripathi and Sabale 2016) (Sabale et al. 2021).

After Math Pimpri, another quaternary geoarchaeological excavation at Panch Pimpla (Jamkhed Tehsil, Ahmednagar District) in upper reaches (IAR), Khairav (Madha Tehsil, Solapur district) (IAR-) and Virvade (Mohol Tehsil, Solapur District) (IAR) in middle reaches, and Rajur (South Solapur) (IAR, 2020) in lower reaches were carried out in the Sina River basin to understand the man–land relationship of the basin.

Aims and Objectives of the Work

The prime aim of this site catchment is to study the man–land relation of the site. Therefore, this analysis was to serve the following purposes, viz., i. Get a chance to understand the details of environmental variability in relation to archaeological sites; ii. Inferences about economic development relied on the environmental conditions within the sites; iii. Occurrences of often local variation that is of human interest; iv. Locations of the economic practices carried out at individual sites, and which suggest new lines of research and relevant test data; v. Study of faunal and floral remains from archaeological deposits, and more careful analysis of the differential processes by which such materials were collected, transported, discarded, and incorporated in archaeological sediments and ultimately preserved or destroyed; vi. Site record to see how far actual subsistence practices and site functions correspond to their environmental setting, leading to more subtle interpretations of the relationship between environment and economy sites in unexpected locations, or having distinctive archaeological features and catchments compared to other sites in similar locations; vii. This may indicate the use for purposes other than subsistence, such as for defence, procurement of valuable raw materials, the control of trade routes or markets, and for social aggregation; viii. Early agriculture has identified the use of satellite camps to collect or partially process specific resources; viii. The use of a landscape for specific purpose and their daily subsistence activities; ix. The study of rural agriculture abundant and immobile resources such as plant foods; x. To study the extensive grazing basins circumscribed by topographic barriers and their relationship with the hunter–gatherers; xi. To understand the site exploitation territory; xii. Function of individual sites or locations and patterns of selectivity in the choice of some locations and environmental resources in preference to others; xiii. To understand the climate and vegetation change, soil-sediments formed by erosion from hill slopes and River valleys and re-deposited in the valley; xiii. Topography of hills and valleys and its impact on humans; xiv. Analyses of lake and river sediments and pollen sequences for the palaeoenvironmental and palaeoclimate study; and xv. Analysis of scope for agriculture.

Geology and Topography

From a geological point of view, the study area is a part of Deccan Trap formations of the central-western part of India and in the basaltic flows shows varieties of basaltic rocks, like compact, amygdaloidal, vesicular, volcanic breccias, and tachylitic basalt. In these, the amygdaloidal flows are predominant than the compact basalt. As you know the amygdaloidal basalt is rich in gas infillings/zeolitic mineral in the form of gas cavities. Therefore, it acts as a good source of varieties of silicate raw materials. As a result, the site catchment area is resourceful for raw material required for Palaeolithic tools and Mesolithic tools (Sabale and Kshirsagar 2014).

From a topographic point of view, the region is a well-distinguished feature in the succession of plateau of flat-topped hills that are so uniformly horizontal and resemble like present almost table land. The isolated hills and forming backbone of the ridges between the streams, though they do not rise to any great heights than the general level of the plateau, locally they form prominent features (Fig. 3).

Towards the channel side of the river, the elevations become less marked, the slopes are more even, and the country land is a rolling upland plateau gently dipping eastwards; at some sections, the plateau is dotted with numerous isolated hills, a result of differential erosion in a semi-arid tract. Around the channel side also the terrain has become much more rugged which shows low-laying plateau summits at a uniform height of about 500 m. At some areas, a few peaks stand out prominently and form a large elevated tract known as *hill plateau and* such tracts of rugged ground are covered with boulders, a rolling upland, interspersed with hill country land which is locally known as the *munja* (Ahmednagar District Gazetteer 1976). The surfaces are almost covered with large proportions with rock boulders and with practically scanty or no vegetal cover. The entire plateau is a reflection of the erratic and unreliable nature of the monsoon rainfall which is precariously marginal (Fig. 3).

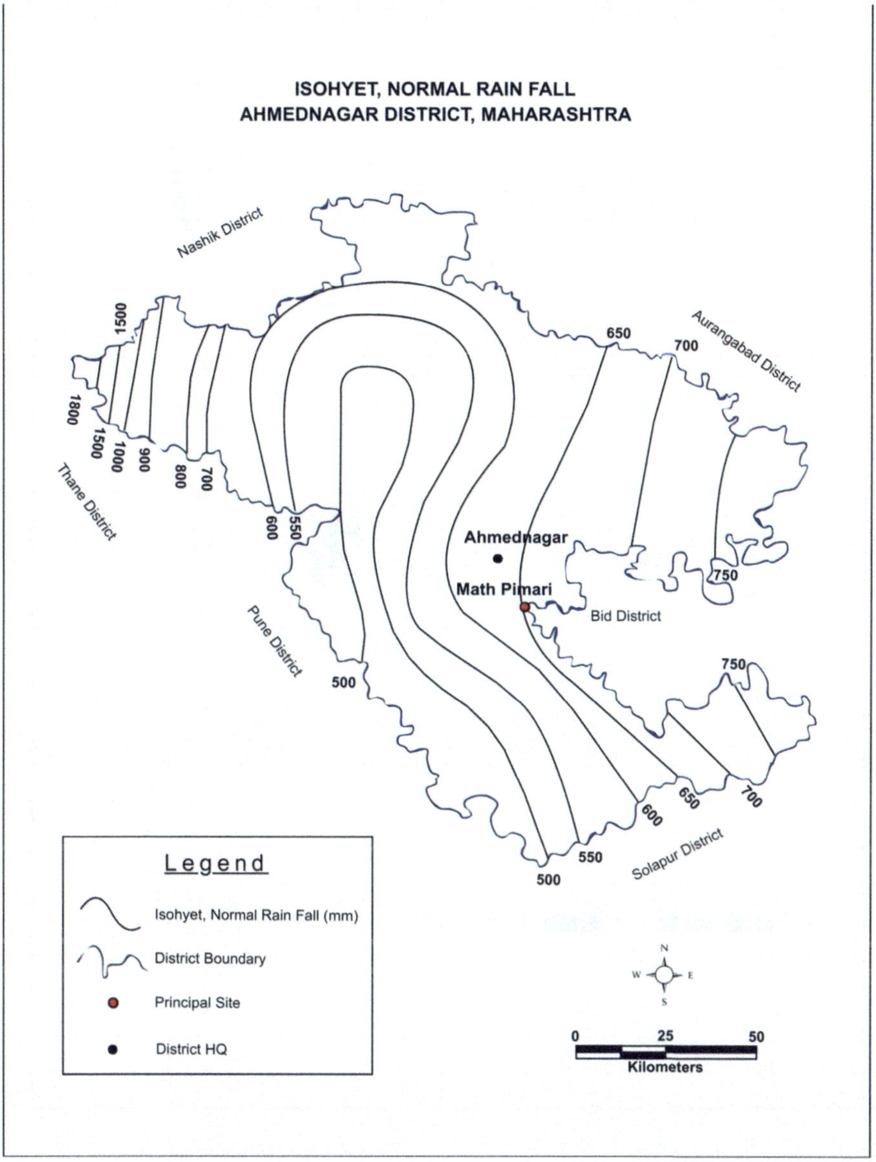

Fig. 3 Rainfall map of Ahmednagar district (Modified after, CGWB, CR, 2018–19)

The whole of this region is a chronic, facing scarcity of water and farming needs to be supplemented by artificial means of water supply. At places, due to the presence of minor irrigation schemes inclusive of construction of numerous wells, farming in this region may have a face-lift since the soils of the region are inherently productive (Fig. 4) (CGWB 2014).

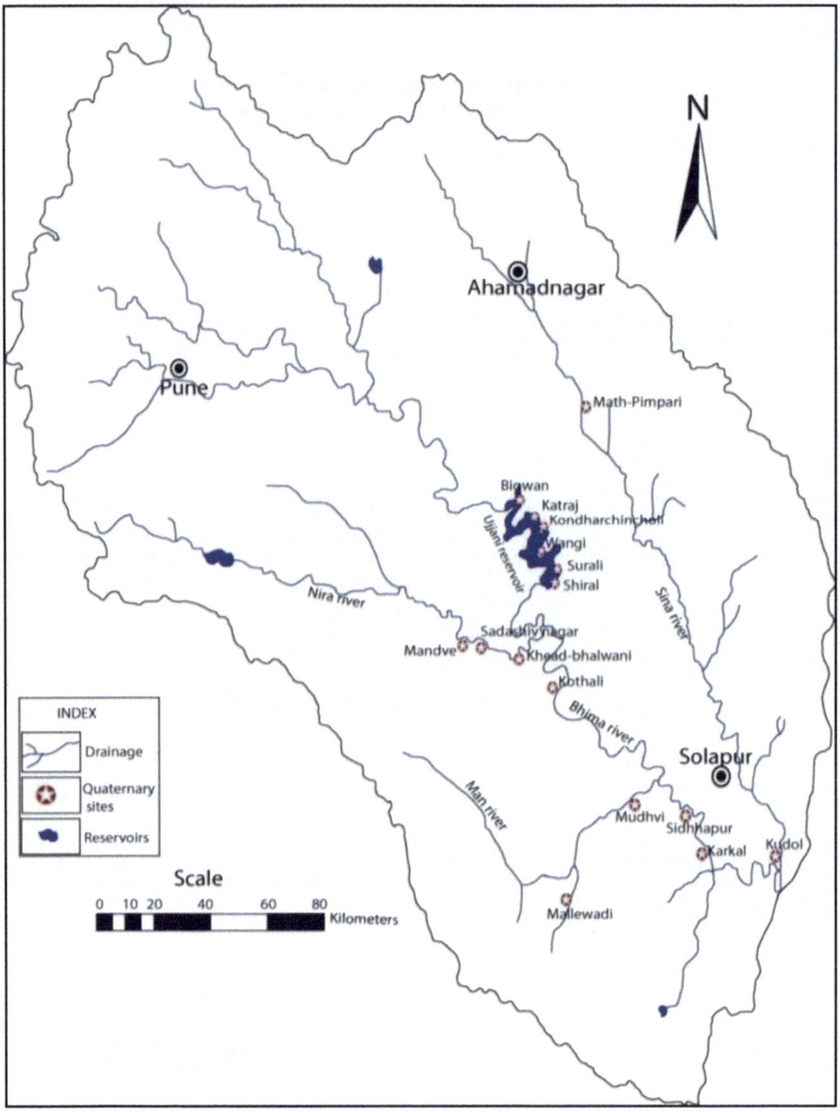

Fig. 4 Drainage map and water bodies present in the Sina River in Bhima Basin

Landforms of Sina River Basin

The drainage system on the earth's surface develops a characteristic type of topography according to its different reaches. Flowing channels shape the earth's surface through the processes of erosion, transportation, and deposition of sediments. The main processes responsible for modifying such landscapes are weathering and erosion. As sediments are eroded from source areas and deposited in low-lying regions, streams continuously reshape the landscape. The sediments deposited during the Quaternary period are particularly important for archaeology because human development occurred contemporaneously.

Major rivers flowing through diverse geological and geomorphological settings are significant for understanding sediment transport characteristics, stratigraphic development, flood episodes, associated hazards, and their impacts on human cultures.

It is clearly observed that in the upper (youth), middle (mature), and lower (old) reaches of the Sina River, the rates of weathering, surface dissection, topography, landforms, and drainage characteristics change accordingly. The hilly and rocky terrain in the upper source region becomes undulating to gently plain in the middle reaches, and finally transforms into a broad plain in the lower reaches. Due to the rocky surfaces in the upper reaches, agricultural activity is difficult. In contrast, the middle reaches contain rocky masses surrounded by depressions filled with soil, which are widely used for agriculture. The lower plains contain thick, fertile alluvial soils that support extensive cultivation. Archaeological and geological evidence together can reveal how past societies responded to climate variability, while the present Math Pimpri site belongs to the upper reaches of Sina river, in South Ahmednagar district.

Interpretation of Landscape Characters (Based on Field Investigations)

The Sina river is a major left-bank tributary of the Bhima river in western Maharashtra. It is clearly observed that the youth, mature, and old stages represented in the upper, middle, and lower reaches of the Sina correspond to significant changes in the rate of weathering, surface dissection, topography, landforms, and drainage characteristics (Fig. 4). The hilly and rocky topography of the upper source region gradually transforms into an undulating to gently plain surface in the middle reaches, and further into a broad plain in the lower reaches. Due to the rocky terrain in the upper reaches, agricultural activities are limited. In contrast, the middle reaches contain rocky masses surrounded by depressions filled with soil, making them suitable for agriculture. The lower reaches contain thick and fertile alluvial deposits, which are extensively used for cultivation.

Archaeological and geological evidence reveals how societies in this region have responded to past climatic variability. These records clearly show that societal responses could not have been predicted in advance and that successful adaptation to new climatic conditions depended largely on the choices—either effective or ineffective—made by the communities.

Field Investigations

The present site has different sorts of catchments. The economic catchment may turn out to be smaller than the hypothetical exploitation territory, or to comprise different-sized catchments for different food resources.

In the present study area, on the basis of characters of the landscape, it has been divided into several types of factors:

1. Economic.
2. Scrub.
3. Cultural.
4. Urban.

The economic land is a province which is having more resources of various raw materials. Such resources found in the catchment area are listed as follows.

a. Geological: Since most parts of the study area are rocky, it shows availability of the rock material on the surface as well as in the form of open cast mines. The other sorts of catchments for example 'geological' catchments comprise sources of raw materials for making stone artefacts or ceramics.
b. River and drainage: The characteristics of the main river and its tributaries are analysed.
c. Land use pattern: This shows the observation about the land used for the different purposes.
d. Topography and slope: It includes the different features in connection to the topography and its inclination with reference character of the landscape.
e. Forest: In the hilly terrain especially, a moderate to scanty vegetation is observed.
f. Ethnobotonical plants: Several ethnomedicinal plants from the catchment were recorded.
g. Fertile soil: Major part of the study area is covered with thick, moisture rich, fertile, and well-textured rich soil.
h. Agriculture: Various agricultural patterns are recorded.
i. Biological: Study area is rich in biodiversity.
j. River alluvium: It is spread along the bank of the main river and its trunk tributaries. It is thick, highly extensive, highly fertile tracts, and therefore are used extensively for the agricultural activities. The thickness of this material increases towards the channel and decreases in peripheral areas. At places, it is filled with the depressions along the channels and shows great thickness while at some rocky cliffs and banks, it shows very few curtains of alluvium.
k. Quaternary deposit: Pleistocene and Holocene deposits are present below the upper soil strata.

In addition to these, the Scrub land is a barren landscape in the study area which is free from crop. Similarly, the urban landscape is the area in which residential structures and other anthropogenic activities are found in the base map (Fig. 5). The above-mentioned various types of landscapes observed in the study area are described as follows.

River/Drainage Network

The development of human settlements in the vicinity of the water channel since immemorial time. Because drainages are the source of water supply and water is the life of plant, animals, and human beings, as rivers are the vital source for existence, most of the big and small settlements are established along the channels of different rivers (Higgs 1972, 1975).

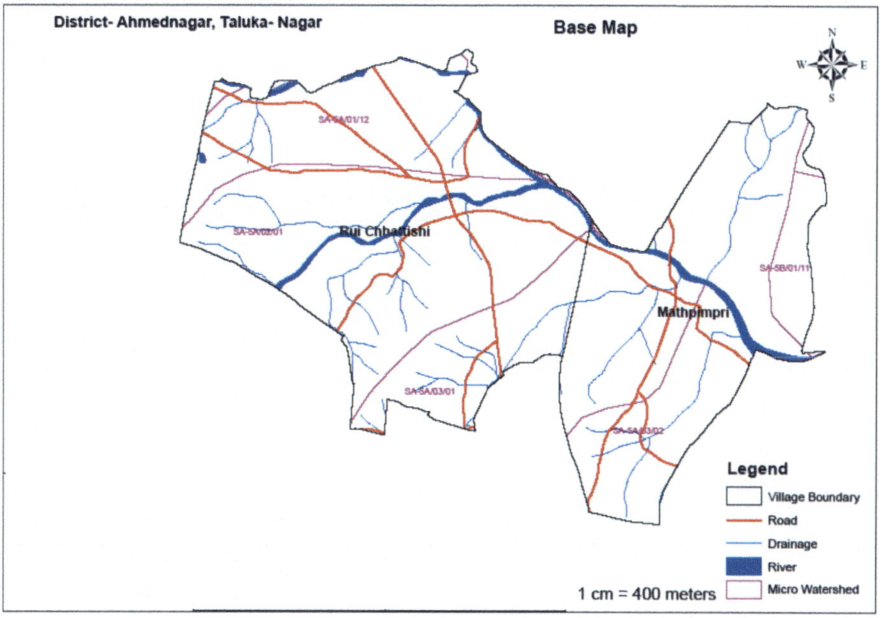

Fig. 5 Base map of the study area

On the same line, in the present context the settlement is present on the island developed in between the present and palaeochannel of Sina River. The supporting settlements to this main are present in the catchment of sub-tributaries of the Sina in the vicinity of the sites. In Fig. 6, it is clearly observed that the drainage density is high on the right-bank channel than on the left. With reference to that more settlements are also found along these particular catchment areas. Overall, the drainage pattern observed in the catchment is a dendritic to sub-dendritic type. These drainages basically originated from the low elevated, deeply weathered, locally developed, circum-denuded hills of undulating to plain topographic land. These drainages show shallow to moderate depth with partially filled immature sediments of the close source region of the Sina River basin.

Utilization of Land and Pattern

Land use and land cover (LULC) change has been one of the most immense and perceptible transformations of the earth's surface (Alam et al. 2020). Since the beginning of human life, mainly activities have accelerated the changes in Land Use and Land Cover. These alterations in the landscape due to exploitation may not seem significant at present, but in the long run they would cause substantial changes in the environment and socio-economic standards. Such important activities of any resourceful region to considerable impact on the forest lands, waste lands and surface water bodies. It is observed more in the Rui Chattisi area along the Ahmednagar to Solapur National Highway belt (Lambin 1997).

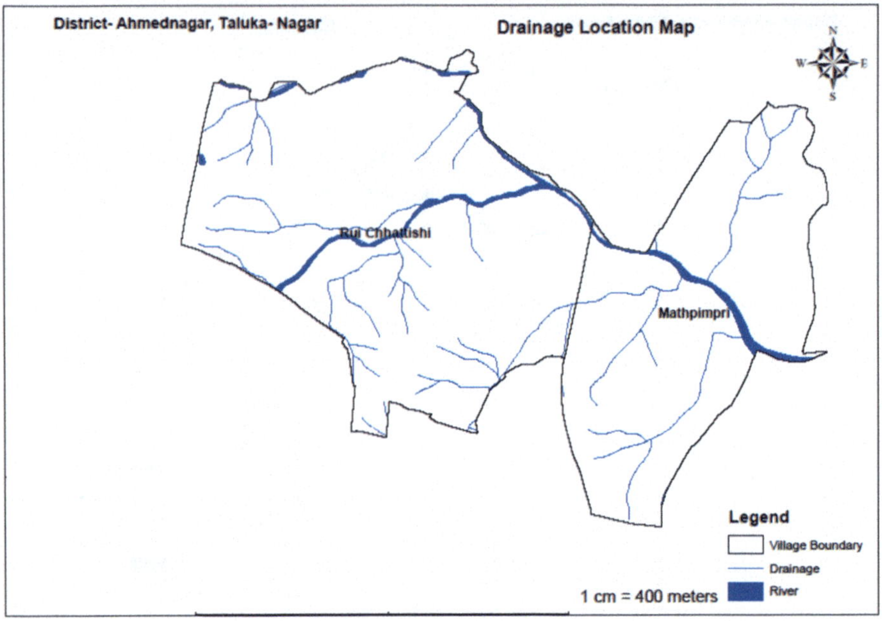

Fig. 6 Drainage map of the study area

Due to the continuous exploitation of land, the long-term indirect impacts occur slowly, but the final results of the same are quite remarkable in nature. We can have various examples of such indirect impacts such as high rate of depletion of water resources, agriculture, forest cover, mineral, and other natural resources which affect local weather, soil quality, crop quality (and quantity), and hence the livelihood of a large mass of people of the region.

It is a process by which anthropogenic activities transform on the earth's natural surface, show use of land, usually showing the functional role of land for economic activities. These changes are often nonlinear and might trigger feedbacks to the system, stress living conditions, and threaten people with vulnerability (Paul and Rashid 2017).

'Land use' refers to 'human activities on and in relation to the land', and are most of the time not directly visible from the satellite imagery (Lo 1986), while on the other hand 'land cover' describes the vegetation and artificial constructions on the land surface (Burley 1961). This factor is an intrinsic element of the landscape and is having direct and indirect links with various earth processes and anthropogenic activities. On the surface of the earth, the rate and scale of human alterations mostly in the form of factors are unprecedented and so invasive that they enormously transformed a large proportion of the planet's land surface, which affect important aspects of earth systems (Foley et al. 2005).

In the present context, the land in the catchment area, on the basis of the above factors, has been divided into mainly four types, i.e. i. agriculture; ii. wasteland,

iii. Built-up, and iv. water bodies. In this land use classification, most of the land (2371.26 ha) shown with green colour comes under agricultural activities, while some land especially in the rocky terrain shown with yellow colour is wasteland, which occupies a significant percentage (102.22 ha). The area under settlements shown with red colour is used for the construction of residents and is the built-up area, which occupies a very low (14.94) proportion of the total area. Only stream and river channel area is covered with the water bodies, shown with blue colour (Fig. 7).

Topography and Slope

Slope is the angle of inclination, steepness, or gradient of a straight line and is often used to describe the steepness of the ground's surface. The terrain slope is an important factor in evaluating land suitability for the establishment of archaeological sites on the surface of the earth.

Especially, the basin has a mean elevation of 650 m with a slope towards the southeast indicated by the direction of the Sina River, and its minor branches shoot out giving a varied and rugged topography. This plateau region along the basin area is dotted with hills all over but the country presents a more rugged appearance. Most of the time, the stream gradients vary, and show the terrain over which they flow in every stage of their journey, in the case of most of the large rivers, while the small tributaries often show steep slopes (at gradients of 60–100%).

In the study area, the landscape is divided into three categories, i.e. i. gentle; ii. Medium, and iii. Steep slopes. Most of the area is occupied by shallow slope

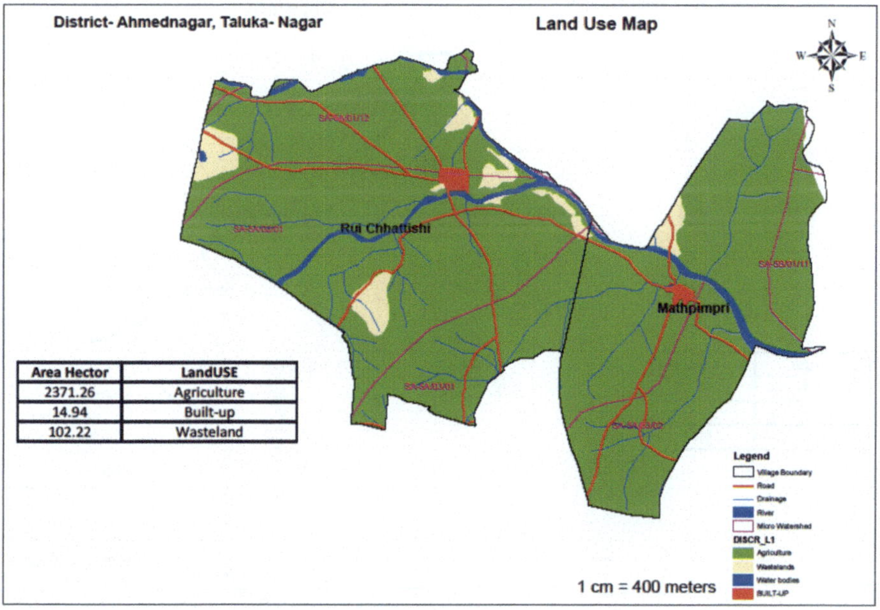

Fig. 7 Land use cover of the study area

(1461.94 ha) and very few (255.1 ha) areas are occupied by steep slope. Since the major part comes under gentle slope category, most parts of the landscape are suitable for the establishment of archaeological sites, in these catchment areas (Fig. 8).

The word *slope* used in a *general* sense refers to any geometric element of the earth's solid surface; the entire landscape is made up of slope elements, and thus the science of geomorphology is largely concerned with the study of slopes. This term is often used to refer to the angle which any part of the earth's surface makes with a horizontal datum. Better terms for this angle include inclination, slope angle, and slope gradient. Most of the time, on slopes with saprolite and other regolith, *soil creep* may normally occur. The rate of creep shows variation with season to season, more values occurring when the soil is unfrozen or when the soil is saturated with moisture.

Soil Characteristics

In the study region, the soil varies considerably depending on the terrain and slope conditions. On the rocky surfaces, the soils are shallow and mixed with calcrete (lime). The soils on the plain terraces are of relatively inferior quality, while the hill slopes consist of stony soils that are generally unsuitable for crop production. In contrast, the low-lying areas, bordered by undulating and broken ground, show a significant amount of deposition of black cotton soils, making these zones fairly suitable for agriculture.

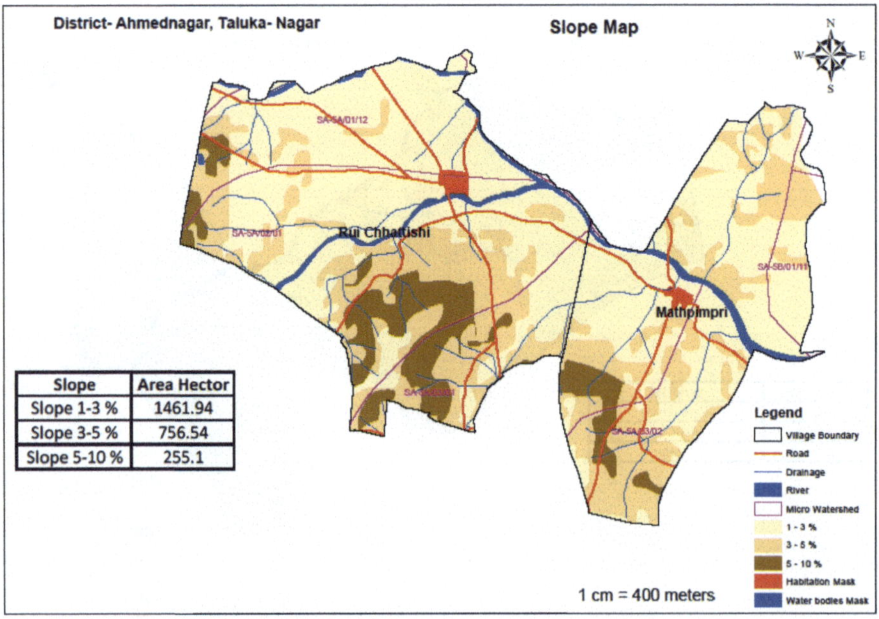

Slope	Area Hector
Slope 1-3 %	1461.94
Slope 3-5 %	756.54
Slope 5-10 %	255.1

Fig. 8 Slope map of the study area

Soil Types

The main soils of this region are *kali* or black cotton soil with a variety of gradations depending upon the local terrain conditions and slope. They are derived from weathering of basalts under tropical semi-arid climatic conditions, are *regurs* (black cotton soil), a type of clay loam that is moisture-retentive. Therefore, such soil in this region is more suited to wheat than to cotton, excessively sticky and hard to work during rains and full of cracks in the hot weather. Being of alluvial and friable nature, the soil needs less moisture and is more easily worked than the stiffer clayey loams on plains.

The crop cultivation in this area extends almost up to the banks of the river except where the banks being stripped off their surface soils by gully erosion present a desolate barren aspect. At places, farmers cultivated up to the edge, i.e. at the border of the steep and high banks. In the dry season when the streams occupy only small portions of water in its channels, in many places the crops of wheat and vegetables are raised on the alluvial deposits within the bed of the river while the gravelly stony soils are generally planted with melons.

At places, the plains are skirted by low hills with flat summits while large tracts are covered with boulders and gravels with level tracts in which soils are comparatively deeper and richer. In general, the land slopes towards the rivers.

As far as economic conditions of local peoples are concerned, the study region is the most backward area of the district and records lower densities of rural population. The settlements are large sized and compact, at a considerable distance from each other, invariably dependent upon well-water supply along the seasonal stream lines. Therefore, very few market centres are developed due to the low agricultural production (Ahmednagar District Gazetteer 1976).

Soil Thickness

In the basin area, the soil is formed from the basaltic rock, mainly by physical and chemical weathering especially of compact and amygdaloidal (predominantly); and tachylitic and volcanic breccias (rarely) by fluvial, wind, and thermal atmospheric agents. This is basically a black cotton soil, predominantly made up of mafic minerals from the source rock. In general, the thickness of this soil increases towards the river channel and decreases towards the ridge lines.

On the basis of thickness of soil in the catchment area, we can divide it into 5 types, i.e. i. Very deep; (>100 m), ii. Deep (50–100 cm); iii. Moderately deep (25–50 cm); iv. Shallow (10 to 25 cm); v. Shallow to very shallow (<25 cm). The details of the distribution of this soil are given in the index table of Fig. 9. On the basis of their area of coverage in hector, shallow to very shallow soils occupy the largest areas (2146.64), while moderately deep soil covers very small areas (36.21) of the site catchment. The whole Math Pimpri settlement is present on the deep to shallow category of the soil depth (Fig. 9).

Soil Texture

This is an important character that influences storm water infiltration rates. Such well-drained characteristic of soils contains air that is similar to atmospheric air, which is conducive to healthy root growth, and thus a healthy crop. This will also

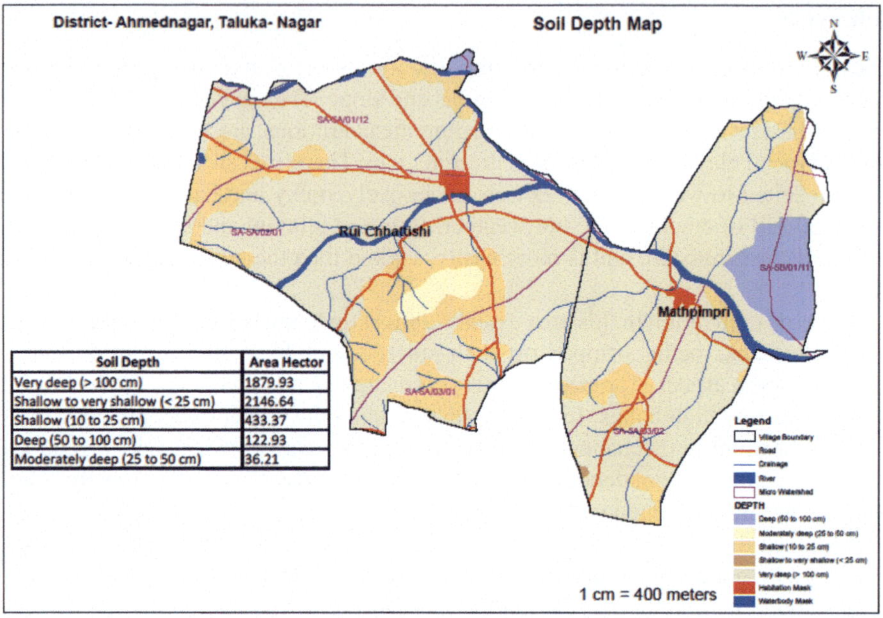

Fig. 9 Soil depth map of the study area

help to improve drainage and reduce soil *erosion* resulting due to the excess surface run-off.

As we know, human use and management of soil and water resources have shaped the development, persistence, decline, and regeneration of human civilizations that are sustained by agriculture (Harlan 1992; Hillel 1992). Therefore, soil and water are essential natural resources for our domesticated animal- and plant-based food production systems. During the last ten millennia, in agriculturally based societies, humans have developed complex, urban civilizations that have cycled through periods of increasing complexity, awe-inspiring intellectual achievement, persistence for millennia, and, in some instances, perplexing decline (Trigger 2003). Many times, the stressed, declining civilizations adapted, or reemerged, into new or similar complex cultures. With such fluctuations, we have remained dependent on a relatively small number of crop and animal species for food, and on integrated soil–water systems that are essential for their production. As far as our social evolution is concerned, it started since the agricultural revolution, and took place synergistically with human biological evolution, as we have become dependent on domesticated plants and animals for human development with the help of soil–water systems (Parikh 2012).

On the basis of the soil textural characters, the soil in the catchment area has been divided into six types, i.e. i. Clay loam; ii. Clayey; iii. Gravely clay loam; iv. Gravely sandy loam; v. Sandy clay loam; and vi. River Island. In this case, the gravely sandy loam occupies large areas (392.49) while River Island only few (0.1) areas (Fig. 10).

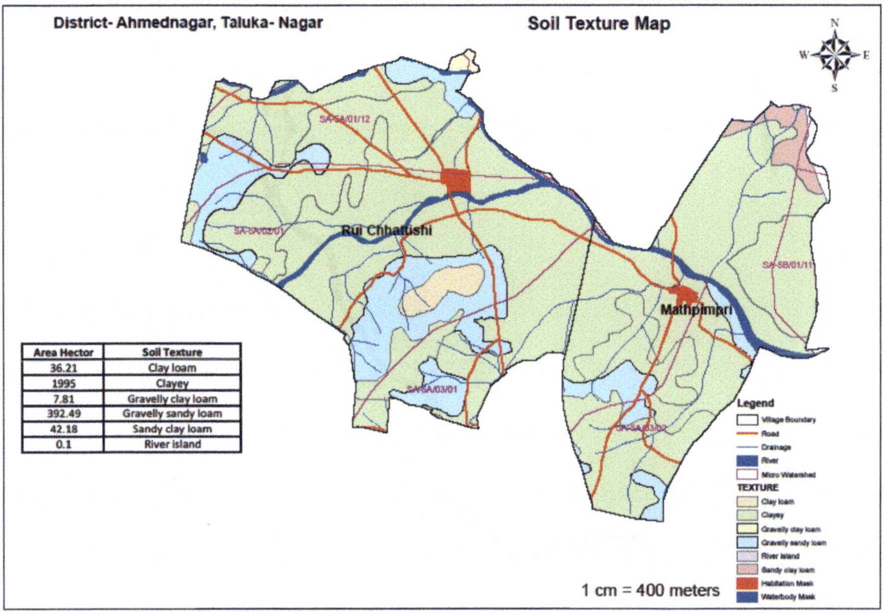

Fig. 10 Soil texture map of the study area

River Alluvium

It spreads along the bank of the main river and its trunk tributaries. It is thick, highly extensive, high fertile tracts, and therefore is used extensively for the agricultural activities. The thickness of this material increases towards the channel and decreases in peripheral areas. At places, it filled the depressions along the channels and shows great thickness while at some rocky cliffs and banks, it shows very few curtains of alluvium.

Quaternary Deposit

At places, numerous exposed sections of quaternary deposits are present throughout the channel of Sina River. Out of this, the Pleistocene deposit is rarely exposed but the Holocene are largely exposed and observed just below the upper soil layer. At the location of broad channel of River, such deposit is generally covered with soil but along the deep channel, where all the layers of quaternary strata are easily exposed. The Quaternary deposit observed along the channel of Sina River is explored, measured, and documented properly, and a contour map of the same was prepared (Fig. 11). Each contour has 1 m thickness of Quaternary deposits. It is clearly observed that the thickness and its stiffness vary with the lithology, terrain topography, rate of weathering and erosion, fluvial geomorphology, and water velocity.

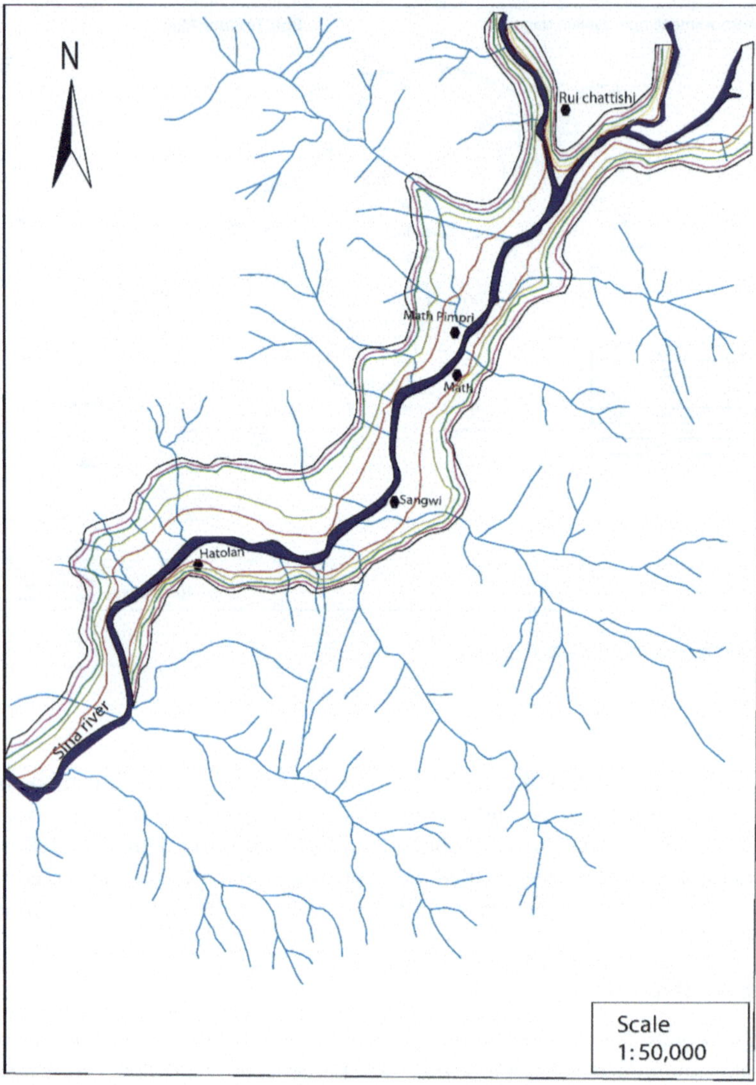

Fig. 11 Thickness of quaternary deposits along the Sina River channel

Agriculture

As we know, the vegetation type is directly linked with the soil setting and therefore, soil has become a highly potential element for the study. Therefore, the economic value is then related to the soil series. So, the soil series are categorized into different groups, based on their economic value based on vegetation. In the case of archaeological sites, it provides an interpretable view.

The hills in this plateau region consist of the soils of friable and mixed characters with gravels, while at the foot hill region and minor valleys, deep reddish soils are observed. In short, in the Sina valley the soils are not different in texture and are less gravelly; they are lighter in weight and poor in fertility.

Vegetation

All over the globe, the native vegetation is under threat under the present different developmental activities such as overgrazing, invasive species, increasing soil and water salinity, ongoing clearance, altered fire regimes, poor land management, and other factors, which results in degradation of land.

The site survey-based methods have long been used to understand the compositional, structural, and functional attributes as indicators of vegetation condition, and these methods continue to be used widely today. In the present day, remote sensing technologies are being employed increasingly for monitoring a range of remotely detectable properties of vegetation, and there is now a growing demand to integrate the two approaches to survey and study the vegetation condition across a range of scales (Lawley et al. 2016).

Locally grown vegetation resources could have been used to identify the populations of site, and hence it is of fundamental importance in settlement pattern investigations. The ethno-botanical information provides the details on the edibility, medicinal and normal, toxic and non-toxic, and other characters of plants. Such sources of information are used to aid in the interpretation of vegetation data collected in the catchment and at the archaeological site in the study area (Vita-Finzi 1977).

Along the summits and slopes of the hills of the plateau, uniformly bare of trees, the stunted bushes of *khair* (*Acacia catechu*) trees and prickly cactus which at a distance can hardly be distinguished from the basalt boulders are strewn over the hill sides, while some parts are well wooded with mango and tamarind groves. In short, the vegetation owes slope and partly due to the geological forms of the hills, the general aspect of the open country of the plateau is decide. During the rainy season, only after heavy precipitations, the hills become green and quickly turn to yellow as the thin surface soils get parched and weather under the scorching sun. During the hot summer, the scenario is very different; after the last harvest, the country is as bleak and barren for whole areas (SCA (by Higgs and Vita-Finzi) 2006).

Agricultural Production (Crop Production)

The thickness decides the productivity and crops, as the deeper the soils, the higher the production. The general crops on such soil are *bajri* grown during the *kharif* season in the western parts, *jowar* over this plateau is almost exclusively a *rabi* crop, while the pulses, mainly *tur* and gram, during the *kharif* and *rabi* seasons, respectively are grown often as mixed crops.

Based on the above category, several plant species were studied during the site catchment analysis and recorded in the field book. The score is given on the basis of medicinal, edible, edible and medicinal grade. This data helps in understanding the importance and use of these plants since the establishment of settlements in the basin area.

But in general, the area shows vegetation, waste land (scrub land), and built-up area (land under construction).

Ethnomedicinal plants

At places, in the study area, the ethnomedicinal plants were observed during the catchment analysis. These plants include mainly Allium sativum Linn. (Liliaceae) *'Lahsun'*; Aloe vera Linn. (Liliaceae) *'Korphad'*; Azardirachta indica A. Juss. (Meliaceae) *'Neem'*; Luffa echinata Roxb. (Cucurbitaceae) *'Dangar'*; Emblica Officinalis Phyllanthus (Euphorbiaceae) *Awalla*; Tamarindus indica L. (Caesalpiniaceae) *'Chinch'*; Vitex negundo L (Verbenaceae), *'Nirgudi'*; Coriandrum sativum Linn. (Umbelliferae) *'Dhane'*; and Thespesia populnea (Malvaceae) and *'Parshi Pimple'* (Kachare et al. 2010). These wild species plants are used for the treatment of various diseases in human settlements. People are aware regarding such traditional medicinal plants and their use for different purpose, in this area.

Biodiversity

This area is noted for rich biodiversity. Several faunal and floral remains were collected from Quaternary excavations carried out in upper, middle, and lower reaches of the basin (Sabale et al. 2021). This clearly indicates that the area is rich in biodiversity since the beginning. In the present condition also, diverse animals were observed in this area.

2 Conclusions

Optimal utilization of land and water resources is essential for sustainable development. In this site catchment analysis, it is observed that the use of maximum natural resources from human beings and the present area as an organizing unit has proven to be an effective scale for natural resource management. Many different natural and artificial activities affect the landscape system, and promote greater integration and collaboration among those actions.

It is observed that the satellite-borne data has played an important role right from the planning, actual survey, and conclusion phases of this site. Such remote sensing and its image processing technology provide access to spatial and temporal information of this region.

Use of Remote and GIS in site catchment analysis becomes a perfect match of technology and application. This tool gives the possibility to extend the catchment method beyond its initial form to a more sophisticated level of analysis. This technology offers a variety of tools to the exploration archaeologist; still its full potential has not been realized. Such data have proved to be indispensable tools in the land use and land cover study of each object on the surface in the Sina basin.

The local and regional topographical and environmental monitoring relies heavily on remote sensing satellite and sensors which are capable of collecting the information of large-extent entities on the earth's surface. The satellite data combined with field survey data can provide a unique and hybrid database for optimal planning and detail field site analysis (Khadri and Kanak 2013). This space-borne data technology provides a unique tool for spatial, multi-spectral, and repetitive information for detail site catchment analysis of different cultural period sites (Legg 2008; Roper 1979a, b; Samarasinghea et al. 2010; Sarker and Sivertun 2011; Tianhong et al. 2003; Uddin et al. 2013).

In the present study, the benefits of remote sensing and GIS technologies in site catchment planning, survey and classification, management, and conservation are found to be very important. On the basis of these images, it was becoming very easy to study the landform characters of each and every corner of the site, which includes soil, land use land cover, slope, geomorphological parameters, drainage density, and drainage pattern. The morphometric analyses of the study area were also easily understood with the help of this data. Still a detailed site catchment analysis by using remote sensing and GIS is required to understand the potential of the study area (Ismail and Opeluwa 2013; Khadri and Kanak 2013; Seeni Mohd and Mansor 2000).

Acknowledgement I thanks the authorities of Deccan College Deemed University, Pune, for providing support during the Math Pimpri excavation. Similarly, I thank the authorities of Central Ground Water Board (CGWB), to avail the online climate data of Sina basin, of South Ahmednagar tehsil.

References

Ahmednagar District Gazetteer, 1976.

Akhtar Alam, M. Sultan Bhat and M. Maheen. 2020. Using Landsat satellite data for assessing the land use and land cover change in Kashmir valley. *Geo Journal* volume 85, Pp.1529–1543.

Bailey, G.N 2005. Site catchment analysis. ARCHAEOLOGY-The Key Concepts. Edited by Colin Renfrew and Paul Bahn. Routledge New York.

Bimal Kanti Paul, Harun Rashid. 2017. Climatic Hazards in Coastal Bangladesh -Non-Structural and Structural Solutions. Science Direct Publ. Pp. 121–152.

Burley, T. M. (1961). Land use or land utilization? *Professional Geographer*, *14*(5), 18–20. economies in Europe.

CGWB. 2014. Ground Water Information, Ahmednagar District (Maharashtra). Ministry of Water Resources. Central Ground Water Board. Govt of India.

CGWB. CR, 2018–19. Ground Water Information, Ahmednagar District (Maharashtra). Ministry of Water Resources. Central Ground Water Board. Govt of India.

Foley, J. A., et al. 2005. Global consequences of land use. *Science,309*(5734), 570.

Harlan, J. R. 1992. *Crops and Man.* Am. Soc. Agron. and Soil Sci. Soc. Am.

Hen J, Parker A, Riverson J. 2005. A new approach for a windows-based watershed modelling system based on a database-supporting architecture. Environ Modelling Softw. 20:1127–1138.

Higgs, E.S. (ed.) 1972. Papers in Economic Prehistory. Cambridge: Cambridge University Press.

Higgs, E.S. (ed.) 1975. Palaeoeconomy. Cambridge: Cambridge University Press. A series of

Hillel, D. *Out of the Earth: Civilization and the Life of the Soil.* University of California Press, 1992.

Ismail M, Opeluwa Saanyol I. 2013. Application of remote sensing (RS) and geographic information systems (GIS) in flood vulnerability mapping: case study of River Kaduna. Int J Geomat Geosci. 3:618–627.

Kachare, S V; Suryavanshi S R and Raut K S.2010. Ethnomedicinal pants of Beed and Nanded District from Marathwada. International Journal of Current Research. Vol.5, Pp.14–16.

Khadri, S.F.R. and Kanak N. Moharir. 2013. Estimation of groundwater prospectus zone mapping and morphometric analysis of WRD-4 watershed in Wardha District of Maharashtra, India using Remote Sensing and GIS Techniques. International Journal of Advancements in Research & Technology, Volume 2, Issue 5.

Lambin, E. F. 1997. Modelling and monitoring land-cover change processes in tropical regions. *Progress in Physical Geography,21*(3), 375–393.

Lawley, V., Lewis, M. Clarke K. and Ostendorf, B. 2016. Site-based and remote sensing methods for monitoring indicators of vegetation condition: An Australian review. Ecological Indicators, Volume 60, Pp.1273–1283.

Legg, R.J.2008. Sites Catchment Analysis. In Encyclopedia of Archaeology, Daborah, MP, Ed; Academic Press New York (USA). Pp. 2002-2004.

Lo, C. P. 1986. *Applied remote sensing*. New York: Longman Inc. of soils (by Webley) (p.227).

Sabale, P. D. and Kshirsagar, S. D. 2013. A preliminary report on geoarchaeological explorations in upper reaches of Sina River basin: A case study of Ahmednagar, Bid, Osmanabad and Solapur districts of M.S. *Bulletin of the Deccan College* 7: 127–138.

Sabale, P. D., Deshmukh, J. D., Kshirsagar, S. D., Pokharkar, D.M., Sarwade, K. G., Kamble, P. V. and Nagraja Rao. 2014. Preliminary report on Quaternary geoarchaeological excavation at Math Pimpri, District Ahmednagar. Bulletin of the Deccan College Research Institute, Pune. ISSN No 980. *Bulletin of the Deccan College, Vol. 74.*

Sabale, P. D., Badam, G. L. and Kshirsagar, S. D. 2021. Quaternary Faunal Records from Upper Reaches of Sina Basin: A Case Study of Math-Pimpri. Quaternary Climate Change cover the Indian Subcontinent. Springer. Pp. 345–256.

Ropar, D.C. 1979. The method and theory of site catchment analysis: A review. Adv. Archeol. Method Theory. 2. Pp. 119–140.

Roper Donna C. 1979a. The Method and Theory of Site Catchment Analysis: A Review. Advances in Archaeological Method and Theory. Vol. 2 (1979), Pp. 119–140.

Roper, D. 1979b. The Method and Theory of Site Catchment Analysis: A Review, pp.119 –40 in

Samarasinghea S, Nandalalb H, Weliwitiyac D, Fowzed J, Hazarikad M, Samarakoond L. 2010. Application of remote sensing and GIS for flood risk analysis: a case study at Kalu-Ganga River, Sri Lanka. Int Arch Photogramm. Remote Sens Spatial Inform Sci. 38:110–115.

Sanjai J. Parikh. 2012. Soil: The Foundation of Agriculture. Nature Education Knowledge 3(10):2.

Sarker M.Z., Sivertun Å. 2011. GIS and RS combined analysis for flood prediction mapping—a case study of Dhaka City Corporation, Bangladesh. J Electr Control Eng. 1:250–257.

SCA (by Higgs and Vita-Finzi), 2006. Definition of the extended territory concept (by Sturdy), and use Schwartz, G. M. & Nichols, J. J. *After Collapse: The Regeneration of Complex Societies*. The University of Arizona Press.

Schiffer, Michael B. 1975. Behavioral chain analysis: Activities, organization, and the use of space. In chapters in the prehistory of Eastern Arizona, IV. Fieldiana: Anthropology 65: 103–119.

Seeni Mohd I, Mansor MA. 2000. Application of remote sensing and hydrological modelling in flood prediction studies. Malaysian J Remote Sens GIS. 1:91–98.

Tianhong L, Yanxin S, An X. 2003. Integration of large-scale fertilizing models with GIS using minimum unit. Environ Modelling Softw. 18:221–229.

Trigger, B. G. *Understanding Early Civilizations: A Comparative Study*. Cambridge University Press, 2003.

Tripathi and Sabale 2016. Note on human skeletal finds belonging to modern time from Math Pimpri. *Bulletin of the Deccan College, Vol. 74.*

Uddin K, Gurung DR, Giriraj A, Shrestha B. 2013. Application of remote sensing and GIS for flood hazard management: a case study from Sindh Province, Pakistan. Am J Geogr. Inform Syst. 2:1–5.

Vita-Finzi, C. 1977. Archaeological Sites in their Setting. London: Thames and Hudson.

Chapter 10
Geotourism in South Bihar, India: Sustainability of a Unique Geological and Historical Heritage Amid Apparent Imprints of Climate Change

Akhouri Bishwapriya, Asaad Ahsan Raza, Ringphami Awungshi, Abhishek Kumar Chaurasia, and Sanjay Kumar Dutta

1 Introduction

India is a remarkable land—dynamic, restless, and constantly shaped by natural forces. Its ever-evolving landscapes bear the unmistakable marks of geological mysteries carved over millions of years. One such mystery lies across the Nalanda, Gaya, and Jehanabad districts of Bihar, where nature has crafted a rich geological setting preserved on the vast canvas of geological time. Civilization arrived much later, adding an equally rich archeological heritage that stands firmly rooted in this geological foundation. Together, these elements create a unique and inseparable identity for the region today. Recognizing the immense potential it holds, geologists are continually working to uncover fresh evidence of how these two rapidly developing fields—geology and archeology—are opening exciting opportunities for Geotourism enthusiasts.

It is rare to find a geological setting so perfectly aligned with the historical landscape as this region. Undoubtedly, it serves as a major attraction for tourists from across the country and around the globe. Its rich geoarcheological significance adds a unique and unmatched dimension, enhancing its aesthetic appeal and earning it recognition as an important Geotourism center. Geotourism represents a modern, sustainable form of tourism—broader and more inclusive than earlier niche approaches. As

A. Bishwapriya (✉) · A. A. Raza
Geological Survey of India, Ranchi, Jharkhand, India
e-mail: akhouri.bishwapriya@gsi.gov.in

R. Awungshi
Geological Survey of India, Itanagar, Arunachal Pradesh, India

A. K. Chaurasia · S. K. Dutta
Geological Survey of India, Patna, Bihar, India

N. Khare (ed.), *Changing Climate over India*,
https://doi.org/10.1007/978-981-92-1098-5_10

more people than ever before visit such destinations globally, the cumulative impact on Geotourism sites is increasing manifold.

The Rajgir hills, along with the surrounding peneplain and pediplain surfaces, possess immense geoarcheological potential. Geotourism acts as a powerful engine that drives key activities such as conservation, education, and sustainable development within the framework. This region offers a journey through a remarkable range of artifact, all preserved within a natural landscape in the historically rich state of Bihar.

The majestic northeast–southwest-trending Rajgir hills are home to the geoarcheological treasures of the ancient land known as Rajgriha. Geologically, this region lies along the northern margin of the Chhotanagpur Gneissic Complex (CGC), a vast ENE–WSW-trending high-grade metamorphic terrain forming a significant part of the eastern Central Indian Tectonic Zone. The geology of the northern CGC margin comprises the Bihar Mica Belt, the Munger Group, the Rajgir and Gaya metasediments, several sub-basins, and the Barabar–Nagarjuni Magmatic Complex. Based on lithological and structural attributes, the Munger and Rajgir–Gaya metasediments were previously correlated with the Mahakoshal Group of the Son–Narmada Graben in Central India (Ghosh and Mukherjee 2000). Later, Chatterjee and Ghosh (2011) proposed that the East Patna Fault and its conjugate, the Munger–Saharsa Ridge Fault, collectively define a graben that may represent the northeastern extension of the Son–Narmada Graben (Fig. 1).

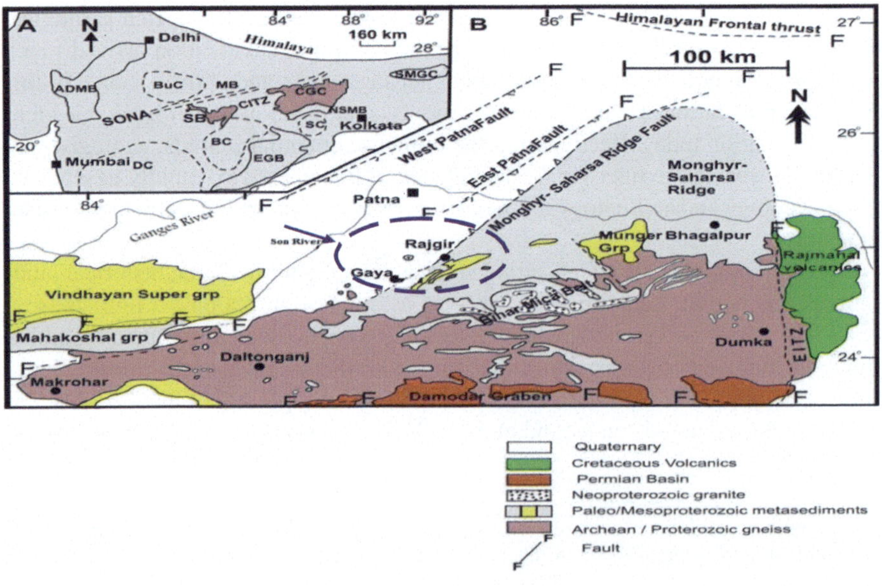

Fig. 1 Regional geological map showing Chhotanagpur Gneissic complex (CGC), Satpura Belt (SB), Mahakoshal Belt (MB), SONA and its eastern continuation represented by East Patna fault and Monghyr–Saharsa ridge fault. (Chatterjee and Ghose 2011, as cited in Asaad Ahsan Raza et al. 2020)

The geotourism transects covering the Gaya–Rajgir volcano-sedimentary belt and the associated magmatic complex represent a unique lithostratigraphic assemblage located along the northern fringe of the Chhotanagpur Gneissic Complex.

Along the northern face of Rajgir, amid its many antiquities, lies the recently explored volcano-sedimentary sequence identified by the Geological Survey of India (GSI). This site, known for its rare and significant geological features, has been proposed for preservation as a National Geological Monument (Ahmad and Wanjari 2009).

This region showcases some of the finest exposures of pillow basalts and their associated geological assemblages. Pyroclastic rocks and Banded Iron Formations further enhance the geoscientific value of the area (Ahmad and Wanjari 2009). In addition, several hot springs near Rajgir—each supported by plausible geological explanations—serve as major attractions for visitors. The chariot marks preserved within phyllite carry strong mythological significance, yet they are equally intriguing from a geological perspective. The Ghora Katora Lake, located at the scenic core of the antiformal closure of the Rajgir fold, stands out as another remarkable geological feature. Encircled by low hills, the lake offers a picturesque setting and provides an excellent vantage point for observing textbook-quality structural characteristics. Outstanding exposures of pillow basalts occur at Churi, Gulni, and Jagarnathpur. However, the preservation challenges faced at the Churi hillock—due to human interference stemming from limited awareness and vested interests—underscore the urgent need for a time-sensitive conservation strategy to protect such rare geological relics.

For better understanding, the sector is divided into the Northern, Central, and Southern domains. The Northern Domain (ND) consists of granitoids containing xenoliths of anorthosite and mafic rocks, occurring as intrusions within the volcano-sedimentary sequence of the central litho-domain. The granitoids are prominently exposed at Bathani, Nimchak, Patharkati, Mahadeo Hill, and Satgarha Hill. The magmatic suite exposed across the Barabar–Mahadeo–Nagarjuni hill ranges is collectively referred to as the Barabar–Nagarjuni Magmatic Complex (BNMC). Titaniferous magnetite ore bodies in this area are either associated with gabbro–anorthosite units or occur as massive xenoliths within the porphyritic biotite granite of the BNMC.

The Central Litho-Domain (CD) is composed of volcanic rocks such as basalt, rhyolite, andesite, and tuff, along with minor occurrences of chemogenic Banded Iron Formation (BIF) and carbonate. The association of pillow lavas, rhyolites, explosive volcanic fragments, and chemogenic sediments indicates the eruption of mafic to felsic magmas, accompanied by the ejection of pyroclasts in a subaqueous environment.

The Southern Litho-Domain (SD) is represented by predominantly arenaceous rocks, locally transitioning into argillaceous varieties. This domain exhibits well-preserved sedimentary characteristics, ranging from quartzite to orthoquartzite, with distinct primary structures.

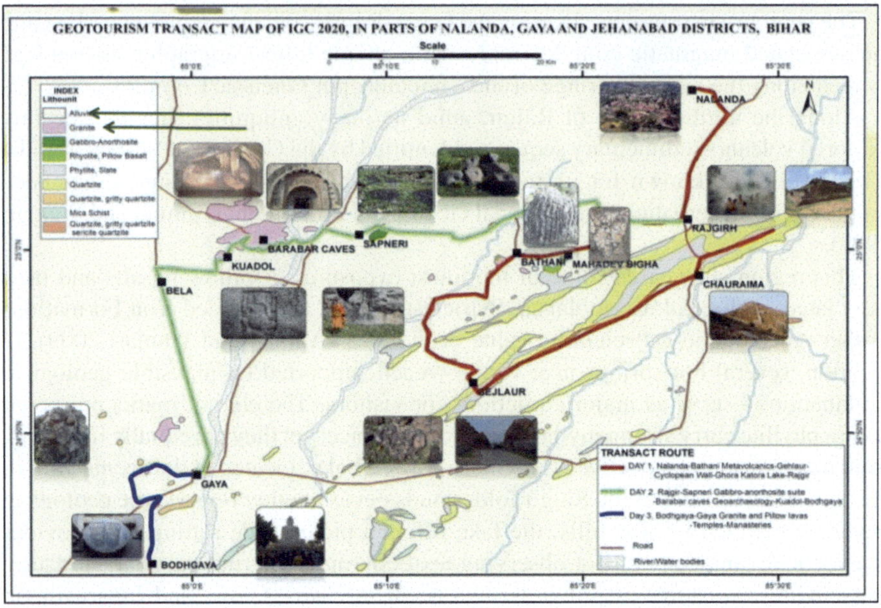

Fig. 2 Geotourism transact map of international geological congress, 2020, in parts of Nalanda, Gaya, and Jehanabad districts, Bihar (*Source* Asaad Ahsan Raza et al. 2020)

A Glimpse of the Geotourism Sites in the Sector:

Figure 2 shows the index map of Geotourism artifacts preserved in the Nalanda-Gaya-Jehanabad districts in Bihar and has the distinction of being selected as a transact for the International Geological Congress delegates for the IGC 2020. Their proximity to each other renders the transact as a compact one intricately interwoven with the archeological heritage thus adding an added dimension of Geotourism for increased footfall from those visiting the sector.

Each of these sites is dealt within the following pages giving due cognizance to their geological uniqueness, understanding in light of geological explanation, and other facets which bring unparalleled novelty to the sector and deserves to be given further visibility for national and international awareness and appreciation.

Amidst challenges posed by short- and long-term climate change and their gradual, though at times evident impact on such natural marvels with a bearing on their longevity, there is a growing need to view this sector through a holistic perspective that aligns with the core principles of geotourism—conservation and education, as long term sustainable development goals for geo-environment.

(a) **The Ruins of Nalanda University at Baragaon**:

Francis Buchanan was the first to explore the ruins at Baragaon which is now the celebrated archeological site of Nalanda. He found a number of Buddhist images at site which reminded him of Bodh Gaya. The digging at Nalanda revealed shrines, several rock sculptures, artifacts, stupas, and other antiquities, and paintings depicting human figurines and floral designs were also discovered. Nalanda was an ancient mahavihara, a large and revered Buddhist monastery in the ancient kingdom of Magadha which is now called Vihar. It is characterized as one of early Indian universities.

At its peak, it attracted students and scholars from near and far. At its zenith, it attracted students and scholars from distant regions. The travelogue of the Chinese pilgrim Hiuen Tsang, later translated and interpreted by subsequent scholars, highlights the exceptional significance of this place.

The Nalanda site comprises the archeological remains of a monastic and scholastic institution dating from the third century BC to the thirteenth century AD. It includes *stupas*, shrines, *viharas* (residential and educational buildings), and important art works in stucco, stone, and metal. Nalanda stands out as the most ancient university of the Indian Subcontinent and counts among the world heritage sites (Figs. 3, 4 and 5).

Excavated from beneath the peneplain–pedeplain surfaces to the north of Rajgir hills, the unique archeological excavation rests on a geological platform intricately

Fig. 3 Aerial view of Nalanda University (*Source* Asaad Ahsan Raza et al. 2020)

Fig. 4 Different components of university campus (*Source* Asaad Ahsan Raza et al. 2020)

Fig. 5 Different sections of university campus (*Source* Asaad Ahsan Raza et al. 2020)

connoting the two sciences which unearth marvels. The unique geological heritage that resides alongside gives it a valued stature of Geotourism.

(b) **The Hot Springs at Rajgir**:

Rajgir is also famous for its hot water springs, locally known as Brahmakund. The thermal springs of Rajgir, Tapovan, and Agnikund are located at the base of Rajgir-Jethian hills (Fig. 6). The rocks are tightly folded and traversed by ENE-WSW faults. Of greatest significance is the hot spring at Brahmakund. This has a geological origin and is a tourist's delight. At the foot of the Vaibhava hills, there is a stair case which leads to various temples. Separate bathing places are earmarked for men and women and the water comes through spouts from Saptdhara, believed to have found their source beneath the Saptarni caves up in the hills.

It is postulated that the percolating meteoric water gets heated on meeting some superheated steam and carbon dioxide from some younger source, thus accounting for rise in temperature and acidity (Raza et al. 2020). The base of the Rajgir Jethian hills is dotted with number of hot springs which bear immense geological significance. Here the rocks mainly quartzite are tightly folded and are traversed by ENE-WSW faults (Fig. 7). The meteoric water through various fractures and joints percolates into the ground and on meeting some heated juvenile source resurfaces in the form of natural springs. Tapovan, Agnikund, and Brahmakund are just a few of them and Brahmakund in particular holds a lot of religious significance.

Rajgir has thus come up as a major health and winter resort due to its natural warm water ponds. These ponds are said to contain medicinal properties, which help to cure many skin ailments and are a perpetual source of attraction for millions of devotees thronging the sector.

(c) **Swarn Bhandar**: The great Jain sage Vairadevaji got the twin caves constructed around fourth century AD meant for meditation. These caves were believed to be king Bimbisara's treasury and hence they were also known as Swarn Bhandar (Fig. 08). The eastern cave has two story height with statues of Jain arhals installed in it. The fine image of Lord Vishnu recovered from the cave has been installed in Archeological Museum at Nalanda.

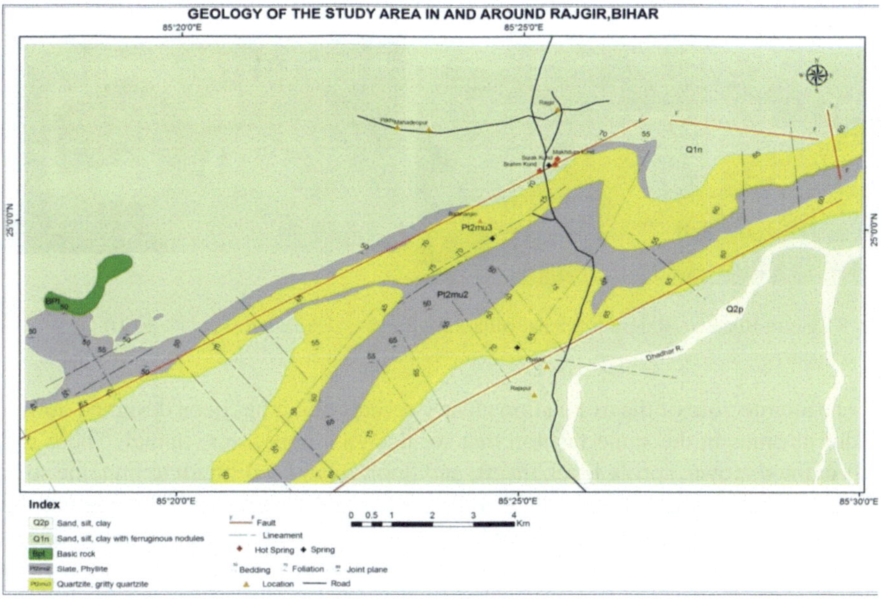

Fig. 6 Geological section along with hot springs, Rajgir, Nalanda, Bihar (*Source* Asaad Ahsan Raza et al. 2020

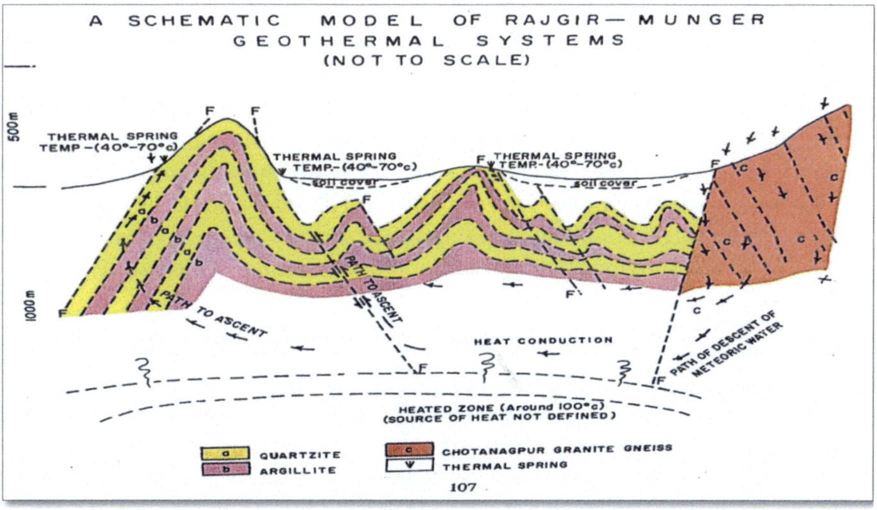

Fig. 7 A schematic geological cross-section showing geothermal systems of Rajgir (Geothermal Atlas of India; GSI 1991, as cited in Asaad Ahsan Raza et al. 2020)

Fig. 8 Swarn Bhandar, Rajgir, Nalanda, Bihar (*Source* Asaad Ahsan Raza et al. 2020)

The architecture of the excavation clearly reveals that it has been designed during medieval times in the same fashion that we design a horse shoe tunnel. There is a well-defined crown, spring level, invert, and floor. It is very difficult to imagine such a huge structure within fractured or jointed quartzite which exists outside but the makers have still found out a massive body which is free of geological infirmities as far as the rock is concerned. There are cracks which have been retreated and have withstood the trials of time which can be seen as a geological marvel in rock (Fig. 8).

(d) **The Chariot Mark**: The Chariot Mark within phyllites exhibits an excellent geoarcheological spot depicting pounding marks along the grooves which points toward anthropogenic activities in the historical past. Believed to be made by chariot's wheel during Mahabharat times, two parallel channel-like depressions (grooves) runs parallel in NE–SW direction with a width of 1ft. The grooves are 10–20 cm deep. The structure lies over phyllite, light purple in color. The foliation trends N55°E-S55°W dipping 60° toward SE. Numerous Z-shaped folds, plunging at angles of 15°–20° towards the southwest, indicate synformal closure towards the eastern side and antiformal closure towards the western side of the linear ridge. Pounding marks along the grooves are observed which point toward human activities (mineral exploration) sometime in the past (Raza et al. 2020).

There is an interesting revelation in the area which also offers a geological explanation to this phenomenon. The two parallel running groves also coincide with two regionally occurring master joints in the area which have a broken continuity but run for kilometers. They trend slightly askew to the orientation of the foliation. The spacing between master joints coincides with the width of a chariot wheel (~1m) and has been observed at several places in the strike continuity (Figs. 9 and 10). It occurs within quartzite and within phyllite; however, its occurrences within the less competent member (phyllite) are more pronounced as observed at this location. The grooves of this joint may have been deepened by anthropogenic activity and may have further been smoothened by loss of detrital grains or by fluvial action behaving as a drain over the period of time. With deepening, it gives the impression of a groove created by movement of a Chariot (Fig. 11).

Fig. 9 View of two parallel channel-like marks within phyllite (*Source* Asaad Ahsan Raza et al. 2020)

Fig. 10 Closer view of the channel-like depression (*Source* Asaad Ahsan Raza et al. 2020)

(e) **Ghora Katora Lake/Rajgir Fold**: The Ghora Katora Lake situated in the picturesque core of antiformal closure large F1 fold is another geological spectacle in Rajgir. Flanked by converging quartzite hill sequences forming the nose of an anticline, the core preserves a ponded water body within the phyllite sequence. Surrounded by small hills, the lake looks picturesque and makes for an ideal spot for witnessing the textbook-type structural features. The Intraformational conglomerate boulders Recovered from the bed of the lake offers delectable admixture of clasts. The feature of geological origin now offers a sought-after site for tourists who enjoy the serenity within and also indulge in water sports, boating spread around a giant statue of Lord Buddha (70 ft in height) in the middle built out of pink sandstone (Figs. 12 and 13).

(f) **The Cyclopean Wall:** A continuous and bastioned dry fortification wall with a circuit of about 40 km runs over the top of flanks of hills occupied by quartzite, and there is subsidiary partition wall defensive in character cutting across the valley. In the widest part of the valley there is also an inner fortification 8 km

Fig. 11 Example of similar feature at ancient gold mining site, Egypt (Dietrich Klemm et al. (2001) as cited in Asaad Ahsan Raza et al. 2020)

Fig. 12 Ghora Katora Lake: Panoramic view of the lake (*Source* Asaad Ahsan Raza et al. 2020)

in perimeter of heaped up earth and rubble at places in ruins but at places rising up to 10 m in height. Founded largely in large and competent quartzite mainly for the protection purposes, it provides a natural fortified look to Rajgir and is a beautiful signature of human effort during those times.

Fig. 13 Closer view of the Buddha Statue situated in the middle of the lake (*Source* Asaad Ahsan Raza et. al. 2020)

Fig. 14 View of Cyclopean wall (*Source* Asaad Ahsan Raza et al. 2020)

The Cyclopean Wall constructed over Rajgir hill and plains comprising several gates and fortifications are a geoarcheological marvel of the Mauryan period. The wall is built along the trend of Rajgir hill, a meta-sedimentary sequence of quartzites and phyllites. It was meant to protect the Rajgriha state against external invaders and enemies (Figs. 14 and 15).

(g) **The Rajgir-Gaya volcano meta-sedimentary sequence**: The recently reported volcano sedimentary sequence in juxtaposition with Rajgir metasediments is extended over known 40 km aerial distance from Bathani to Churi Jagannathpur in Gaya district of Bihar (Fig. 16). The occurrence of volcano sedimentary sequence of Bathani in unison with Rajgir metasediments presents the Munger

Fig. 15 Closer view of the wall (*Source* Asaad Ahsan Raza et al. 2020)

Rajgir belt in an entirely novel litho stratigraphic perspective. Emplacement of diverse mafic and felsic rocks together justifies for multistage and multi-source magmatism in the area. The presence of pillow lava marks the eruption of these rocks in subaqueous environment (Ahmad et al. 2012, Ahmad et al. 2013, Ahmad et al. 2014).

The central lithodomain is essentially represented by volcanic rocks mainly basalt, rhyolite, andesite, tuff, and a little amount of chemogenic BIF and carbonate (Figs. 18

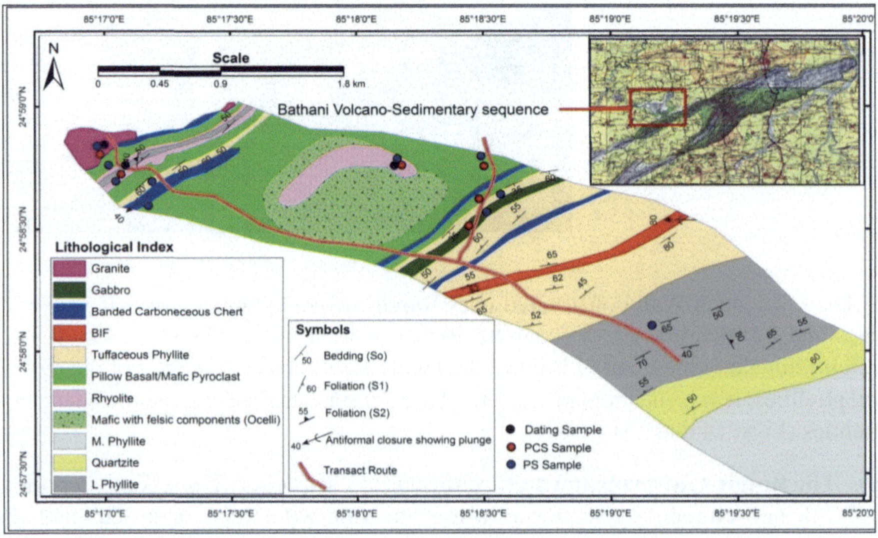

Fig. 16 The Bathani volcano sedimentary sequence package in Gaya district, Bihar (*Source* Asaad Ahsan Raza et al. 2020)

and 19). Volcanic rocks of central domain display megascopic as well as microscopic evidences of mixing and mingling of mafic and felsic magmas ultimately forming hybrid rock like basaltic andesite and andesite. The association of pillow lava, rhyolite, explosive fragment, and chemogenic sediments suggest eruption of mafic and felsic magmas and ejection of pyroclasts in subaqueous condition. Subaqueous emplacement of mafic lavas associated with pyroclasts is a characteristic feature of subduction (Sakia et al. 2014; Ashima et al. 2017).

The pillow basalt exposed in Mahadev Bigha and Churi Jagannathpur has requisite attributes and eligibility criteria to qualify as a National Geological Monument, and an initiative has already been taken in that direction (Fig. 17).

Fig. 17 **a** & **b** Exposed section of pillow basalt in nala section near Mahadev Bigha. **c** **d** and **f** Exposure of pyroclastic rock with various volcanic bombs and lapillis west of Gulelwa Hill (*Source* Asaad Ahsan Raza et al. 2020)

Fig. 18 Photographs showing **a** sectional view of BIF, **b** planar view of brecciated BIF—fragments of chert and tuff are embedded in fine-grained ferruginous matrix, **c** planar view of fragile felsic tuff showing reddish color, **d** transverse sectional view of tuffaceous phyllite, **e** sectional view of BIF bands in felsic tuff, and **f** sectional view of contact relation of tuff and BIF (*Source* Asaad Ahsan Raza et al. 2020)

The litho-units of volcano-sedimentary sequence have been classified into (i) mafic volcanics, comprising pillow basalt and mafic pyroclasts, (ii) acid volcanics, comprising the rhyolite and the alkali rhyolite, (iii) felsic ocelli within mafic groundmass, (iv) volcanogenic sediments/sedimentary sequence comprises tuff, banded chert, banded iron formation (BIF) and siliceous dolomite, and (v) concordant intrusion of gabbro/hornblende gabbro.

Fig. 19 **a** & **b** Banded chert showing typical weathering pattern and F1 fold closure near Gulelwa hill. **c** Botryoidal iron exposure. **d** Exposure of ferruginous body (part of BIF) within phyllite (*Source* Asaad Ahsan Raza et al. 2020)

(h) **Gehlaur: The Mountain Man**: In the district of Gaya also lies an interesting Geotourism destination at Gehlaur, the native village of Dasarath Manjhi, the man who realized a near impossible task to excavate a mountain of quartzite and phyllite in an untiring effort for 22 long years with a hammer, chisel, and homemade tools, and make way for the other side of the mountain which otherwise required a longer path of several kilometers to avail basic amenities of life (Fig. 20a, b, and c). It is a glorifying example of how will can give way to something impossible. In the pursuit, he developed all that was required to achieve the ominous task. Hurt profusely by the sudden death of his beloved wife while maneuvering the mountain, he dedicated this herculean feat to his late wife. We salute the Mountain Man.

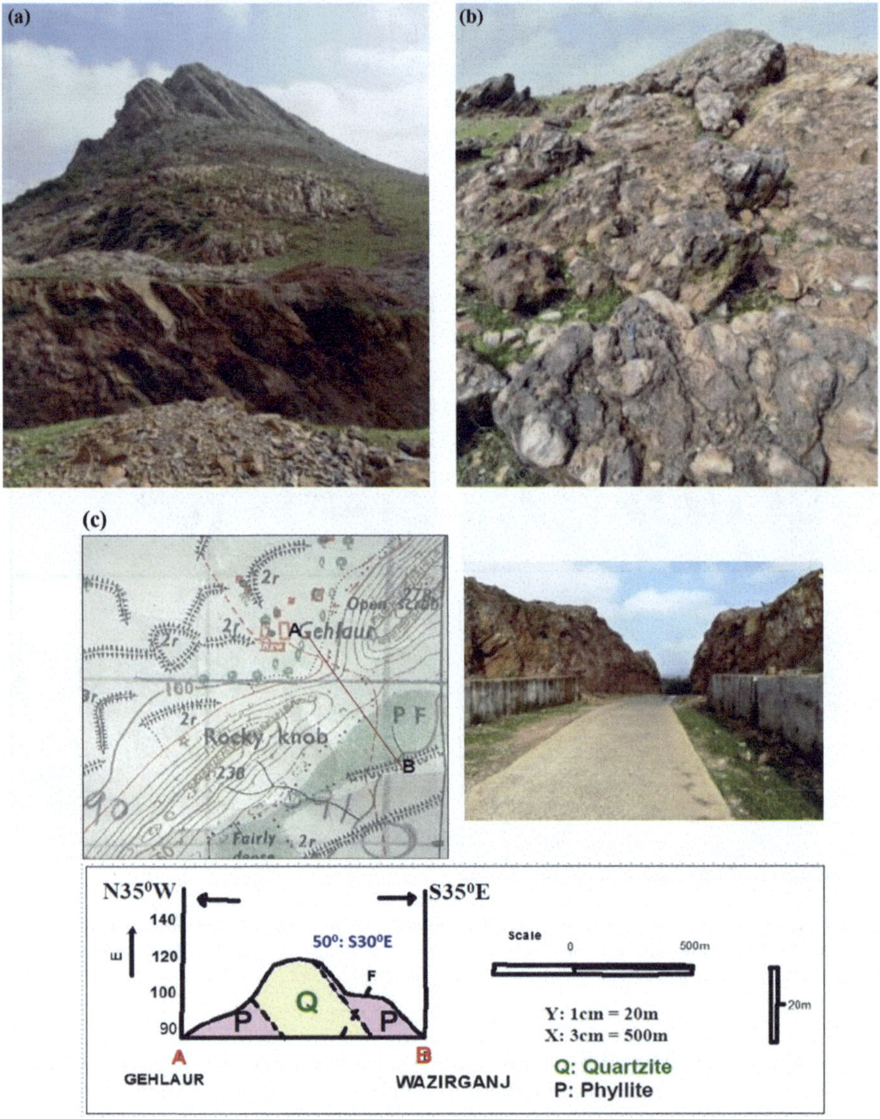

Fig. 20 **a** Sectional view of quartzite ridge (*Source* Asaad Ahsan Raza et al. 2020). **b** Fault Breccia along the ridge (*Source* Asaad Ahsan Raza et al. 2020). **c** Geological Section along the excavated road (*Source* Asaad Ahsan Raza et al. 2020)

The meta-sedimentary sequence of quartzite and phyllite is exposed along the section. The quartzite is generally massive, compact, and silicified in nature and exhibits foliation trending N60°E and 55° dips toward S30° E. Mineralogically, they are composed of quartz, sericite, muscovite, feldspar, and opaque ore minerals. Texture is phyllitic (Raza et al., 2020).

Fault breccia as manifestation of brittle faulting is observed along the section with associated slickenside which is a not-to-be missed feature for those visiting the place. He excavated the 360-foot road in quartzite and phyllite in 1982. Tearing straight through the mountain, the shorter route not only shortened ~25 miles of travel for inhabitants, but it made commutation to the village on the other side of the hill, much safer and easily accessible.

(i) **Barabar and Nagarjuni Hills**: The Barabar and Nagarjuni hills hosting man-made caves are the glorifying examples of Indian rock-cut architecture and one of its kinds in the country. The caves have been fashioned with the finest architecture prevailing during those times and carry antiquities, relics, and inscriptions of the rich historical past engraved in the country rock. They date back to Mauryan empire, some with Ashokan inscriptions located in the Makhdumpur region of Jehanabad district, 24 km north of Gaya. Housed in the twin hill of Barabar with four caves and Nagarjuni with three caves, they are magnificent excavations within massive granite (Figs. 21, 22a and b).

From the geological perspective, it is observed that special care was taken to locate these caves within massive granite which are free of any oblique or basal joints. It

Fig. 21 Panoramic view of the elongated granite body where twin caves are excavated (*Source* Asaad Ahsan Raza et al. 2020)

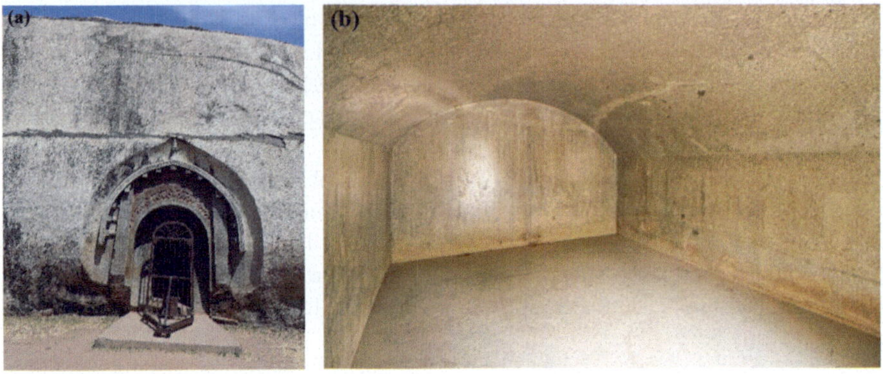

Fig. 22 **a** Entrance of Vapiyaka Caves (*Source* Asaad Ahsan Raza et al. 2020). **b** Interior of Gopiyaka caves (*Source* Asaad Ahsan Raza et al. 2020)

is a difficult proposition, but this approach helped these caves to remain intact for a length of time that it has seen. The Barabar Group of rocks is represented by coarse-grained biotite granite, medium-grained hornblende granite, and porphyritic biotite granite. Apart from this granite, highly silicified sheared and fine-grained granite is also recorded in the northern part of the Barabar hills which corresponds with a prominent regional shear zone along the north-east south-west trend.

Most caves at Barabar consist of two chambers carved entirely out of granite with a highly polished internal surface and exciting echo effect. The first chamber was meant for worshippers to congregate in a large rectangular hall and the second, a small circular dome chamber for worship. The inner chamber probably had a small stupa-like structure at some point, though it is now empty, with a character of the inscription as triangular-headed and having the paleographical features seen in the records of Mahanadi region. Architecturally, Loma Rishi cave is the most attractive cave, but it does not have any donative Ashokan inscription. The magmatic suite of rocks exposed at Barabar and Nagarjuni hill ranges are referred to as Barabar–Nagarjuni Magmatic Complex. Geologically, the area exposes two distinct suites of rocks, namely gabbro anorthosite suite and the granitic suite of rocks. Small pods and lenses of gabbro anorthosite occur as enclaves within the granitoid country rocks. Titaniferous magnetite ore bodies are associated with anorthosite pockets or as a large pocket within porphyritic biotite granite. The occurrences at Barabar and Sapneri have triggered positive hopes of mineralization within a layered complex and are being thoroughly investigated (Raza et al., 2020).

(j) **Kuadol; Ancient Town Site, Buddhist Ruins, and Rock-Cut Sculptures**: The rock engravings in granite at Kuadol are some of the best-preserved arts of those times. The deities, the life styles, culture, religion, worship, and much more reflect the sculptor's brilliance in depicting stories in stone. Kuadol is detached with the rest of Barabar hills to the south-west. A continuous stretch of 1.5 km along the periphery of the granite hills exhibits numerous rock carvings and ruin stupas. The statue of Lord Buddha in basalt and the pillars doing the premises

Fig. 23 View of the Majestic Kuadol granite hill (*Source* Asaad Ahsan Raza et al. 2020)

are rich remnants of the bygone era. The sculptures are carved out in granite, having steep scarp on its northern face and sloping at an angle of 45° on the southern side (Raza et al. 2020).

The Geotourism potential in Rajgir-Gaya and Jehanabad sector sustains and enhances the identity of its historically important center along with its geological environment, culture aesthetics, heritage, and the well-being of its residents (Figs. 23 and 24).

(k) **Churi, Gulni, and Jagannathpur Hills**: Textbook-type exposures of pillows basalt at Churi and Jagarnathpur village are observed (Figs. 25 and 26). The isolated outcrops of the basalt, pillow basalt, pillow breccia, and agglomerate are situated around Gaya. Excellent exposures of pillow basalt are at Churi, Gulni, and Jagarnathpur. The entire hill at Churi comprises pillow lava which is best preserved. Detailed studies on the morphology of pillow, nature of outcrop, pillow shape, sizes, chilled margin, vesicle size and percentage, nature of packing, inter pillow space, pillow breccia, radial cracks, convexity, etc., in the Gaya-Rajgir belt have been carried out and are also in progress. The exposures found in this belt at various locations are rare in context with the geology of Bihar and not found elsewhere in the state.

Fig. 24 Rock carvings and sculptures in Kuadol (*Source* Asaad Ahsan Raza et al. 2020)

Fig. 25 Textbook-type pillow basalt, Churi (*Source* Asaad Ahsan Raza et al. 2020)

Fig. 26 Jagannathpur quarry (*Source* Asaad Ahsan Raza et al. 2020)

Faint Imprints of Short-Term Climate Change in the Sector Along with Anthropogenic Interference:

In a flattish country, with minimal relief difference between low rocky uplands and pediplain surfaces, the primary indicator of seasonal changes and thus climate over a period of years is the depleting water table and also the depleting tiers of aquifer. A conspicuous decline in annual precipitation in the past three decades in the sector has been observed which has several spikes of annual lows vis-à-vis highs as compared to the average annual precipitation. This establishes the fact that a shift in the climatic regime has been experienced by most districts in Bihar which has undoubtedly affected the surface runoff in major streams and also the groundwater recharge conditions within confined (shallow-deep) aquifers (Bishwapriya et al. 2022).

A significant rainfall reduction in districts of Bihar from 1957 to 2002 during the monsoon season has been observed (Warwade et al. 2018). Overall, a declining trend in rainfall magnitude was found between 1901 and 2002. The resultant annual decrease was −18.92% from 1957 to 2002. Average monsoonal rainfall declined up to 21.84% during 1901–2002 (Bishwapriya et al. 2022).

Now, let us see the apparent influences such aberrations have brought to the geoarcheological sites. The Rajgir hot springs which are a source of hot water with immense medicinal and spiritual value have of late reduced considerably in their discharge, and some of them have also dried during summer seasons. The piezometric surface which gets daylighted in the form of springs has reduced in their discharge because of a drastic fall in the water table. The surface water repositories have also dried down and the natural pondage at Ghora Katora was found water-deficient posing a serious blow to tourism potential.

Investigations were carried out by GSI, Patna, and it was found that along the course of the Saraswati Nala, feeding the channel in front of Brahmakund, most of the springs had dried down or miserably clogged by anthropogenic wastes. The Government of Bihar pondered on this critical issue and contemplated lifting water from the Ganga River in the reverse gradient to replenish the shortage of water and rejuvenate the water table.

In addition to this, increasing spells of heat waves possibly due to climatic shift are also affecting the surface polish of granites within caves and causing increase in gaping joints within quartzites.

In addition to this, the area is rampant to quarrying activities and in the absence of clear-cut guidelines, quarrying activities are also affecting geological sites of importance which possess lithology which are rare to the region. It cannot be overemphasized that the pillow lava features around Mahadeo Bigha and Churi require urgent preservation against anthropogenic activities, as the consequences would be far-reaching and catastrophic. An appeal was thus raised to declare the site as a National Geological Monument Site.

The sculptures in granites at Kuadol which are the finest examples of engravings in granites are being lost in the absence of clear-cut preservation guidelines. The people need to be sensitized about the importance of these structures, and there needs to be a conscious effort from governing agencies to preserve and protect them against odds.

2 Conclusions

Geological and geoarcheological studies were taken up in parts of Nalanda–Gaya–Jehanabad areas as a part of IGC-2020 transect. The work included traverse mapping and collection of samples for petrography, chemical analysis, and dating. This study highlights "Geotourism" in the state of Bihar, which is a perfect example of how geological ground expressions can be clubbed with internationally acclaimed archeological sites.

These are rich historical antiquities that are a reflection of the predominance of Buddhism and Jainism religions and add significantly to the geoarcheological importance of the place. Turning the pages of history, the enquirers have looked for facts and have been delivered with fables; they have looked for historic lights and they unraveled poetic coruscations but the truth that has been buried in times has been slowly unleashed in phases with a concerted effort. This presentation is merely a window to introduce this valued tourist destination to the facet provided by the science of geology which is deeply engrained in its soul.

Earth is ancient now, but she keeps a record of all that has happened with time. In time, off time, she says little and in a language which people have only been able to understand partially. It is at this juncture that a geologist attempts to study the earth at a point where an archeologist leaves it, and he takes it further back into remote antiquity. This home away from home offers all of that in a well-knit package.

Acknowledgements The authors express their sincere thanks and gratitude to Director General, GSI, for facilitating the groundwork for the publication. Thanks are due to Shri P.V. Ramana Murthy, Additional Director General (Retd.), GSI, ER, for his support and encouragement in going ahead with this publication. Dr. Mansoor Ahamad, Director, GSI, and Dr. Nischal Wanjari, Geologist, GSI (Formerly), who are credited with the reporting of the volcano sedimentary sequence in and around Rajgir are the pillars that unraveled the typical geological setup that became the driving factor for other aspects to gain recognition and significance with the passage of time. Their works have been duly referred to here particularly in regards to the volcano sedimentary package of immense geological relevance. Thanks are due to Dr. Rajesh Asthana, Deputy Director General (Retd.), GSI, for being the source of encouragement and mentor for giving shape to this publication. The authors express their gratitude to Shri Niloy Khare, Editor of the series, for giving valued comments and guidance during the peer-review and editing work of this publication. Sincere gratitude to all concerned for the encouragement and support received from them. All the figures and tables which are re-used in the research article are **reproduced with permission from the concerned publisher.**

References

Ahmad, M. and Paul, A. Q. (2013): Tectonic stratigraphic constraints of Bathani volcano sedimentary, volcanic sequence and associated rocks Chotanagpur Granite Gneiss Complex, Gaya district, Bihar. GSI, ER, News V.33 (1&2), pp.13-15.

Ahmad, M., Paul, A.Q. and Haque, M.W. (2014): Electron micro probe dating of monazite in carbonate rock of Bathani volcano sedimentary sequence within Chotanagpur Granite Gneiss Complex. GSI, ER, News V.34 (1&2), pp.15-18.

Ahmad, M. and Dubey, J. (2012): Report on Prospecting for Gold and Silver Mineralization in Munger Rajgir Group of Rocks in Nalanda district, Bihar. Unpubl. Rep. Geol. Surv. India, FS: 2008-09 and 2009-10.

Ahmad, M. and Paul, A. Q. (2012): Final Report on Investigation for Gold mineralization in Bathani area, Munger Rajgir Group of Rocks in Gaya and Nalanda districts, Bihar. Unpubl. Rep. Geol. Surv. India, FS: 2010-12.

Ahmad, M., and Wanjari, N (2009): Volcano-sedimentary sequence in the Munger Rajgir metasedimentary belt, Gaya district, Bihar. Indian Journal of Geoscience, v.63 (4), pp. 351–360.

Akhouri Bishwapriya, Abhishek Kumar Chaurasia, Aneesh Kumar V (2023): *"Anthropogenic Interference and Climatic Changes Influencing Precipitation, Sediment Load and Sand Deposition by the Phalgu River in and Around Jahanabad District, Bihar: A Case Study."* In: Khare, N. (eds.) Science, Policies and Conflicts of Climate Change. Springer Climate. Springer, Cham. pp 131–171. https://doi.org/10.1007/978-3-031-16254-1_7.

Dietrich Klemm, Rosemarie Klemm, Andreas Murr (2001): Gold of the Pharaohs – 6000 years of gold mining in Egypt and Nubia, Journal of African Earth Sciences.

Asaad Ahsan Raza, Ringphami Awungshi, and Akhouri Bishwapriya (2020): Rajgir-Bodh Gaya-Barabar Geotourism Spots: A unique Geological and Historical Heritage of Bihar; Field Trip Guide.

Ashima Saikia, Bibhuti Gogoi, Tatiana Kaulina, Liudmila Lialina, Tamara Bayanova and Mansoor Ahmad (2017): Geochemical and U-Pb zircon age characterization of granites of Bathani Volcano-Sedimentary sequence, Chotanagpur Granite Gneiss Complex, Eastern India: vestiges of Nuna supercontinent in Central Indian Tectonic Zone. Geological Society, London, Special Publications. Volume 457. Pages 233 – 252. https://doi.org/10.1144/SP457.11.

Chatterjee, N. and Ghose, N.C. (2011): Extensive Early Neo-Proterozoic high-grade metamorphism in North Chotanagpur Gneissic Complex of the Central Indian Tectonic Zone, Gondwana Research, 20, 362-379.

Geological Survey of India (1991): Geothermal Atlas of India; Geological Survey of India, Calcutta, India, 1991.

Ghose, N. C., Mukherjee, D. (2000): Chhotanagpur gneiss-granulite complex, Eastern India -A kaleidoscope of global events. In: A.N. Trivedi, B.C. Sarkar, N.C. Ghose, Y.R. Dhar (eds.), Geology and Mineral Resources of Bihar and Jharkhand, Platinum Jubilee Commemoration Volume, Indian School of Mines, Dhanbad, Institute of Geo-exploration and Environment, Monograph 2, pp.33–58, Patna.

Sakia.A. Gogoi, B., Ahmad, M. and Ahmad, T. (2014): Geochemical constrains on the evolution of mafic and felsic rocks in the Bathani volcanic and volcano-sedimentary sequence of Chotanagpur Granite Gneissic complex. Journal of earth science system, v. 123, no. 5, pp. 959-987.

Warwade P, Tiwari S, Ranjan S, Chandniha SK, Adamowski J (2018): Spatio-temporal variation of rainfall over Bihar State, India. J Water Land Dev 36:183–197. https://doi.org/10.2478/jwl d2018-0018.

Part V
Climate Changes on Longer Scales

Chapter 11
A Geochemical Investigation from Tropical India that Reveals the Way the Monsoon Climate Affects Sedimentary Processes at Different Longitudes

Shubhra Singh, Ishfaq Ahmad Mir, Kalyan Pal, and M. Thara

1 Introduction

Understanding various aspects such as catchment area geology, mineralization, environmental conditions, agriculture, and land use, as well as sedimentary processes such as weathering, erosion, sediment sources, depositional mechanisms, and tectonics, is made possible by the analysis of stream and slope wash sediments through geochemical investigations (Mir et al. 2022a; Mir et al. 2022b; Mir and Mir 2019). Under the National Geochemical Mapping (NGCM) program, the Geological Survey of India (GSI) is mapping the geochemistry of the whole nation. The creation of a multielemental database for use in diverse geoscientific applications is one of the key goals of this program. Geochemical mapping at a scale of 1:50,000 was done as part of this program in the SOI toposheet nos. (TS Nos) 57D/1 and 48P/1 in the years 2018–19 and 2019–20, respectively. During these mapping initiatives, the Hassan District and Dakshina Kannada districts of the Karnataka state in the peninsula of India were partially mapped.

The climatic circumstances that lead to weathering and erosion have an impact on sediment composition (Mir and Maria, 2023; Mir et al. 2022c; Pattan et al. 2012 and 2013; Mir et al. 2013). Chemical weathering alters the mineral phases, textural features, and compositions of rocks relative to the original rock through the effects of geology, vegetation, temperature, precipitation, and geomorphological circumstances (Scarciglia et al. 2008). Chemical weathering is significantly influenced by climate and tectonic activity over a longer geological timescale, and it is crucial to the global element cycles (Giovanoli et al. 1988). Labile minerals like feldspar are easily

S. Singh · I. A. Mir (✉) · K. Pal · M. Thara
Geological Survey of India, Bengaluru, Karnataka, India
e-mail: geoishfaq@gmail.com

N. Khare (ed.), *Changing Climate over India*,
https://doi.org/10.1007/978-981-92-1098-5_11

chemically worn in hot, humid climates at lower latitudes, whereas at higher latitudes, physical weathering is more intensive in semi-arid and glacial environments.

Geochemistry of weathered rocks provides important information on the extent of weathering, as indicated by various geochemical proxies such as the Chemical Index of Alteration (CIA; Nesbitt and Young 1982), Chemical Index of Weathering (CIW; Harnois 1988), Plagioclase Index of Alteration (PIA; Fedo et al. 1995), Weathering Index of Parker (WIP; Parker 1970), and ratio of Rb/Sr (McLennan et al. 1993). Proxies like the Index of Composition Variability (ICV; Cox et al. 1995) and the ratio of Si/Al (Potter 1978) are indicators of sediment maturity.

The goal of this study is to better understand how diverse climatic circumstances affect chemical weathering and sediment maturity in two distinct regions of peninsular India that are located at various longitudes and have varying climatic conditions. The study makes use of the geochemical characteristics of sediments from streams and slope wash. The research region, which is situated in Dakshina Kannada (TS no. 48P/1), is characterized by an upland pediplain topography that is peppered with small hills and extends between the Western Ghats and the Arabian Sea. The Karnataka Plateau is a feature of the study area, which is situated in Hassan District (TS no. 57D/1). In the western and eastern flanks of the Western Ghats, respectively, are both areas of study.

2 Study Area

The Dakshina Kannada district of Karnataka, India, is included in the research area on the western flanks of the Western Ghats. The study area is located between latitudes of 12° 45′ 00″ and 13° 00′ 00″ N and longitudes of 75° 00′ 00″ and 75° 15′ 00″ E. Within toposheet number 48P/1, the area of research spans roughly 751 km^2 (Fig. 1). The area is located along the southern Karnataka coast, bordered to the west by the Arabian Sea and to the east by the imposing Western Ghats. With altitudes ranging from 80 m to 300 m (Fig. 2), it is distinguished by a continuous area of sharply angled hills that resemble flat terraces. The River Netravathi and its tributaries, as well as seasonal streams, drain the area. A pattern is consequently followed by the drainage system. The Netravathi River is mature and flowing at a slow rate. The River's straight course points to the existence of a sizable lineament that is moving northwest to southeast.

Parts of the Hassan district in Karnataka, India, are included in the research area on the eastern flanks of the Western Ghats. 12°45′00″–13°00′00″ N latitudes and 76°00′00″–76°15′00″ E longitudes define the study area's boundaries. Within the toposheet with the code 57D/1, the region of research spans around 738 km^2 (Fig. 1). Generally speaking, the area has gently undulating terrain and is located at an average elevation of about 900 m above mean sea level (Fig. 2). The topography typically slopes east and southeast, with rivers and streams serving as drainage in these directions. Small streams eventually converge with the Yagachi and Hemavati Rivers after flowing westward and southwestward. The principal drainage system in the area is

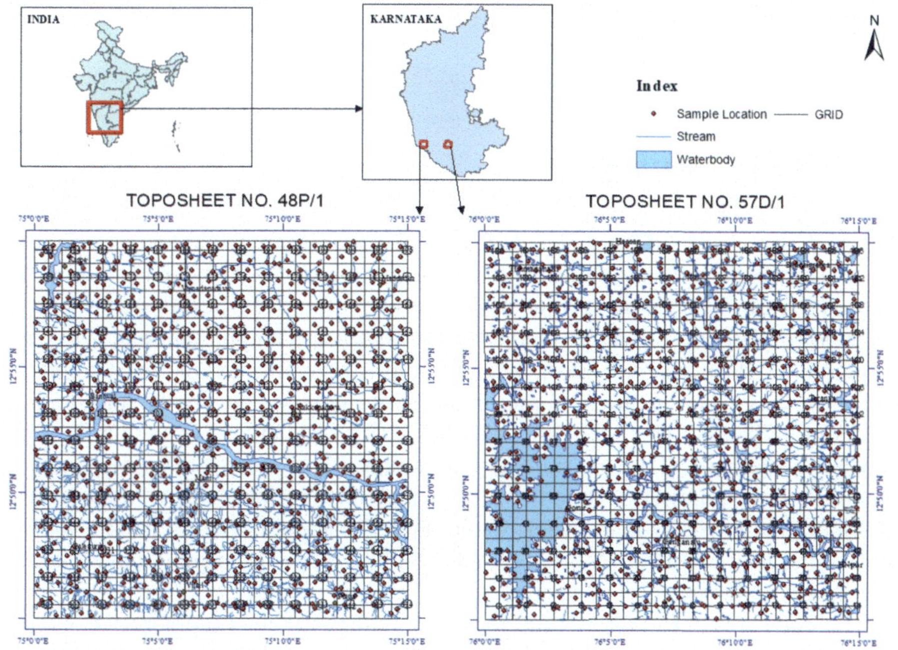

Fig. 1 of the study areas along with sampling points and drainage pattern

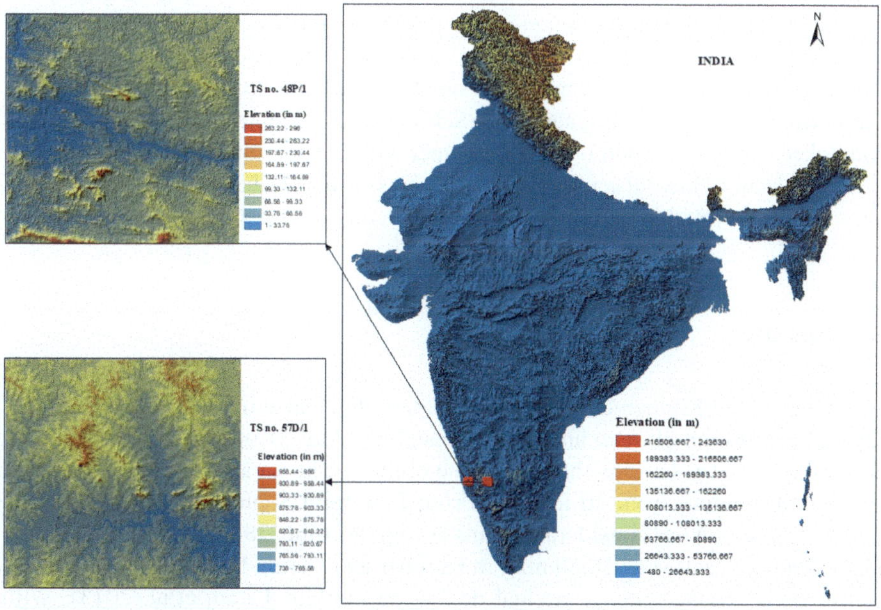

Fig. 2 Digital elevation maps (DEM) of study areas, with reference to India DEM, after USGS SRTM data

made up of the Hemavati River and its tributary, the Yagachi River. Subdendritic drainage is the predominant kind.

3 Geological Setting

The study area in Dakshina Kannada falls in Western Dharwar Craton and consists of rocks of Sargur group, Peninsular Gneissic Complex-I (PGC-I), and Charnockite suite which are traversed by acid and basic intrusives (Table 1). The oldest lithology in the area is of Sargur group and is marked by the presence of amphibolite and pyroxene amphibolite. The major portion of the area comprises Peninsular Gneissic Complex-I and consists of migmatites, biotite-granodiorite gneiss, and hornblende biotite-granodiorite gneiss. The study area also has rocks of Southern Granulite Terrain, consisting of charnockite and pyroxene granulite. Basic dykes are represented by olivine dolerite, porphyritic dolerite, gabbro, and olivine gabbro. Acid intrusives are present in the form of Quartz veins within PGCI. Laterite occurs as cappings over most of the hills as well as plains in the area; both ferruginous and aluminous laterite are present in the area.

The study area in Hassan also has rocks belonging to ancient supracrustal enclave (Sargur group), PGI of Peninsular Gneissic Complex, Bababudan group of Dharwar Supergroup, and Charnockite suite of rocks of Archaean age, which are intruded by both acid and basic intrusives of Paleo-Proterozoic (Table 1). Granite gneiss and migmatite of PGC are the dominant rock types in the area. Rocks belonging to Sargur group and Bababudan group are exposed in the part of the Holenarasipur schist belt. The Sargur group of rocks comprises metapelites, amphibolite, and ultramafic rocks. The northern margin of Holenarasipur Schist belt is occupied by intrusive granite (Mavinkere granite), which occurs in contact with metabasalt of Bababudan group. Several dolerite dykes cut across the granite gneiss and meta-basalt in the area. Quartz veins are present within metabasalt and sporadically in granite gneiss.

4 Climate

According to the Köppen climatic classification, the Dakshina Kannada region has a tropical monsoon climate and is designated as "Am" (Koppen 2011). There are clear rainy and dry seasons that define the climate. The area has high humidity all year round, ranging from 70 to 95%. From June until September is the monsoon season. The region has considerable rains during this time. 2872 mm of precipitation fall on average each year. Beginning in October and lasting through December, the post-monsoon season sees a gradual decline in rainfall (Pattan et al. 2013). With January being the coldest month and an average temperature of 26 °C, the winter season begins in January and lasts through February. The summer months of March

Table 1 Stratigraphic succession of the Study Area in Dakshina Kannada District, T.S. No. 48P/1, after Krishnamurthy et al. (1980), and Hassan District, T.S. No. 57 D/1, after Rao (1924)

			Dakshina Kannada Region		Hassan Region	
Age	Supergroup	Group	Formation	Lithology	Formation	Lithology
Cenozoic				Laterite		
Paleo-proterozoic			Granites	Granite		
		Younger Intrusives	Acidic Intrusives	Quartz vein	Acidic Intrusives	Quartz vein
			Basic Intrusives	Dolerite	Basic Intrusives	Dolerite
Archean	Southern granulite complex	Charnockite	Southern granulite complex	Charnockite		
				Pyroxene granulite		
				Granite gneiss		
				Ultramafite		
	Dharwar	Bababudan (Holenarsipur)			Modalagud-da	Banded magnetite quartzite
						Quartz sericite schist
						Meta basalt
						Quartzite
					Granitoids	Grey Granitoid (Mavina-kere)
	Peninsular gneissic complex	Peninsular gneiss-I	Gneisses	Peninsular gneiss-I	Gneisses	Granite gneiss
		Sargur		Peridotite		Metaultr-amafite
				Ultramafics		Amphibo-lite
				Amphibolite		Kyanite-muscovite schist
		Gorur gneiss				Trondjh-metic gneiss

through May are known as the "summer" months, with April being the hottest with an average temperature of 29 °C (https://en.climate-data.org, Fig. 3).

According to the Köppen climatic classification, the Hassan region has a tropical savannah climate and is categorized as "Aw" (Koppen 2011). The peninsular plateau contains areas with this type of climate. Throughout the year, the region experiences moderate humidity levels, which range from 50 to 70%. Most of the district's rainfall occurs between June and September, which is the monsoon season. 1211 mm of precipitation falls on average each year. Beginning in October to November, the post-monsoon season sees a gradual decline in rainfall. The coldest month, January, with an average temperature of 21 °C, marks the beginning of the winter season, which lasts from December through February. The summer months of March through May have average temperatures of 25 °C, with April being the hottest month (https://en.climate-data.org, Fig. 3).

5 Materials and Methods

The Geological Survey of India (GSI) established guidelines for the sampling process for the National Geochemical Mapping (NGCM) initiative, following the standard operating procedure (SOP 2014), to collect samples from both the study area and the field. 1x1 gridded toposheets were the base maps that were used for sampling. Stream sediment (SS) samples were obtained from first and second order streams, while slope wash (SW) sampling was conducted in areas without streams. From the two study locations, 391 sediment samples in total were taken, including 8 extra samples for cross-check analysis. The samples underwent air drying, powdering, and sieving to a mesh size of 120 μm. Then, they were quartered, coned, and separated into two sets of 500 g each. In order to conduct a chemical analysis, one set was sent to the GSI Chemical Laboratory in Hyderabad and the second set was stored in the sample repository at GSI, SU: Karnataka & Goa, Bengaluru.

The main oxides and a few trace elements were measured using an XRF Spectrometer made by M/S Panalytical called the MAGIX, 2.4 KW. For major elements, the detection limit is 0.1%, whereas for trace elements, it is 1.00 ppm. Perkin Elmer Sciex ELAN 6100 ICP-MS was used to analyze trace and rare earth elements (REE). Analysis was conducted using the fusion method, whose quantification limits range from 0.1 to 1 ppm. A known element concentration Standard Reference Material (SRM) (GSD10) was analyzed for accuracy after each batch of 20 samples, and duplicate samples were analyzed for repeatability after each batch of 10 samples. The measurement's precision and accuracy are both within 5% and 3%, respectively. Processing geochemical data, doing statistical analysis, creating and plotting maps, and other tasks were done using the M/S Excel, Python, ArcGIS, and Origin Pro software.

The Plagioclase Index of Alteration (PIA; Fedo et al. 1995) and the Rb/Sr ratio (McLennan et al. 1993) were computed to assess the degree of chemical weathering in the study area. Sediment maturity was determined using the Index of Composition

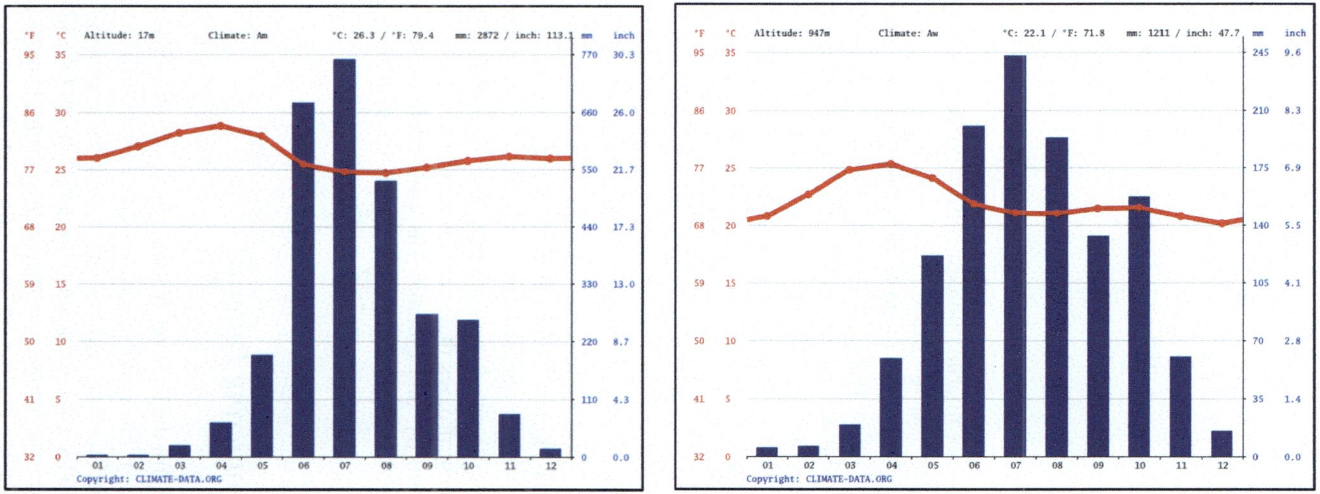

Fig. 3 Annual rainfall (in mm) and temperature (in °C/°F) of Bantwal Station (Dakshina Kannada) on left side and of Hassan Station on right side, data source: https://en.climate-data.org

Variability (ICV; Cox et al. 1995) and the Si/Al ratio (Potter, 1978):

$$\text{PIA} = [(\text{Al}_2\text{O}_3 - \text{K}_2\text{O})/(\text{Al}_2\text{O}_3 + \text{CaO} + \text{Na}_2\text{O} - \text{K}_2\text{O})] \times 100 \quad (1)$$

$$\text{ICV} = \text{Fe}_2\text{O}_3 + \text{K}_2\text{O} + \text{Na}_2\text{O} + \text{CaO}^* + \text{MgO} + \text{TiO}_2/\text{Al}_2\text{O}_3 \quad (2)$$

CaO* represents the amount of CaO added in the silicate fraction of the rock.

6 Results

The descriptive statistics of weathering indices, climate proxies, and sediment maturity indicators used in this study are given in Table 2. In Dakshina Kannada area, PIA varies from 94.00 to 99.69 with a mean of 98.62, Rb/Sr varies from 0.19 to 3.04 with a mean of 1.18, ICV varies from 0.24 to 0.83 with a mean of 0.44, and Si/Al varies from 0.99 to 1.87 with a mean of 1.34. In the Hassan area, PIA varies from 42.43 to 90.42 with a mean of 81.35, Rb/Sr varies from 0.08 to 0.72 with a mean of 0.30, ICV varies from 0.46 to 3.53 with a mean of 0.79, and Si/Al varies from 2.62 to 10.47 with a mean of 3.48.

7 Discussion

A practical way to assess the level of weathering in the source sediment and to interpret the intensity of weathering, sediment maturity, and climatic conditions is geochemical analysis of sediments based on major and trace elements. Table 2 lists the major and trace element proxies that were chosen to assess the research area's weathering, sediment maturity, and climate. The table includes the variables that are used for determining geochemical weathering (PIA and Rb/Sr) and sediment maturity (ICV and Si/Al).

7.1 Changes in Weathering Patterns

The PIA and Rb/Sr studies for the litho-units within the research sites revealed various weathering regimes (Fig. 4). The results of this study's PIA and Rb/Sr measurements (Table 2) for the Dakshina Kannada region support the presence of significant chemical weathering change. PIA values ranged from 94.00 to 99.69 and Rb/Sr values ranged from 0.19 to 3.04. The Rb/Sr and PIA values in the Hassan region, which range from 0.08 to 0.72 and 42.43 to 90.42, respectively show low to moderate levels of chemical weathering. It is worth noting that PIA values are used to evaluate the

Table 2 Descriptive statistics of weathering indices and sediment maturity indicators

Data	Dakshina Kannada region				Hassan region			
	PIA	ICV	Si/Al	Rb/Sr	PIA	ICV	Si/Al	Rb/Sr
Mean	98.62	0.44	1.34	1.18	81.35	0.79	3.48	0.3
Median	98.89	0.43	1.32	1.08	81.95	0.67	3.32	0.27
Standard deviation	0.98	0.11	0.17	0.6	5.47	0.36	0.8	0.14
Sample variance	0.97	0.01	0.03	0.36	29.91	0.13	0.64	0.02
Kurtosis	5.6	2.38	−0.05	−0.34	13.45	21.65	34.17	0.1
Skewness	−2.08	1.18	0.33	0.52	−2.57	3.87	4.8	0.76
Minimum	94	0.24	0.99	0.19	42.43	0.46	2.62	0.08
Maximum	99.69	0.83	1.87	3.04	90.42	3.53	10.47	0.72

degree of weathering of plagioclase minerals in host rocks, with values ranging from 0 to 50 indicating unaltered rocks and values ranging from 50 to 100 indicating altered rocks. Accordingly, the PIA values obtained for sediments suggest better-preserved plagioclase minerals in the Hassan area sediments than in the Dakshina Kannada sediments. Rb/Sr ratios greater than 1.0 are also indicative of strong chemical weathering (McLennan et al. 1993), as are high Rb/Sr ratios. As a result, the Rb/Sr ratios for the sediments from the Dakshina Kannada area (ranging from 0.19 to 3.04) indicate a high level of chemical weathering, while the values for the sediments from the Hassan area (ranging from 0.08 to 0.72) reflect a low level of chemical weathering alteration.

7.2 *Changes in Sediment Maturity*

Utilizing the Index of Compositional Variability (ICV; Cox et al. 1995) and the Si/Al (Potter 1978) ratio, the compositional maturity of the sediments is assessed. The ICV (Equation 2) is a method for evaluating detrital mineralogy. More clay-rich mature sediments have lower ICV values (1.0), and they primarily come from cratonic environments (Cox et al. 1995). The ICV results in the Dakshina Kannada region range from 0.24 to 0.83 and from 0.46 to 3.53 in the Hassan area (Table 2), showing that the Dakshina Kannada region's sediments are mineralogically better developed than the sediments of the Hassan area (Fig. 5). ICV values in this study decrease from east to west longitudes, indicating high clay content and mineralogical maturity. Due to the transformation of feldspars into clay minerals in hot, humid climates, sediments from high rainfall areas have lower ICV values than sediments from low rainfall areas that are rich in non-clay silicate minerals. Mature sediments have low ICV values (1.0) showing several cycles of chemical weathering, erosion, and redeposition, whereas immature sediments have high ICV values (>1.0) indicating a single weathering cycle (Cox et al. 1995). The ratio of Si/Al also reflects the sediments' maturity. The

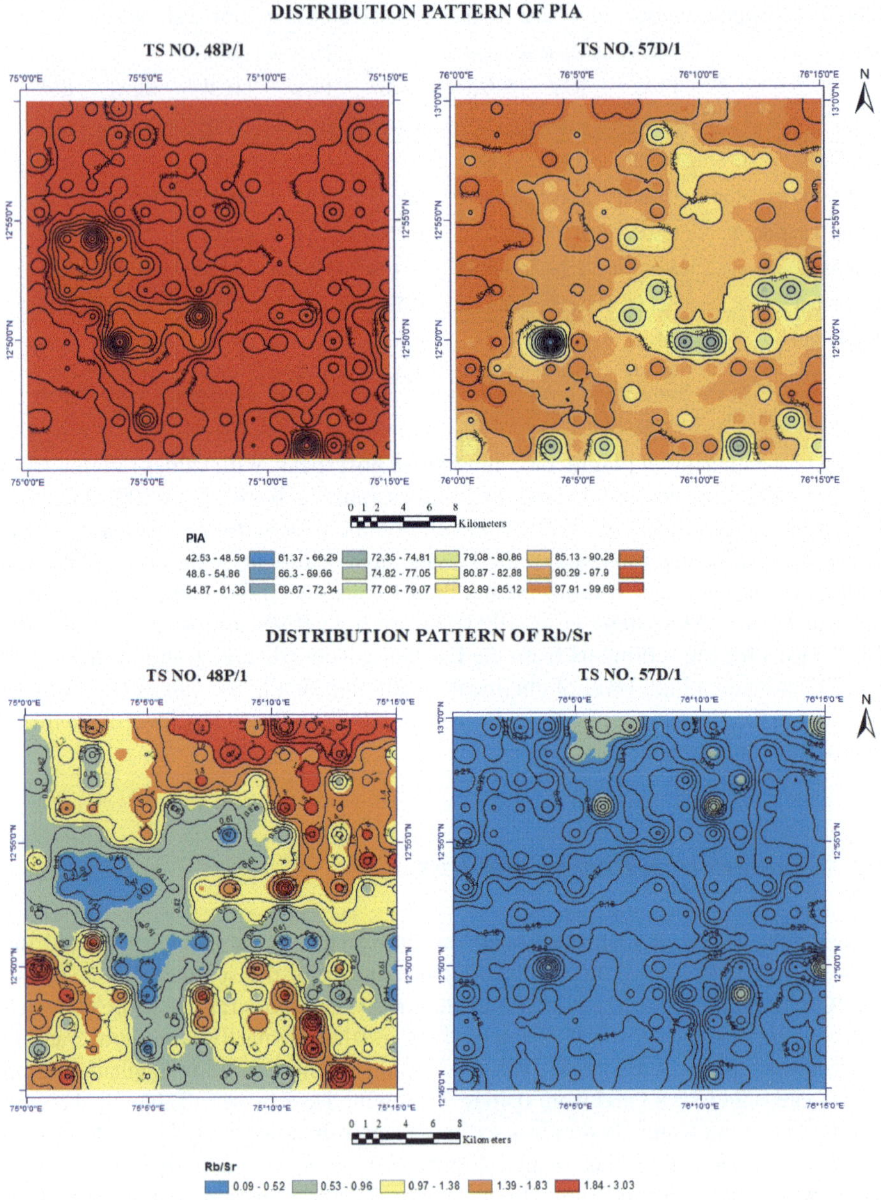

Fig. 4 Indicators of variance in chemical weathering intensity include PIA (top) and Rb/Sr ratio (bottom) from various longitudes in tropical southern India

level of clayness decreases with increasing sediment Si content. Si/Al ratios between high and low imply chemically immature and mineralogically mature sediments, respectively. Si/Al values in the Dakshina Kannada region range from 0.99 to 1.87 (Table 2), showing that the sediments were generated under humid conditions that tended to increase chemical maturity. Si/Al values in the Hassan area range from 2.62 to 10.47 (Table 2), indicating that the sediments were created in semi-humid circumstances with a tendency toward maturing less chemically. According to our research, there is an increase in chemical weathering, humidity, sediment maturity, and clay mineral growth as longitudes go from east to west (Fig. 5).

7.3 Geology, Chemical Weathering, Sediment Maturity, and Climatic Interactions

The changes in the chemistry of sediments are influenced by the extremely humid climate in the Dakshina Kannada region and the moderately humid climate in the Hassan area. While Dakshina Kannada, which was established in a tropical monsoon environment, gets significant chemical weathering, the Hassan area acquires moderate chemical weathering. At greater temperatures and moisture levels, the chemical weathering and sediment maturity processes advance more quickly (Binkley and Fisher 2019). By providing more moisture to mineral surfaces and accelerating the net dissolution reactions forward, high precipitation and temperature in tropical monsoon climate regions stimulate the dissolution of minerals (Ferrier et al. 2012). Thus, a combination of high temperatures and heavy precipitation speeds up chemical weathering. Numerous works from climatically various parts of the world support this phenomena (Rasmussen et al. 2011; Mir et al. 2019). The rate of chemical weathering and erosion is also influenced by other elements such as lithology, physical erosion rates, flora, land use-land cover, and human activity (Ferrier et al. 2012). Chemical weathering indicators (PIA, Rb/Sr, Fig. 4), mean annual precipitation (Fig. 3), and proxy measures of sediment maturity (ICV, Si/Al, Fig. 5) all show a strong correlation, collectively indicating that the degree of chemical weathering and sediment maturity increases with increasing precipitation and temperature.

The primary determinants of weathering rates are the makeup of the rocks and minerals as well as the local climate. Various rock kinds weather at various rates in a region with a consistent climate (Raymo and Ruddiman 1992). Acidic rocks weather more slowly than basic, ultrabasic, and carbonates rocks in a homogeneous environment. Because alkaline rocks are softer, weathering agents can easily alter and modify their structures (Daniel et al. 2016). Although there is little lithological difference between the Dakshina Kannada and Hassan areas in the current study, the Dakshina Kannada terrain is still more worn than the Hassan area. Here, climate is more crucial to the rate of weathering than lithology. Temperature and precipitation both affect a region's climate. High temperatures and precipitation intensify chemical

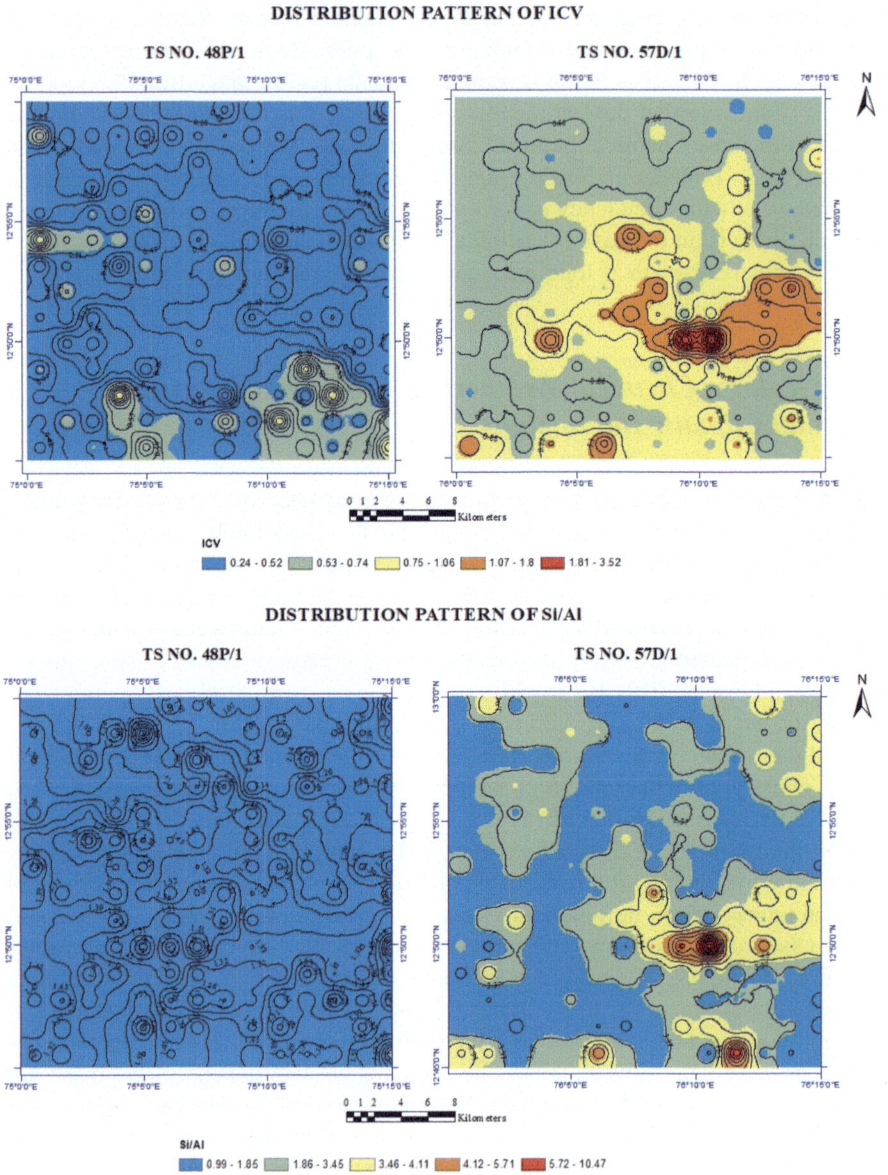

Fig. 5 ICV and Si/Al ratio comparisons from various longitudes in tropical southern India are shown on the top and bottom, respectively, as a representation of the variance in mineralogical maturity and clay mineral formation

reactions and accelerate weathering (White and Blum 1995; Curtis 1990). Additionally, a warmer environment encourages the growth of flora, which aids in weathering. More than the Hassan region, the Dakshina Kannada region's hot, humid temperature and lush evergreen foliage weather the region's acidic terrain. The PIA and Rb/Sr concentration of the study's stream sediments had strong relationships with temperature, precipitation, and longitude, demonstrating the major influence of the southwest monsoon environment on chemical weathering. Instead of using precise and direct proxies, the weathering indices and sediment maturity indices used in this work offer an overall picture of the weathering history in the drainage basins over the longitudinal shift. We cannot dispute the contribution of anthropogenic activities to the acceleration of human-induced weathering in the study area, but more research is required to fully comprehend this contribution.

8 Conclusions

The processes of chemical weathering and sediment maturity are discussed in this work from different longitudes in southern India under differing climatic circumstances. The following is a summary of the key findings of this study:

(a) Variations of PIA and Rb/Sr have been used to characterize chemical weathering conditions. All weathering indices are very similar to one another and differ primarily due to varying environmental circumstances. While rocks in the Hassan area are moderately weathered, those in the Dakshina Kannada region have experienced extensive weathering. High temperatures and rainfall-related climate forcing appear to be the cause of the region's high chemical weathering conditions in Dakshina Kannada.

(b) Using the ICV and Si/Al ratio, the compositional maturity of the sediments is assessed. Overall, the findings show that the sediments in the Dakshina Kannada region are more mineralogically developed than the sediments in the Hassan region. From west to east longitudes, sediment maturity decreases, indicating a high degree of clayness and mineralogical maturity that coincides with monsoon severity.

Acknowledgments The Director-General of GSI is thanked by the authors for developing the National Geochemical Mapping program and providing the tools required to complete this work. The Additional Director General, GSI, Hyderabad (SR) and the Deputy Director-General, State Unit, Karnataka and Goa, are acknowledged for providing logistical and financial support for this study, respectively. The GSI geochemistry laboratory team in Hyderabad was helpful in analyzing the samples.

References

Binkley, D., Fisher, R.F., 2019. Ecology and Management of Forest Soils, 5th edition. WileyBlackwell, pp. 456.

Cox, R., Lowe, D.R., and Cullers, R.L. 1995. The influence of sediment recycling and basement composition on evolution of mud rock chemistry in the southwestern United States: Geochimica et Cosmochimica Acta, 59: 2919–2940.

Curtis, C. D., 1990. Aspects of climatic influence on the clay mineralogy and geochemistry of soils, palaeosols and clastic sedimentary rocks: Journal of the Geological Society, 147, 351–357.

Daniel, E. I., Jeremy, K. C., Seulgi, M., Dana, L. T., Jens, H., Chamberlain, C. P., Kate, M., 2016. Differential weathering of basaltic and granitic catchments from concentration–discharge relationships. Geochimica et Cosmochimica Acta, 190, 265-293.

Fedo, C.M., Nesbitt, H.W., and Young, G.M., 1995. Unraveling the effects of potassium metasomatism in sedimentary rocks and paleosols, with implications for paleoweathering conditions and provenance. Geology, 23: 921–924.

Ferrier, K.L., Kirchner, J.W., Finke, R.C., 2012. Weak influences of climate and mineral supply rates on chemicalerosion rates: Measurements along two altitudinal transectsin the Idaho Batholith. J. Geophys. Res., 117:20-26.

Harnois, L., 1988. The CIW index: a new Chemical Index of Weathering. Sedimentary Geology, 55: 319–322.

Giovanoli, R., Schnoor, J.L., Sigg, L., Stumm, W., Zobrist, J., 1988. Chemical weathering of crystalline rocks in the catchment area of acidic Ticino Lakes, Switzerland. Clay Clay Miner., 36: 521–529.

Koppen, W., 2011. The thermal zones of the earth according to the duration of hot, moderate, and cold 9 periods and to the impact of heat on the organic world. Meteorologische Zeitschrift, 20: 351–360.

Krishnamurthy, K.V., Malhotra, A.K., Awasthi, S.K. (1980) Geology of parts of Bantval and Puttur Taluks, South Kanara District, Karnataka. Geological Survey of India Progress Report for the field season 1975-76.

McLennan, S.M., Hemming, S., McDaniel, D.K., and Hanson, G.N., 1993. Geochemical approaches to sedimentation, provenance, and tectonics. Processes controlling the composition of clastic sediments. Geological Society of America, Special Paper, 284: 21–40.

Mir, I.A., Maria Brenda Luzia Mascarenhasa, 2023. Redox changes during the past 100 kilo years in the deeper eastern Arabian Sea: a study based on trace elements and multivariate statistical analysis. Water, 15, 7, 1252.

Mir, I.A., SreePrabash, M., Sridhar, V., Maruthi, K.V., 2022a Distribution, sources and toxicity of heavy metals in surface sediments of north western Karnataka, south India. Sci. Rep. 12, 15782.

Mir, I.A., Bhat, A.A., Sreeprabash, M. Sridhar, V. Maruthi, K.V., 2022b. Surface sediment geochemistry for understanding the recent sedimentary environment in north western Karnataka, south India. Geosciences Journal, 26, 6, 669–683.

Mir, I.A., MariaBrenda Luzia Mascarenhas, Neloy Khare, 2022c. Geochemistry and granulometry as indicators of paleoclimate, weathering, and provenance of sediments for the past 100000 years in the eastern Arabian Sea. J. of Asian Earth Sci., 105102.

Mir, I.A., Mir, R. A., 2019. Geochemistry of surface sediments in parts of Bandipora-Ganderbal areas, Kashmir valley, western Himalaya: Implications for provenance and weathering. J. Earth Syst. Sci., 128: 223.

Mir, I.A., Pattan J.N., Matta V.M., V.K, Banakar, 2013. Variation of paleo-productivity and terrigenous input in the Eastern Arabian Sea during the past 100 ka. Journal of the Geological Society of India, 81, 647-654.

Nesbitt, H.W. and Young, G.M. 1982. Early Proterozoic Climates and Plate Motions Inferred from Major Element Chemistry of Lutites. Nature, 299, 715-717. http://doi.org/10.1038/299715a0

Parker, A., 1970. An index of weathering for silicate rocks. Geological Magazine, 107: 501–504.

Pattan, J.N., Parthiban, G., Karapurkar, S. G., Matta, V.M., Naidu, P.D., Naqvi, S.W.A., 2013. Coupling between the suboxic condition in sediments of the western Bay of Bengal and southwest monsoon intensification: A geochemical study. Chemical Geology, 343, 55–66.

Pattan, J.N., Parthiban, G., Gupta, S.M., and Mir, I.A., 2012. Fe speciation and Fe/Al ratio in the sediments of southeastern Arabian Sea as an indicator of climate change. Quaternary International, 250: 19–26.

Potter, P.E., 1978. Petrology and chemistry of modern big River sands. The Journal of Geology, 86(4): 423–449.

Rao, R. B., 1924. Report on the revision survey work of parts of Holenarsipur Taluk, Hassan District, Record from Mysore Geology Deptartment, volume 4.10

Rasmussen, C., Brantley, S., Richter, D., Blum, A., Dixon, J., White, A.F., 2011. Strong climate and tectonic control on plagioclase weathering in granitic terrain. Earth Planet. Sci. Lett., 301–530.

Raymo, M. E., Ruddiman, W. F., 1992. Tectonic forcing of late Cenozoic climate: Nature, 359, 117– 122.

Scarciglia, F., De Rosa, R., Vecchio, G., Apollaro, C., Robustelli, G., Terrasi, F., 2008. Volcanic soil formation in Calabria (southern Italy): the Cecita Lake geosol in the late Quaternary geomorphological evolution of the Sila uplands. J. Volcanol. Geotherm. Res., 177: 101–117.

SOP, 2014: (Standard Operation Procedure for National geochemical mapping & quality management. Revised and updated 2014), Mission-I: Baseline Geosciences Data Generation of Geological Survey of India.

White, A. F., Blum, A. E., 1995. Effects of climate on chemical weathering rates in watersheds:Geochimica et Cosmochimica Acta, 59, 1729 –1747.

Chapter 12
Phytolith-Based Palaeoclimate Studies in the Indian Subcontinent

Waseem Qader, Malik Zubair Ahmad, Jehangeer Ahmad Mir, Suhail Hussain Sheikh, Sajad Ahmad, Reyaz Ahmad Dar, and Irfan Rashid

1 Introduction

Numerous studies have been conducted worldwide to reconstruct past climates and assess the variables influencing contemporary climate change (Zalasiewicz and Williams 2009; Mir et al. 2022). Paleoclimate proxies, like sediment colour, magnetic properties, stable isotopes, and phytolith, amongst others (Dar et al. 2015 a,b) obtained from paleoclimate archives, such as ice cores, sediment cores, loess-paleosol sequences, and tree rings, constitute the foundation for these past climatic reconstructions (Shah et al. 2024; Dar and Zeeden 2020). Amongst various palaeoclimate proxies, phytolith studies have made a substantial contribution to our understanding of past climate changes (Rashid et al. 2019). Phytoliths, microscopic silica bodies formed within plants (Zexer et al. 2023), accumulate in soil and other sediments after the death and decay of plants (Fig. 1) and remain preserved for millions of years, thus making them useful proxies for reconstructing past vegetation types and patterns (Qader et al. 2023a, b). Vegetation plays a critical role in the Earth's climate through its effects on surface albedo (Mykleby et al. 2017), atmospheric aerosol (Jin and Wang 2018), greenhouse gas composition, and global carbon cycle (Piao et al. 2009). Also, the current climate trends indicate that the rise in both CO_2 and temperature is expected to continue, attaining levels that will, without any doubt, affect both flora and fauna, including mankind. The reconstruction of the type

W. Qader · M. Z. Ahmad · J. A. Mir · S. H. Sheikh · S. Ahmad · R. A. Dar (✉) · I. Rashid
Department of Earth Sciences, University of Kashmir, Srinagar 190006, India
e-mail: reyazdar@kashmiruniversity.ac.in

W. Qader
Centre for Disaster Risk Reduction, Islamic University of Science and Technology, Kashmir, Awantipora 192122, India

I. Rashid
Department of Botany, University of Kashmir, Srinagar 190006, India

N. Khare (ed.), *Changing Climate over India*,
https://doi.org/10.1007/978-981-92-1098-5_12

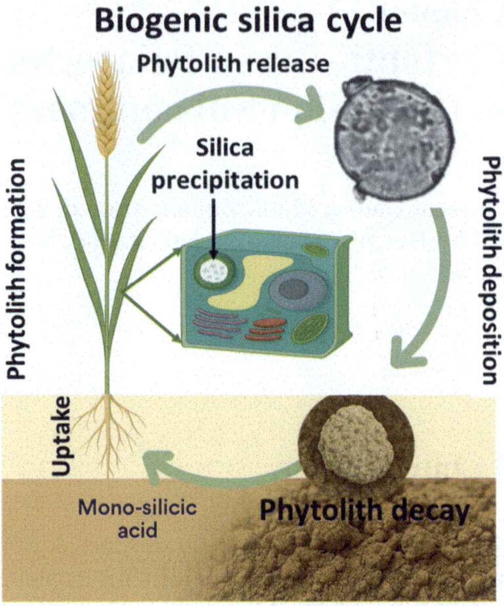

Fig. 1 Conceptual diagram showing the biogenic silica cycle: plants absorb mono-silicic acid from the soil and precipitate silica to form phytoliths in and/or between cells, which later get deposited into the soil upon plant decay, and over time, the phytoliths dissolve back into mono-silicic acid which reinters in to the soil solution and becomes available for plant uptake, completing the cycle

and distribution of past vegetation will, therefore, play an important role in understanding past climate changes, fluctuations in atmospheric pCO2, and numerous other environmental disturbances (Ghosh et al. 2017; Jha et al. 2020; Roy et al. 2020).

The Indian subcontinent, located entirely in the Northern Hemisphere, is characterised by diverse climatic conditions ranging from warm temperate in the northern parts, equatorial rainy belt in the peninsular part, and subtropical arid belt in the central and western part (Chatterjee et al. 2013). However, the voyage of the Indian plate from the south to its current location has a profound effect on the climate, especially the uplift of the Himalayas and Tibetan Plateau (TP), which obstructed the atmospheric flow from Central Asia, resulting in long-term climatic changes (Dar et al., 2025; Hay et al. 2002; Molnar 2005). For instance, the fossil records of both plants and animals show that the early Neogene experienced a warm, humid climate, which subsequently changed to a colder, drier period in the late Neogene. The Indian subcontinent thus experienced a significant shift in both climate and vegetation. The possible link between vegetation and monsoon strength has drawn special attention to understand the C3–C4 vegetation transition in the Indian subcontinent (House et al. 2023). During the Late Oligocene, mixed vegetation (tropical rain forests, humid forests, and dry open areas) followed by temperate broad-leaved vegetation was reported from the Western Pakistan region (De Franceschi et al. 2008). The temperate vegetation of the Oligo-Miocene was followed by tropical evergreen to moist deciduous species from the northern mountainous region (Srivastava et al. 2014 and references therein). The late Early Miocene of Kutch indicates the presence of warm and humid tropical rainforests (Shukla and Mehrotra 2018). The Middle Miocene conditions were warm and humid with a diverse evergreen and some moist deciduous rainforest, as evident from the Siwaliks (Srivastava et al. 2014). Late in the Miocene, at ~ 5.9 Ma, forests declined, giving way to C4 grasslands (Cerling et al. 1997; Hoorn et al. 2000). Whilst C4 grasslands predominated in the Late Miocene and Pliocene, C3 vegetation flourished in lowland areas during the Early and Middle Miocene with warm and humid tropical flora. Various studies show variations in the timing of the late Miocene expansion of C4 vegetation and imply that local climatic conditions played a significant role (e.g. Dunlea et al. 2020; Tauxe and Feakins 2020; House et al. 2023). The C_4 vegetation in the Indian subcontinent expanded to substantially greater abundance during the Middle Pliocene to the Middle Pleistocene (~3.5–1.5 Ma; Tauxe and Feakins 2020). Modern grasslands expanded widely during the Early Pleistocene, likely in response to increasing aridity (House et al. 2023). Further, in the Middle Pleistocene, shifts between C4 and C3 vegetation and the occurrence of mixed C3/C4 vegetation with a gradual increase in C4 vegetation towards the Late Pleistocene were reported (Krishnamurthy et al. 1986; Dar et al. 2015a, b). However, the Indian subcontinent still has very little evidence of Quaternary vegetation, especially before the Last Glacial Maximum, due to the low preservation capabilities of conventional vegetation proxies like pollens (Farooqui et al. 2010). Currently, the region exhibits diverse vegetation zones including tropical forests, thorn forests, and grasslands. In this chapter, we reviewed the phytolith-based studies conducted to understand the palaeoenvironment and monsoon variations in various regions of the Indian subcontinent.

2 Indian Subcontinent and its Geological History

The Indian subcontinent constitutes an important component of the Indian plate spanning across the equator that was a part of the Gondwana land, and fragmented at ~100 Ma and started moving towards the north (Stojanovic et al. 2016; Bhat et al. 2023a). The Indian subcontinent also includes the portions of the Indian Ocean basin, Tibet, and western Indonesia (Wallace 1876). Indus-Tsangpo Suture Zone (ITSZ) marks the boundary between the Indian plate and the Eurasian plate (Bhat et al. 2023b, 2024). The Indian subcontinent comprises the southern part of Asia, which includes countries of India, Pakistan, Nepal, Bhutan, Bangladesh, Sri Lanka, and Myanmar. The physiographic subdivisions of the Indian subcontinent for the present study include the Great Mountains (Himalayas) in its north, the north Indian Plains, the Peninsular Plateau, the Indian Desert, and the Coastal Plains and the Islands (Fig. 2).

The Indian plate was located in the Southern Hemisphere as an island continent for most of the Cretaceous. The climate of the Early Cretaceous was mild, followed by warmer global climates during the Late Cretaceous, with average annual temperatures above 38 °C at the equator. During the Late Cretaceous, the Indian subcontinent had drifted northward to 25°S, lying between the Equatorial Belt and Subtropical Arid Belt, with mean annual temperatures between 17 and 26 °C (Scotese et al. 2021; Goswami 2011). Tropical rainforests evolved and diversified somewhere during

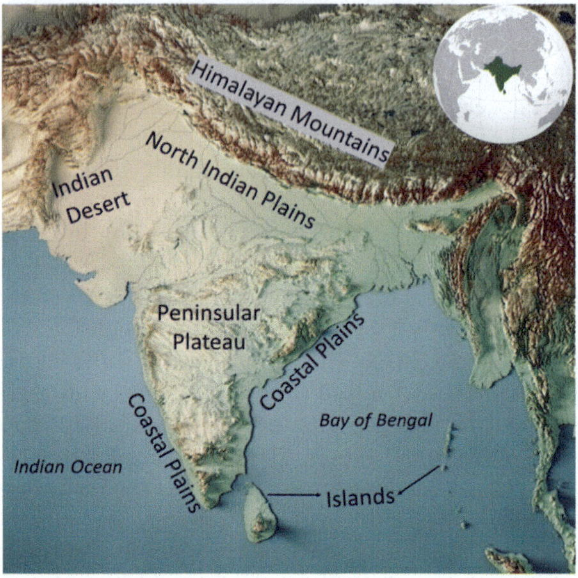

Fig. 2 Physiographic map of the Indian Subcontinent showing the broad divisions discussed in this chapter

the Late Cretaceous/Early Paleogene, influenced by tectonic movements (Givnish and Renner 2004) as revealed by the western Indian fossil records (Shukla and Mehrotra 2018). During the Early Cenozoic Era (Palaeocene and Early Eocene), the Indian plate entered the Equatorial Humid Belt. Cooling started during the Eocene–Oligocene boundary owing to the persistent weathering and erosion of the Deccan Traps that removed much of the CO_2 from the atmosphere (Kent and Muttoni 2008). However, during the Oligocene (~30 Ma), the Indian subcontinent reached close to its present position and was centrally located at 10°N with the southern part in the Equatorial Rainy Belt and the northern part lying in the Subtropical Arid Belt with mean annual temperatures ranging between 27 °C (southern part) and 5 °C (northern part; Scotese et al. 2021). Subsequently, the climatic divide formed due to the uplift of the Himalayan and TP (Hay et al. 2002; Molnar 2005). The uplift disrupted the circulation, bringing the low-pressure area southward over India, resulting in the evolution of the Asian monsoon system, although it is still a matter of debate (Dar and Zeeden 2020). The glaciation in the Himalayas and TP was extensive during 60–32 ka and at other places during the global maximum at ~18 to 20 ka (Benn and Owens 1998; Dar et al. 2017; Murtaza et al. 2021; Joya et al. 2021). However, towards the end of the Pleistocene ice age (18 to 20 ka), oscillations of dry-cold and wet-warm climates have been recorded (Paul et al. 2022; Dar et al. 2022). Drier conditions dominated with reduced fluvial activity during the Last Glacial Maximum (LGM; ~21.5 ka; Sirocko et al. 1993; Thompson et al. 1997; Kale et al. 2003) followed by intensification of summer monsoons during the Early Holocene (~11.5–5.0 ka; Sirocko et al. 1993; Overpeck et al. 1996).

3 Climate and Vegetation History of the Indian Subcontinent

The present-day climate is characterised by warm-wet summers and is governed by the annual latitudinal migration of the Intertropical Convergence Zone (ITCZ), with 70% of annual precipitation falling during late summers and the rest during dry winters (Dixit and Tandon 2016). The reversal of monsoon winds results in large seasonal variations in temperatures, precipitation, water, and sediment transport (Nair et al. 1989; Sirocko and Lange 1991) in the Indian subcontinent. The Indian summer monsoon (ISM) system, an important component of the global climate, has a profound impact on the vegetation pattern of the Indian subcontinent (Kudrass et al. 2001; Gupta et al. 2003). However, the uplift of the Himalayas and TP had played an important role in the climatic variability and palaeo-biogeography of the flora and fauna of the Indian subcontinent. The vegetation shifts from C3 (forests) to C4 (grassland) type were documented during the Late Miocene as revealed by numerous studies carried out across the subcontinent (Flynn and Jacobs 1982; Barry et al. 2002; Quade et al. 1989; Quade et al. 1995; Hoorn et al. 2000; Sanyal et al. 2004; Galy et al. 2008). The shift to C4 vegetation resulted from seasonal droughts that prevailed due to increased aridity (Barry et al. 2002) and seasonality in response to the increasing distance between the ICTZ and the Himalaya, thus lowering the amount of precipitation required to support the forests in the region (Allen and Armstrong 2012). This is also evident from the decreased sedimentation flux in Himalayan basins during the Late Miocene (Burbank et al. 1993; Clift and Plumb 2008) and also from the dietary patterns and habitat shifts of mammals in the Indian subcontinent (Patnaik et al. 2019). Currently, the region can be broadly divided into the following vegetation zones: (1) tropical evergreen forests, found in the Western Ghats, the north-eastern states, and the Andaman and Nicobar Islands, characterised by high rainfall and high humidity; (2) tropical deciduous forests, found in central and eastern India characterised by a dry season and a rainy season; (3) thorn forests and scrubs, found in the semi-arid regions of western and southern India characterised by low rainfall and extreme temperatures; and (4) grasslands, found in the western and northern regions of India, are characterised by low rainfall and high temperatures.

4 History of Phytolith Research in the Indian Subcontinent

Phytolith research in the Indian subcontinent has a long history dating back to the early twentieth century, with the majority of the studies contributed from the Indian region. The first significant study of phytoliths in India was conducted by the British archaeologist Mortimer Wheeler in the 1930s, who used phytolith analysis to study the ancient plant remains at the Indus Valley Civilisation site of Harappa (Wheeler 1936). From the 1950s, Indian scientists began to take an interest in phytolith research. One of the earliest studies was conducted by Ponnaiya (1951), who studied the distribution of epidermal silica bodies in Sorghum and described irregular and dumbbell-shaped silica bodies. Later, Thapar (1955) used phytolith analysis to study the ancient plant remains at the Harappan site of Mohenjo-daro. Thapar's work was significant in demonstrating the potential of phytolith analysis for reconstructing past plant use and subsistence patterns in ancient societies. Mehra and Sharma (1965) accounted for conical silica bodies from sedges, Govindrajalu (1968, 1975) described silica bodies from Eleocharis, Rhynchospora and Scleria, and Sharma (1972) described the morphology and distribution of silica structures in Scirpus squarrosus. In the 1970s and 1980s, Indian scientists continued to develop phytolith research, with a focus on investigating the use of phytoliths as indicators of past environmental change. One of the notable studies was conducted by Bisht (1976), who used phytolith analysis to reconstruct past vegetation patterns in the Western Ghats region of India. Besides, Srivastava (1978) described the location and morphology of silica bodies in Digitaria. Vaughan et al. (1987) found silica of plant origin on gloss that appeared on sickles used for harvesting the sorghum crop in Baluchistan (Pakistan). Weber (1989) studied the archaeobotanical remains, including phytoliths, at the site of Rojdi, in north-west India, during their study on Plants and Harappan subsistence. In the 1990s, Indian scientists began to focus on using phytoliths as indicators of past human activity. One of the most notable studies from this period was conducted by Chauhan (1994), who used phytolith analysis to investigate ancient land-use practices in the Deccan Plateau region of India.

From 2000 onwards, Indian scientists continued to make significant contributions to phytolith research. For example, Krishnan et al. (2000) studied phytolith size and morphology of modern grasses from Tamil Nadu, India, to identify extant and extinct grasses for exploring past vegetation and climate. Kajale and Eksambekar (2001) investigated phytoliths at Balathal, Rajasthan, India, indicating the presence of oryzoid cereals or panicoid millets that were used by the ancient farmers of Balathal. Saijo and Tanaka (2002) used phytoliths from palaeosols from the Thakkhola basin, central Nepal, to understand their palaeoclimatic significance, suggesting stronger summer monsoons, increased rainfall, and dense vegetation at the time of their formation. Ghosh et al. (2006) studied phytoliths from ^{14}C-dated sedimentary layers to understand vegetation patterns in an archaeological site of Pakhanna district, Bankura, West Bengal. Prasad et al. (2005) found phytoliths preserved in Dinosaur coprolites that showed five taxa from extant grass (Poaceae) subclades from the Indian subcontinent during the late Cretaceous. Saxena et al. (2006) studied phytoliths from Lahuradewa Lake deposits of the Ganga Plain, finding evidence of rice-based agriculture during the last 10 ka. Phytolith studies were also carried out by Singh et al. (2007) at Itola, Dhadhar River basin, to indicate monsoonal variabilities during the Holocene in western India. Alam et al. (2009) used phytolith from archaeological soils from the Somapura Mahavihara site, Bangladesh, suggesting five different climate zones marked by alternatively cool and temperate events during the late Holocene.

Similarly, Ghosh et al. (2011) studied phytolith for reconstructing deltaic environments in the Indian Sundarbans. Tripathi et al. (2013) investigated the patterns of phytolith to trace the history of cultivation of *Sorghum* in Allahabad, UP, India. Thomas et al. (2012) analysed an archaeological net from Shahi Tump, Baluchistan,

Pakistan, which suggested the use of fibres from the leaves of a local palm species: the desert palm or *Nannorrhops ritchieana* (Griff.) Aitch. In addition, the study discussed past and present uses of the desert palm in the arid regions of the Middle East, as well as the possible utilisation (fishing, carrying, etc.) of the protohistoric net. Saxena et al. (2013), whilst using the phytolith index (Iph) of phytolith assemblages from lake sediments of the Ganga Plain, found alternating wet and dry phases during the past 10,600 BP. Farooqui et al. (2013) studied phytoliths to address the climate–culture relationship between Kaj and Kanjetar, mid-Saurashtra coast, Gujarat. Prasad et al. (2014) carried out a multiproxy study, which also included phytoliths on lake sediments to reconstruct climatic fluctuations of the mid-Holocene. Similarly, Raj et al. (2015) used phytoliths to understand Holocene climatic fluctuations from the alluvial plains of western India. An et al. (2015) used surface soil phytoliths as vegetation and altitude indicators from the southern Himalaya, which suggested a good correlation between phytolith assemblages and vegetation zones. Sakai et al. (2016) used phytoliths to assess the lowering of lake water caused by possible earthquakes in the Paleo-Kathmandu Lake, central Nepal Himalaya. Patil et al. (2017) used phytolith analysis to investigate the use of plants in ancient human burials in the Western Ghats region. Nogué et al. (2017) in their study used phytoliths in sediment sequences from the Western Ghats of India to assess environmental changes in the region. From the humidity-aridity index (Iph), they suggested changes in phytolith percentages during the last 1000 years. Premathilake and Hunt (2018) used phytolith records from Fahien Rock Shelter, Sri Lanka, which indicated the interactions of anatomically modern humans with the lowland rainforests of south-western Sri Lanka when the Rock Shelter occupants extracted their livelihood from several wild plants, including bananas, rice, breadfruits, durians, *canarium,* and species of palm and bamboo. Pillai et al. (2018) used phytoliths as evidence for the Mid-to-Late Holocene shift in climate and vegetation of the Banni Grasslands of Western India.

Sarkar et al. (2020) suggested climate-culture co-evolution during 3100–900 years B.P. in the Thar Desert, India, by using phytoliths along with other proxies. They also suggested that the early withdrawal of the Late Holocene ITCZ and declining monsoon accounted for the drying of the fluvial system. In core sediments of Tarna-Satighat Lake, Central India, Humane et al. (2020) used phytoliths and diatoms to study their implications as indicators of climatic variability and trophic status. Recently, Biswas et al. (2021) proposed a revised climatic index (rIc) and estimated the Late Holocene mean annual temperature variations in the Himalayas. They suggested that in the Himalayas, with irregular precipitation, phytolith distribution is a function of temperature. Premathilake et al. (2022) investigated the taphonomy of phytoliths at the Palaeolithic site of Attirampakkam, South India. Kumar et al. (2022) used multiproxy analysis, including phytoliths from lacustrine sediments, to determine the drivers of Late Quaternary C_3/C_4 plant variability in the northern Gangetic Plain. Recently, Quaternary vegetation and climate based on phytoliths have been documented in the Kashmir Himalaya (Fig. 3; Rehman et al. 2024; Qader et al. 2024; Sheikh et al. 2025).

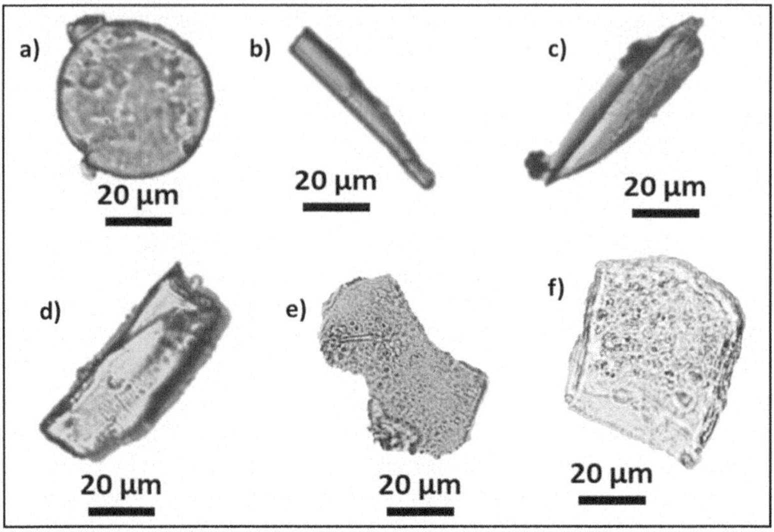

Fig. 3 Microphotographs of some phytolith morphotypes; **a** spheroid psilate; **b** elongate entire; **c** acute bulbous; **d** tracheary; **e** bilobate; **f** spheroid ornate; identified from loess/palaeosols sequences of Kashmir Valley, India (Unpublished PhD thesis data, Department of Earth Sciences, University of Kashmir)

Overall, the history of phytolith research in the Indian subcontinent demonstrates the potential of this technique for reconstructing past plant use and subsistence patterns, as well as investigating past human activity and environmental changes.

5 Phytoliths as Regional Indicators of Palaeoenvironment

Phytoliths have been used regionally to reconstruct past climate and vegetation patterns. The presence and abundance of different types of phytoliths in sediment samples can provide information on the types of plants that were present in a particular area, and thus, the climate and vegetation conditions of that area. Based on the studies carried out on phytoliths, the Indian subcontinent has been divided into many broad zones for a better understanding of the regional climatic scenarios due to its complicated terrain and the diverse climatic conditions seen across its length and breadth. These zones, based on physiographic subdivisions, are the Himalayan Mountains, the North Indian Plains, the Peninsular Plateau, the Indian Desert, and the Coastal Plains and the Islands.

5.1 The Himalayan Mountains

Previous research on vegetation and climate records suggests long-term past climatic variability in the Himalayas (Bali et al. 2017; Pillai et al. 2018; Kumar et al. 2021). However, very few studies have used phytoliths for such vegetation reconstructions in the Himalayas. Very recently, Sheikh et al. (2025) used phytoliths from palaeosols to understand climate dynamics in the Pliocene Karewa deposits of the Kashmir Himalaya. Ganjoo and Shaker (2007) studied phytoliths in Ramnagar Member palaeosols of Lower Siwalik at Udhampur, Jammu and Kashmir state, finding well-preserved phytoliths reflecting climatic seasonality, indicating the influence of ISM, which suggests wet and humid climatic conditions. Recently, Biswas et al. (2021) inferred that phytolith spectra variability in C3/C4 grass distributions and irregular precipitation distribution are a primary function of temperature in mountainous regions like the Himalayas. Similarly, Biswas et al. (2016) found a significant relation between grass silica short cell phytoliths and rising elevation to interpret vegetation and climate in Darjeeling and Arunachal Pradesh. The grass silica short cell phytolith assemblages, such as bilobate, cross, saddle, and rondel from modern grasses, showed significant variability and were found to be reliable in discriminating C3/C4 grasses in high-altitude mountainous regions. Mukherjee et al. (2019) distinguished different vegetation–climate zones along the elevation gradient by using phytolith spectra of some eastern Himalayan pteridophytes. They suggested that phytoliths can be successfully employed to discriminate tropical and subtropical from temperate vegetation and for palaeoclimate reconstructions. Tripathi et al. (2019) studied phytoliths in the dung of endangered Sangai (Rucervus eldii eldii) and Hog deer (Axis porcinus) from Manipur, India, to understand their implications for palaeo-herbivory and palaeoecology of the region. The study revealed significant variation in vegetation. For example, *Sapotaceae,* and marshy and aquatic taxa such as *Polygonum, Impatiens, Cyperaceae, and Nymphaea,* are indicative of the tropical mixed deciduous forest and perennial waterlogged condition in the region.

Given the paucity of palaeovegetation studies, more detailed phytolith-based studies on vegetation and climatic fluctuations of the Himalayas are needed to extend our understanding beyond the instrumental era through natural archives for future climate change predictability.

5.2 The Northern Plains

The northern plains, also known as the Indo-Gangetic plains, serve as potential archives for understanding the regional climatic history (Misra et al. 2020). For example, Saxena et al. (2013) based on a phytolith study reconstructed the Holocene climate from a lake-fill sequence of the Ganga Plain. The dominant phytoliths reported from the sequence are Panicoideae, indicating a warm and humid climate, Chloridoideae, indicating a warm and dry climate, and Festucoidae, indicating a cool and moist climate. The authors have identified five wet phases alternating with four dry phases during the past 10.6 ka. The four dry phases are: (1) 10,300–9200 cal yrs. BP, (2) 5300–4100 cal yrs. BP, (3) 1650–1200 cal yrs. BP, and (4) 950–700 cal yrs. BP. Out of these, the dry phase from 5300 to 4100 cal yrs. BP was the most prominent. Similarly, Misra et al. (2020) studied millennial-scale vegetation and climatic changes from an Early to Mid-Holocene lacustrine sequence in Central Ganga Plains using multiple biotic proxies, including phytoliths, to develop an understanding of the role of local and/or regional climate forcing. Between 13.2 and 4.8 ka, five major biotic zones were delineated; Zone I (~13–12 ka) showed dry tropical woody flora along with tall C4 grasses indicating a moderately dry climate; Zone II (12–9 ka) indicating the presence of stagnant, lime-rich inland waters under relatively colder climatic conditions; Zone III (9–8 ka) records millennial-scale drying with a complete absence of aquatic pollen between 8.9 and 7.7 ka BP; Zone IV (8–7 ka) indicates drier conditions; and Zone V (7 –~5 ka) represents a strengthened ISM and a subsequent lake expansion (Misra et al. 2020).

5.3 The Peninsular Plateau

The palaeovegetation archives in peninsular India are scarce due to high levels of rock decomposition (von Lengerke 1977) although few sites, such as shola grasslands and Sandynallah valley (Nilgiris, Western Ghats), have facilitated Late Quaternary vegetation and climate studies (Vasanthy 1988). Veena et al. (2014) found phytoliths from Pookode Lake, Kerala, South India, during the reconstruction of Late Holocene monsoon variability. The results suggest that warm and dry conditions prevailed during 6200–420 BP; however, short, intense wet phases between ~3900 and ~1900, ~1400–~760, and ~420–~140 BP resulted from the strengthening of the southwest monsoon (SWM). Premathilake et al. (2017) used phytoliths as indicators of Quaternary vegetation at the Palaeolithic site of Attirampakkam, south India, finding that this region was arid during the earlier stages of the early Pleistocene, lacking woody plants. The later phase of the Early Pleistocene was wooded, suggesting a shift from a drier to a relatively more humid climate, and the transition towards the Middle Pleistocene was marked by a moderately humid climate, with a shift towards aridity during the Late Pleistocene.

5.4 The Great Indian Desert and Adjoining Areas

Multiproxy studies, including phytolith analysis, from various sedimentary sequences in western India have been instrumental in reconstructing Mid-to-Late Holocene climatic shifts. Research in the Kothiyakhad sedimentary sequence (Mahi estuary) identified three distinct climatic phases: i.e. Phase I (3660—~3400 yr. BP), an initial period of weakening SW monsoon, overlapped by enhanced winter precipitation; Phase II (~3400—~3000 yr. BP), severe drought-like conditions due to monsoon decline, and Phase III (~3000—2850 yr. BP), fluctuating but generally warmer and wetter conditions, akin to present (Prasad et al. 2007). Similarly, a study in the Dhadhar River basin noted a Mid-to-Late Holocene weakening of the SW monsoon, with a period of cool and moist climatic conditions sustained by winter precipitation before a brief phase of arid conditions, followed by a late revival of monsoon strength (Singh et al. 2007). A multiproxy record of Wadhwana Lake in Gujarat traced climate from ~7500 to 5560 cal yr. BP, suggesting cool and moist climatic conditions initially. This was later gradually replaced by a transition to a seasonally drier climate, suggesting the presence of evergreen to deciduous pollen taxa. A documented period of low lake levels, and a return to a stronger summer monsoon after ~3500 cal yr. BP, interspersed with a short dry pulse before stabilising to present-like conditions (Prasad et al. 2014). The archive from Pariej Lake further extends this narrative, indicating a very humid onset of the Holocene. Phytolith evidence shows a very wet phase at ~11 ka with moisture-loving grasses, followed by reduced phytolith yields between ~9000 and 8000 cal yr. BP, indicating lake contraction. Around 7630 cal yr. BP, moderate phytolith production reflects fluctuating precipitation, whilst very low phytolith counts between 5864 and 4680 cal yr. BP mark a pronounced dry spell. After ~4680 cal yr. BP, high phytolith density and diversity point to a return to wetter conditions, in agreement with other records from western India (Raj et al. 2015). Pillai et al. (2018) analysed phytoliths along with other proxy records to track woodland–grassland shifts in the Banni grassland, western India, during the Mid-to-Late Holocene. Their study shows that 4600–2500 cal yr. BP was a high-precipitation phase, marked by abundant pollen and diverse phytolith morphotypes indicative of woody savanna and mesic herbaceous vegetation. After 2500 cal yr. BP, the region experienced reduced rainfall, a decline in mesic taxa, and a rise in grass phytoliths, reflecting a transition towards more open, drier grassland conditions. Beyond terrestrial conditions, multiproxy studies have also documented the interplay between climate and sea level. Recently, Raj et al. (2021) reconstructed palaeoclimate and sea-level changes from the last deglaciation to the Late Holocene using a multiproxy dataset including phytoliths and identified four episodes of marginal-marine conditions during this interval. The first marine-influenced deposition occurred after the LGM, with core records capturing both the dry conditions of the Older and Younger Dryas and the warm, humid climate of the Bølling–Allerød. A second sea-level rise occurred between ~11,210 and ~9795 cal yr. BP, corresponding to a period of rapid sea-level increase along the western Indian coast. A third rise is documented at ~5747 cal yr. BP, aligning with a regional high sea-level phase. Finally, the reappearance of sediments with strong marine signatures from ~2650 ± 180 to ~1530 ± 350 cal yr. BP indicates a return to marine to marginal-marine conditions in the region.

5.5 The Coastal Plains and the Islands

Phytolith analysis has proven highly effective for reconstructing palaeoenvironments in the dynamic coastal and island regions surrounding India, including the Sundarbans, the coast of Kerala, and the islands. In the Indian Sundarbans, studies have established a strong correlation between specific phytolith morphotypes and soil salinity. For instance, bilobate and rondel forms are associated with high-salinity zones, whilst cross and polylobate types dominate in low-salinity areas. This relationship provides a powerful tool for identifying distinct phytoecological zones and interpreting past depositional environments in deltaic settings (Das et al. 2013; Naskar et al. 2018). Further refining this approach, research on mangrove aerial roots has shown that subtle size differences in blocky phytoliths can help distinguish true mangrove environments from other deltaic sub-environments (Das et al. 2014). Applied to a long-term sedimentary record, the presence of mangrove-indicator grass phytoliths suggests a stable mangrove environment in the Sundarbans from ~13.6 to 3.9 ka, which then disappeared before recolonising between ~2.2 and 0.8 ka (Naskar et al. 2022). Similar work on the south-west coast of India (Thrissur, Kerala) indicated a peak in monsoon-driven mangrove expansion between 7270 and 3530 cal BP (Misra and Bhattacharyya 2014). The discovery of the diagnostic phytolith of *Porteresia coarctata*, a robust grass, in Late Quaternary sediments of the east coast serves as a clear indicator of historical coastal swamp or intertidal mangrove habitats (Naskar et al. 2021). Beyond mangroves, phytolith assemblages from Sal (*Shorea robusta*) in the Bengal Basin are being used to understand their implications for reconstructing broader tropical deciduous and semi-evergreen forest histories (Saikia et al. 2022). Collectively, these phytolith-based investigations provide a robust foundation for regional palaeovegetation and palaeoenvironmental reconstructions across India's diverse coastal landscapes.

6 Summary and Way Forward

This chapter reviews the geological history, climate, vegetation, and phytolith research in the Indian subcontinent. It highlights the use of phytolith records as regional indicators of vegetation and climate change. The Indian subcontinent has experienced significant climatic and environmental changes throughout its geological history, influenced by tectonic movements, CO_2 levels, and the evolution of the monsoon system. These changes have had a profound impact on the region's vegetation and other sectors. The ISM has a significant impact on the region's socio-economic structure and is responsible for seasonal fluctuations in precipitation and the run-off of major river systems. The closure of the Tethys Ocean has played a crucial role in the climatic variability and palaeo-biogeography of flora and fauna. The vegetation of the Indian subcontinent has undergone significant changes, with a shift from C3 vegetation dominance in the early and middle Miocene to C4 grasslands in the late Miocene and Pliocene. Currently, the region exhibits a diverse range of vegetation zones, including tropical evergreen forests, tropical deciduous forests, thorn forests and scrubs, and grasslands. Phytolith studies in different regions of the Indian subcontinent have contributed to our understanding of past climate and vegetation dynamics. These studies have provided valuable insights into the impact of the ISM, changes in precipitation patterns, and shifts in vegetation composition over time. These studies highlight the importance of using palaeoclimate proxies, such as phytoliths, to extend our understanding of past environments beyond the instrumental era. The use of phytolith analysis in archaeo-anthropological investigations (not included in this review) in the Indian subcontinent has also provided valuable insights into past human activities and environmental changes. Phytolith records have revealed information about crop domestication, agriculture, and palaeodietary patterns in ancient societies such as the Harappan Civilisation. Some studies have also explored the use of phytoliths to assess human utilisation of rainforest plant resources in prehistoric Sri Lanka to reconstruct past land cover and assess plant resources in various archaeological sites in India. Phytolith analysis has helped interpret archaeological records, identify agricultural practices, and understand the cultural and ecological landscapes of different regions and periods.

Moving forward, there are several avenues for further research and exploration based on the findings and implications presented in this chapter. Some potential directions for future studies should include refining phytolith analysis techniques to enhance the accuracy and precision of phytolith records. High-resolution phytolith studies across various regions within the Indian subcontinent can provide a broader understanding of regional vegetation and climate dynamics. Combining phytolith analysis with other palaeoenvironmental proxies, such as pollen, charcoal, isotopes, and sedimentological data, etc., can enhance the reliability of palaeoenvironmental reconstructions (Qader et al. 2023b). Further exploration of the application of phytolith analysis in archaeological and anthropological investigations can shed light on various aspects of human history and culture. Investigating the relationship between phytolith assemblages and human activities, such as agriculture, land use, and dietary patterns, can deepen our understanding of past societies and their interactions with the environment. Integrating geological, climatic, ecological, and anthropological data in a holistic framework can provide a comprehensive picture of the Indian subcontinent's environmental history. The knowledge gained from phytolith research can have practical implications for conservation efforts in the Indian subcontinent. By understanding past vegetation dynamics and their response to climatic changes, conservation strategies and management plans can be enhanced for preserving the region's diverse ecosystems and protecting its vulnerable flora and fauna.

By pursuing these avenues of research, scientists, archaeologists, and conservationists can expand the understanding of the geological history, climate, vegetation, and human–environment interactions in the Indian subcontinent. This knowledge will not only contribute to our understanding of the past but also inform decision-making processes related to climate change adaptation, biodiversity conservation, and sustainable land-use practices in the region.

Acknowledgement This study was supported by the Ministry of Earth Sciences (MoES), Government of India, under grant number MoES/CCR/Paleo-8/2019. The authors are grateful to the editor for careful review of the chapter that greatly improved its content and quality.

References

Alam, A. M., Xie, S., & Wallis, L. A. (2009). Reconstructing late Holocene palaeoenvironments in Bangladesh: phytolith analysis of archaeological soils from Somapura Mahavihara site in the Paharpur area, Badalgacchi Upazila, Naogaon District, Bangladesh. *Journal of Archaeological Science, 36*(2), 504–512.

Allen, M. B., & Armstrong, H. A. (2012). Reconciling the Intertropical Convergence Zone, Himalayan/Tibetan tectonics, and the onset of the Asian monsoon system. *Journal of Asian Earth Sciences, 44*, 36–47.

An, X., Lu, H., & Chu, G. (2015). Surface soil phytoliths as vegetation and altitude indicators: a study from the southern Himalaya. *Scientific Reports, 5*(1), 1–13.

Bali R, Chauhan MS, Mishra AK, Ali SN, Tomar A, Khan I, Singh DS, Srivastava P. (2017). Vegetation and climate change in the temperate-subalpine belt of Himachal Pradesh since 6300 cal. yrs. BP, inferred from pollen evidence of Triloknath palaeolake. Quaternary International. 15; 444:11–23.

Barry, J. C., Morgan, M. E., Flynn, L. J., Pilbeam, D., Behrensmeyer, A. K., Raza, S. M.,... & Kelley, J. (2002). Faunal and environmental change in the late Miocene Siwaliks of northern Pakistan. *Paleobiology, 28*(S2), 1–71.

Benn, D. I., & Owen, L. A. (1998). The role of the Indian summer monsoon and the mid-latitude westerlies in Himalayan glaciation: review and speculative discussion. *Journal of the Geological Society, 155*(2), 353–363.

Bhat, I.M., Chauhan, H., Ahmad, T. and Dar, R.A., (2023a). Geochemistry and petrogenesis of ophiolitic rocks from the Indus Suture Zone (ISZ), Ladakh Himalaya: insights for depleted mantle beneath an intra-oceanic Island arc complex. *International Geology Review, 65*(21), pp.3329–3347.

Bhat, I.M., Chauhan, H., Ahmad, T., Tanaka, T., Bickle, M., Asahara, Y., Chapman, H. and Dar, R.A., (2024). Fate of an oceanic plate in the Neo-Tethys intra-oceanic subduction system: Evidence from elemental and Rb/Sr–Sm/Nd isotopic systematics. *Gondwana Research, 125*, pp.266–283.

Bhat, I.M., Chauhan, H., Dar, R.A. and Ahmad, T., (2023b). Ladakh Himalayan Ophiolites (LHO): A Geological Heritage of Northwestern India. *Geoheritage, 15*(1), p.2.

Bisht, R. S. (1976). Phytoliths from the Deccan Intertrappean beds and their palaeoecological implications. Palaeogeography, Palaeoclimatology, Palaeoecology, 22(3), 259–268.

Biswas, O., Ghosh, R., Agrawal, S., Morthekai, P., Paruya, D. K., Mukherjee, B.,... & Bera, S. (2021). A comprehensive calibrated phytolith based climatic index from the Himalaya and its application in palaeotemperature reconstruction. *Science of the total Environment, 750*, 142280.

Biswas, O., Ghosh, R., Paruya, D. K., Mukherjee, B., Thapa, K. K., & Bera, S. (2016). Can grass phytoliths and indices be relied on during vegetation and climate interpretations in the eastern Himalayas? Studies from Darjeeling and Arunachal Pradesh, India. *Quaternary Science Reviews, 134*, 114–132.

Burbank, D. W., Derry, L. A., & France-Lanord, C. (1993). Reduced Himalayan sediment production 8 Myr ago despite an intensified monsoon. *Nature, 364*(6432), 48–50.

Cerling, T.E., Harris, J.M., MacFadden, B.J., Leakey, M.G., Quade, J., Eisenmann, V., Ehleringer, J.R., (1997). Global vegetation changes through the Miocene/Pliocene boundary. Nature 389, 153–158.

Chatterjee, S., Goswami, A., & Scotese, C. R. (2013). The longest voyage: Tectonic, magmatic, and paleoclimatic evolution of the Indian plate during its northward flight from Gondwana to Asia. *Gondwana Research, 23*(1), 238–267.

Chauhan, P. R. (1994). Phytolith analysis of ancient land use practices in the Deccan Plateau, India. Journal of Archaeological Science, 21(5), 535–542.

Clift, P. D., & Plumb, R. A. (2008). *The Asian monsoon: causes, history and effects* (Vol. 288). Cambridge: Cambridge University Press.

Dar, R.A. and Zeeden, C., (2020). Loess-palaeosol sequences in the Kashmir Valley, NW Himalayas: a review. *Frontiers in Earth Science, 8*, p.113.

Dar, R.A., Chandra, R., Romshoo, S.A. and Kowser, N., (2015b). Micromorphological investigations of the Late Quaternary loess–paleosol sequences of the Kashmir Valley, India. *Journal of Asian Earth Sciences, 111*, pp.328–338.

Dar, R.A., Chandra, R., Romshoo, S.A., Lone, M.A. and Ahmad, S.M., (2015a). Isotopic and micromorphological studies of Late Quaternary loess–paleosol sequences of the Karewa Group: inferences for palaeoclimate of Kashmir Valley. *Quaternary International, 371*, pp.122–134.

Dar, R.A., Jaan, O., Murtaza, K.O. and Romshoo, S.A., (2017). Glacial-geomorphic study of the Thajwas glacier valley, Kashmir Himalayas, India. *Quaternary International, 444*, pp.157–171.

Dar, R.A., Murtaza, K.O., Paul, O.J., Nisa, A.U., Akhter, N., Dar, F.A. and Mir, R.A., (2022). River response to melting cryosphere since late quaternary in the pir panjal range of NW Himalaya. *Frontiers in Water, 4*, p. 879001.

Dar, R. A., Qader, W., Rashid, I., & Mir, J. A. (2025). Multi-Proxy Records of Climate Change over the Indian Subcontinent. *Climate Dynamics Over the Indian Subcontinent: Geological Perspectives* (2025), 22–59.

Das, S., Ghosh, R., & Bera, S. (2013). Application of non-grass phytoliths in reconstructing deltaic environments: A study from the Indian Sunderbans. *Palaeogeography, Palaeoclimatology, Palaeoecology, 376*, 48–65.

Das, S., Ghosh, R., Paruya, D. K., Yao, Y. F., Li, C. S., & Bera, S. (2014). Phytolith spectra in respiratory aerial roots of some mangrove plants of the Indian Sunderbans and its efficacy in ancient deltaic environment reconstruction. *Quaternary International, 325*, 179–196.

De Franceschi, D., Hoorn, C., Antoine, P.-O., Cheema, I.U., Flynn, L.J., Lindsay, E.H., Marivaux, L., Metais, G., Rajpar, A.R., Welcomme, J.-L., (2008). Floral Data from the Mid-Cenozoic of central Pakistan. Rev. Palaeobot. Palynol. 150, 115–129.

Dixit, Y., & Tandon, S. K. (2016). Hydroclimatic variability on the Indian subcontinent in the past millennium: review and assessment. *Earth-Science Reviews, 161*, 1–15.

Dunlea, A. G., Giosan, L., & Huang, Y. (2020). Pliocene expansion of C< sub> 4</sub> vegetation in the Core Monsoon Zone on the Indian Peninsula. *Climate of the Past, 16*(6), 2533–2546.

Farooqui, A., Gaur, A. S., & Prasad, V. (2013). Climate, vegetation and ecology during Harappan period: Excavations at Kanjetar and Kaj, mid-Saurashtra coast, Gujarat. *Journal of Archaeological Science, 40*(6), 2631–2647.

Farooqui, A., Ray, J. G., Farooqui, S. A., Tiwari, R. K., & Khan, Z. A. (2010). *Tropical rainforest vegetation, climate and sea level during the Pleistocene in Kerala, India. Quaternary International, 213(1–2), 2–11.* doi:https://doi.org/10.1016/j.quaint.2009.09.024

Flynn, L. J., & Jacobs, L. L. (1982). Effects of changing environments on Siwalik rodent faunas of northern Pakistan. *Palaeogeography, Palaeoclimatology, Palaeoecology, 38*(1–2), 129–138.

Galy, V., François, L., France-Lanord, C., Faure, P., Kudrass, H., Palhol, F., & Singh, S. K. (2008). C4 plants decline in the Himalayan basin since the Last Glacial Maximum. *Quaternary Science Reviews, 27*(13–14), 1396–1409.

Ganjoo, R. K., & Shaker, S. (2007). Middle Miocene pedological record of monsoonal climate from NW Himalaya (Jammu & Kashmir state), India. *Journal of Asian Earth Sciences, 29*(5-6), 704–714.

Ghosh, R., Bera, S., D'Rozario, A., Banerjee, M., & Chakraborty, S. (2006). Plant Remains from an Archaeological Site as Indicators of Vegetation and Agricultural Practice Between (3 320±400) and (2 080±80) yr BP in Gangetic West Bengal, India. *Journal of Integrative Plant Biology, 48*(6), 628–641.

Ghosh, R., Naskar, M., & Bera, S. (2011). Phytolith assemblages of grasses from the Sunderbans, India and their implications for the reconstruction of deltaic environments. *Palaeogeography, Palaeoclimatology, Palaeoecology, 311*(1–2), 93–102.

Ghosh, S., Sanyal, P., Kumar, R., (2017). Evolution of C4 plants and controlling factors: insight from n-alkane isotopic values of NW Indian Siwalik paleosols. Org. Geochem. 110, 110–121.

Givnish, T. J., & Renner, S. S. (2004). Tropical intercontinental disjunctions: Gondwana breakup, immigration from the boreotropics, and transoceanic dispersal. *International Journal of Plant Sciences, 165*(S4), S1–S6.

Goswami, A. (2011). *Predicting the geographic distribution of ancient soils with special reference to the Cretaceous.* Ph.D. Thesis, Department of Earth and Environmental Sciences, University of Texas at Arlington, Arlington, Texas, 339 p.

Govindarajalu, E. (1968). Further contribution to the anatomy of the marantaceae: Schumannianthus virgatus (Roxb.) rolfe. In *Proceedings/Indian Academy of Sciences* (Vol. 68, No. 5, pp. 250–260). New Delhi: Springer India.

Govindarajalu, E. (1975). Studies in Cyperaceae. XIV. Endomorphic evidences for placing Cyperus hyalinus under the new subgenus Queenslandiella. *Reinwardtia, 9*(2), 187–195.

Gupta, A. K., Anderson, D. M., & Overpeck, J. T. (2003). Abrupt changes in the Asian southwest monsoon during the Holocene and their links to the North Atlantic Ocean. *Nature, 421*(6921), 354–357.

Hay, W. W., Soeding, E., DeConto, R. M., & Wold, C. N. (2002). The Late Cenozoic uplift–climate change paradox. *International Journal of Earth Sciences, 91*, 746–774.

Hoorn, C., Ohja, T., Quade, J., (2000). Palynological evidence for vegetation development and climatic change in the Sub-Himalayan Zone (Neogene, central Nepal). Palaeogeogr., Palaeoclimatol., Palaeoecol. 163,133–161

House B, Pickering KT, Norris R. (2023). Multi-phase ecological change on Indian subcontinent from the late Miocene to Pleistocene recorded in the Nicobar Fan. *Geological Magazine*;160(7):1428-1440. doi:https://doi.org/10.1017/S0016756823000481

Humane, S. S., Humane, S. K., & Borkar, N. (2020). Diatoms and Phytoliths Records from Sediment Core of Tarna-Satighat Lake, Central India: Implications as Indicators of Changing Wet/Dry Climatic Phases and Trophic Status. *Earth Science India, 13*(2).

Jha, D.K., Sanyal, P., Philippe, A., (2020). Multi-proxy evidence of Late Quaternary climate and vegetational history of north-central India: implication for the Paleolithic to Neolithic phases. Quat. Sci. Rev. 229, 106–121.

Jin, Q., Wang, C., (2018). The greening of Northwest Indian subcontinent and reduction of dust abundance resulting from Indian summer monsoon revival. Sci. Rep. 8.

Joya, E., Bromand, M.T., Murtaza, K.O. and Dar, R.A., (2021). Current glacier status and ELA changes since the Late Pleistocene in the Hindu Kush Mountains of Afghanistan. *Journal of Asian Earth Sciences, 219*, p.104897.

Kajale, M. D., & Eksambekar, S. P. (2001). Phytolith approach for investigating ancient occupations at Balathal, Rajasthan, India. *Phytoliths: Application in earth sciences and human history*, 199–212.

Kale, V. S., Gupta, A., Singhvi, A. K. (2003). 13 late pleistocene–holocenehydrology of Monsoon Asia. *Hydrology: Understanding Global Change*, 213.

Kent, D. V., & Muttoni, G. (2008). Equatorial convergence of India and early Cenozoic climate trends. *Proceedings of the National Academy of Sciences, 105*(42), 16065–16070.

Krishnamurthy, R.V., Bhattacharya, S.K., Kusumgar, S., (1986). Palaeoclimatic changes deduced from $^{13}C/^{12}C$ and C/N ratios of Karewa lake sediments, India. Nature. 323(6084), 150–152. https://doi.org/10.1038/323150a0.

Krishnan, S., Samson, N. P., Ravichandran, P., Narasimhan, D., & Dayanandan, P. (2000). Phytoliths of Indian grasses and their potential use in identification. *Botanical Journal of the Linnean Society, 132*(3), 241–252.

Kudrass, H. R., Hofmann, A., Doose, H., Emeis, K., & Erlenkeuser, H. (2001). Modulation and amplification of climatic changes in the Northern Hemisphere by the Indian summer monsoon during the past 80 ky. *Geology, 29*(1), 63–66.

Kumar Das, S., Gangopadhyay, K., Ghosh, A., Biswas, O., Bera, S., Ghosh, P.,... & Yoshida, K. (2021). Organic geochemical and palaeobotanical reconstruction of a late-Holocene archaeological settlement in coastal eastern India. The Holocene, 31(10), 1511–1524.

Kumar, M., Saikia, K., Agrawal, S., Ghosh, R., Ali, S. N., Arif, M.,... & Bajpai, S. (2022). Climatic control on the C3 and C4 plant abundance during the late Pleistocene–Holocene in the northern Gangetic Plain, India. *Palaeogeography, Palaeoclimatology, Palaeoecology, 591*, 110890.

Lengerke, H. J. V. (1977). The Nilgiris: weather and climate of a mountain area in south India. *(No Title).*

Mehra, P. N., & Sharma, O. P. (1965). Epidermal silica cells in the Cyperaceae. *Botanical Gazette, 126*(1), 53–58.

Mir, J.A., Dar, R.A., Vinnepand, M., Laag, C., Rolf, C. and Zeeden, C., (2022). Environmental reconstruction potentials of Loess-Paleosol-Sequences in Kashmir through high-resolution proxy data. *Palaeogeography, Palaeoclimatology, Palaeoecology, 601*, p.111100.

Misra, P., Farooqui, A., Sinha, R., Khanolkar, S., & Tandon, S. K. (2020). Millennial-scale vegetation and climatic changes from an Early to Mid-Holocene lacustrine archive in Central Ganga Plains using multiple biotic proxies. *Quaternary Science Reviews, 243*, 106474.

Misra, S., & Bhattacharyya, A. (2014). Analysis of the late Holocene climate vis-a-vis vegetation dynamics along the southwest coast of India: Thrissur (Kerala). *Quaternary international, 325*, 150–161.

Molnar, P. (2005). Mio-Pliocene growth of the Tibetan Plateau and evolution of East Asian climate. *Palaeontologia Electronica, 8*(1), 1–23.

Mukherjee, B., Biswas, O., Ghosh, R., Paruya, D. K., & Bera, S. (2019). Phytolith spectra of some eastern Himalayan pteridophytes: Assessing potential for distinguishing different vegetation–climate zones along the elevation gradient. *Flora, 254*, 95–112.

Murtaza, K.O., Dar, R.A., Paul, O.J., Bhat, N.A. and Romshoo, S.A., (2021). Glacial geomorphology and recent glacial recession of the Harmukh Range, NW Himalaya. *Quaternary International, 575*, pp.236–248.

Mykleby, P.M., Snyder, P.K., Twine, T.E., (2017). Quantifying the trade-off between carbon sequestration and albedo in midlatitude and high-latitude North American forests. Geophys. Res. Lett. 44 (5), 2493–2501. https://doi.org/10.1002/2016gl071459

Nair, R. R., Ittekkot, V., Manganini, S. J., Ramaswamy, V., Haake, B., Degens, E. T.,... & Honjo, S. (1989). Increased particle flux to the deep ocean related to monsoons. *Nature, 338*(6218), 749–751.

Naskar, M., Blinnikov, M., Ghosh, R., Das, S., Paruya, D. K., Majumdar, S., & Bera, S. (2021). A diagnostic phytolith morphotype found in Porteresia coarctata (Roxb.) Tateoka indicates coastal swampy mangrove environments: A case study from the Indian east coast. *Flora, 282*, 151884.

Naskar, M., Ghosh, R., & Bera, S. (2018). Variability in phytolith spectra of some Panicoid grasses from different soil salinity zones of the deltaic West Bengal, India: implications in understanding depositional environments. *Botany Letters, 165*(2), 255–264.

Naskar, M., Ghosh, R., Das, S., Paruya, D. K., Saradar, B., Yadava, M. G., & Bera, S. (2022). Grass phytoliths in surface sediments of the Sunderbans, India and their implications in reconstructing past deltaic environmental changes. *The Holocene, 32*(11), 1193–1208.

Nogué, S., Whicher, K., Baker, A. G., Bhagwat, S. A., & Willis, K. J. (2017). Phytolith analysis reveals the intensity of past land use change in the Western Ghats biodiversity hotspot. *Quaternary International, 437*, 82–89.

Overpeck, J., Anderson, D., Trumbore, S., & Prell, W. (1996). The southwest Indian Monsoon over the last 18 000 years. *Climate Dynamics, 12*, 213–225.

Patil, S. J., Prasad, S., & Kar, S. (2017). Investigating the use of plants in ancient human burials in the Western Ghats, India: A phytolith analysis. Journal of Archaeological Science, 81, 1–9.

Patnaik, R., Singh, N. P., Paul, D., & Sukumar, R. (2019). Dietary and habitat shifts in relation to climate of Neogene-Quaternary proboscideans and associated mammals of the Indian subcontinent. *Quaternary Science Reviews*, *224*, 105968.

Paul, O.J., Romshoo, S.A., Dar, R.A., Kumar, P., Dhal, S.P. and Chopra, S., (2022). Paleoglacial reconstruction of the Thajwas glacier in the Kashmir Himalaya using 10Be cosmogenic radionuclide dating. *Geoscience Frontiers*, *13*(6), p.101432.

Piao, S., Fang, J., Ciais, P., Peylin, P., Huang, Y., Sitch, S., et al., (2009). The carbon balance of terrestrial ecosystems in China. Nature 458 (7241), 1009–1082.

Pillai, A. A., Anoop, A., Prasad, V., Manoj, M. C., Varghese, S., Sankaran, M., & Ratnam, J. (2018). Multi-proxy evidence for an arid shift in the climate and vegetation of the Banni grasslands of western India during the mid-to late-Holocene. *The Holocene*, *28*(7), 1057–1070.

Ponnaiya BWX (1951). Studies on the genus Sorghum. II. The cause of resistance in sorghum to the insect pest *Atherigona indica* M. Madras Univ J. 1951;21:203–17.

Prasad, V., Farooqui, A., Sharma, A., Phartiyal, B., Chakraborty, S., Bhandari, S.,... & Singh, A. (2014). Mid–late Holocene monsoonal variations from mainland Gujarat, India: A multiproxy study for evaluating climate culture relationship. *Palaeogeography, Palaeoclimatology, Palaeoecology*, *397*, 38–51.

Prasad, V., Phartiyal, B., & Sharma, A. (2007). Evidence of enhanced winter precipitation and the prevalence of a cool and dry climate during the mid to late Holocene in mainland Gujarat, India. *The Holocene*, *17*(7), 889–896.

Prasad, V., Stromberg, C. A., Alimohammadian, H., & Sahni, A. (2005). Dinosaur coprolites and the early evolution of grasses and grazers. *Science*, *310*(5751), 1177–1180.

Premathilake, R., & Hunt, C. O. (2018). Late Pleistocene humans in Sri Lanka used plant resources: A phytolith record from Fahien rock shelter. *Palaeogeography, Palaeoclimatology, Palaeoecology*, *505*, 1–17.

Premathilake, R., Akhilesh, K., Anupama, K., Pappu, S., Prasad, S., Gunnell, Y., & Orukaimani, G. (2017). Phytoliths as indicators of Quaternary vegetation at the Paleolithic site of Attirampakkam, India. *Journal of Archaeological Science: Reports*, *14*, 479–499.

Premathilake, R., Akhilesh, K., Anupama, K., Prasad, S., Gunnell, Y., Orukaimani, G., & Pappu, S. (2022). Issues of phytolith taphonomy at Palaeolithic sites: Investigation and results from Attirampakkam, India. *Journal of Archaeological Science: Reports*, *42*, 103357.

Qader, W., Dar, R.A. and Rashid, I., (2023a). Phytolith particulate matter and its potential human and environmental effects. *Environmental Pollution*, *327*, p.121541.

Qader, W., Mir, S. H., Meister, J., Dar, R. A., Madella, M., Rashid, I. (2023b). Sedimentological perspective on phytolith analysis in palaeoecological reconstruction. *Earth-Science Reviews*, 104549.

Qader, W., Dar, R. A., Rehman, I. U., Rashid, I., & Sheikh, S. H. (2024). Assessing phytolith preservation in a Late Quaternary loess-paleosol sequence from the Kashmir Valley, Northwest Himalaya, India. *Quaternary Science Advances*, *16*, 100238.

Quade, J., & Cerling, T. E. (1995). Expansion of C4 grasses in the late Miocene of northern Pakistan: evidence from stable isotopes in paleosols. *Palaeogeography, Palaeoclimatology, Palaeoecology*, *115*(1–4), 91–116.

Quade, J., Cerling, T. E., & Bowman, J. R. (1989). Development of Asian monsoon revealed by marked ecological shift during the latest Miocene in northern Pakistan. *Nature*, *342*(6246), 163–166.

Raj, R., Chamyal, L. S., Prasad, V., Sharma, A., Tripathi, J. K., & Verma, P. (2015). Holocene climatic fluctuations in the Gujarat Alluvial Plains based on a multiproxy study of the Pariyaj Lake archive, western India. *Palaeogeography, Palaeoclimatology, Palaeoecology*, *421*, 60–74.

Raj, R., Tripathi, J. K., Kumar, P., Singh, S. K., Phartiyal, B., Sharma, A.,... & Chamyal, L. S. (2021). Palaeoclimatic and sea-level fluctuations from the last deglaciation to late Holocene from western India: Evidence from multiproxy studies. *Journal of Asian Earth Sciences, 214*, 104777.

Rashid, I., Mir, S.H., Zurro, D., Dar, R.A. and Reshi, Z.A., (2019). Phytoliths as proxies of the past. *Earth-Science Reviews, 194*, pp.234–250.

Rehman, I. U., Qader, W., Dar, R. A., Rashid, I., & Shah, R. A. (2024). Phytolith based paleoecological reconstruction from a loess-paleosol sequence in the Kashmir Himalaya, India. *Catena, 245*, 108318.

Roy, B., Ghosh, S., Sanyal, P., (2020). Morpho-tectonic control on the distribution of C3-C4 plants in the central Himalayan Siwaliks during Late Plio-Pleistocene. Earth Planet. Sci. Lett. 535, 116–119.

Saijo, K., & Tanaka, S. (2002). Paleosols of middle Holocene age in the Thakkhola basin, central Nepal, and their paleoclimatic significance. *Journal of Asian Earth Sciences, 21*(3), 323–329.

Saikia, K., Paruya, D. K., Ghosh, R., & Bera, S. (2022). Phytolith assemblages of Shorea robusta Gaertn.(Sal) from West Bengal part of the Bengal Basin, India and their implications in palaeoclimate reconstructions. *Botany Letters, 169*(1), 127–139.

Sakai, H., Fujii, R., Sugimoto, M., Setoguchi, R., & Paudel, M. R. (2016). Two times lowering of lake water at around 48 and 38 ka, caused by possible earthquakes, recorded in the Paleo-Kathmandu Lake, central Nepal Himalaya. *Earth, Planets and Space, 68*(1), 1–10.

Sanyal P, Bhattacharya SK, Kumar R, Ghosh SK, Sangode SJ. (2004) Mio–Pliocene monsoonal record from Himalayan foreland basin (Indian Siwalik) and its relation to vegetational change. Palaeogeography, Palaeoclimatology, Palaeoecology. 20;205(1–2):23–41.

Sarkar, A., Mukherjee, A. D., Sharma, S., Sengupta, T., Ram, F., Bera, M. K.,... & Juyal, N. (2020). New evidence of early Iron Age to Medieval settlements from the southern fringe of Thar Desert (western Great Rann of Kachchh), India: Implications to climate-culture co-evolution. *Archaeological Research in Asia, 21*, 100163.

Saxena, A., Prasad, V., & Singh, I. B. (2013). Holocene palaeoclimate reconstruction from the phytoliths of the lake-fill sequence of Ganga Plain. *Current science*, 1054–1062.

Saxena, A., Prasad, V., Singh, I. B., Chauhan, M. S., & Hasan, R. (2006). On the Holocene record of phytoliths of wild and cultivated rice from Ganga Plain: evidence for rice-based agriculture. *Current Science*, 1547–1552.

Scotese, C. R., Song, H., Mills, B. J., & van der Meer, D. G. (2021). Phanerozoic paleotemperatures: The earth's changing climate during the last 540 Ma. *Earth-Science Reviews, 215*, 103503.

Shah, R.A., Dar, R.A. and Romshoo, S.A., 2024. Paleoclimatic reconstruction of the Karewa deposits of Kashmir Valley, northwest Himalaya: A review. *Quaternary International.*

Sharma, O. P. (1972). ANATOMY OF SCIRPUS SQUARROSUS L. *Current Science, 41*(13), 494–497.

Sheikh, S. H., Ahmad, M. Z., Qader, W., Rehman, I. U., Rashid, I., Mir, S. H.,... & Dar, R. A. (2025). Phytolith insights into palaeosol formation and climate dynamics in the Pliocene Karewa deposits, Kashmir Himalaya, India. *Proceedings of the Geologists' Association*, 101154.

Shukla, A., & Mehrotra, R. C. (2018). Early Eocene plant megafossil assemblage of western India: Paleoclimatic and paleobiogeographic implications. *Review of palaeobotany and palynology, 258*, 123–132.

Singh, V., Prasad, V., & Chakraborty, S. (2007). Phytoliths as indicators of monsoonal variability during mid–late Holocene in mainland Gujarat, western India. *Current Science*, 1754–1759.

Sirocko, F., & Lange, H. (1991). Clay-mineral accumulation rates in the Arabian Sea during the late Quaternary. *Marine Geology, 97*(1-2), 105–119.

Sirocko, F., Sarnthein, M., Erlenkeuser, H., Lange, H., Arnold, M., & Duplessy, J. C. (1993). Century-scale events in monsoonal climate over the past 24,000 years. *Nature, 364*(6435), 322–324.

Srivastava AK (1978). Study of leaf epidermis in the genus Digitaria Rich (Gramineae), J. Indian Bot. Soc., 37: 155–160.

Srivastava, Gaurav, Mehrotra, R.C., Shukla, A., Tiwari, R.P., (2014). Miocene Vegetation And Climate In Extra Peninsular India: Megafossil Evidences. Spec. Publ. Palaeontol. Soc. India 5, 283–290.

Stojanovic, D., Aitchison, J.C., Ali, J.R., Ahmad, T. and Dar, R.A., (2016). Paleomagnetic investigation of the Early Permian Panjal Traps of NW India; regional tectonic implications. *Journal of Asian Earth Sciences*, *115*, pp.114–123.

Tauxe, L., & Feakins, S. J. (2020). A reassessment of the chronostratigraphy of late Miocene C3–C4 transitions. *Paleoceanography and Paleoclimatology*, *35*(7), e2020PA003857.

Thapar, B. K. (1955). Plant remains from Mohenjo-daro. Ancient India, 11, 1–10.

Thomas, R., Tengberg, M., Moulhérat, C., Marcon, V., & Besenval, R. (2012). Analysis of a protohistoric net from Shahi Tump, Baluchistan (Pakistan). *Archaeological and Anthropological Sciences*, *4*(1), 15–23.

Thompson, L. O., Yao, T., Davis, M. E., Henderson, K. A., Mosley-Thompson, E., Lin, P. N.,... & Bolzan, J. F. (1997). Tropical climate instability: The last glacial cycle from a Qinghai-Tibetan ice core. *science*, *276*(5320), 1821–1825.

Tripathi, D. K., Mishra, S., Chauhan, D. K., Tiwari, S. P., & Kumar, C. (2013). Typological and frequency-based study of opaline silica (phytolith) deposition in two common Indian Sorghum L. species. *Proceedings of the National Academy of Sciences, India Section B: Biological Sciences*, *83*, 97–104.

Tripathi, S., Basumatary, S. K., Singh, Y. R., McDonald, H. G., Tripathi, D., & Singh, L. J. (2019). Multiproxy studies on dung of endangered Sangai (*Rucervus eldii eldii*) and Hog deer (Axis porcinus) from Manipur, India: Implications for paleoherbivory and paleoecology. *Review of Palaeobotany and Palynology*, *263*, 85–103.

Vasanthy, G. (1988). Pollen analysis of Late Quaternary sediments: evolution of upland savanna in Sandynallah (Nilgiris, South India). *Review of Palaeobotany and Palynology*, *55*(1-3), 175–192.

Vaughan, P., Jarrige, C., & Anderson-Gerfaud, P. (1987). Sickles and Harvesting Motions in Baluchistan (Pakistan). *MOM Éditions*, *15*(1), 311–318.

Veena, M. P., Achyuthan, H., Eastoe, C., & Farooqui, A. (2014). A multi-proxy reconstruction of monsoon variability in the late Holocene, South India. *Quaternary International*, *325*, 63–73.

Wallace, A.R., (1876). The Geographical Distribution of Animals. Harper & Brothers, New York, 607 p.

Weber, S. A. (1989). *Plants and Harappan subsistence: An example of stability and change from Rojdi*. University of Pennsylvania.

Wheeler, M. (1936). Harappa: City of the Indus. London: John Murray.

Zalasiewicz, J., & Williams, M. (2009). A geological history of climate change. In *Climate change* (pp. 127–142). Elsevier.

Zexer, N., Kumar, S., & Elbaum, R. (2023). Silica deposition in plants: Scaffolding the mineralization. *Annals of Botany*, *131*(6), 897–908.

Chapter 13
Unravelling the Climate During the Meghalaya Stage from the Mudflats of the Southern Saurashtra Coast, Western India

Upasana Swaroop Banerji

1 Introduction

The ongoing interglacial period, the Holocene Epoch (11.8 cal. ka BP to present), has been evidently characterised by pervasive millennial-scale climate perturbations and abrupt climatic events (Mayewski et al. 2004; Wanner et al. 2008). Various chronostratigraphic methods have been applied to characterise the Holocene Epoch; however, there remain considerable inconsistencies in correlating global climatic events (Wanner et al. 2008). Recently, a proposal for dividing the Holocene Epoch (Greenlandian, Northgrippian and Meghalayan Stages) was initiated by the Subcommission on Quaternary Stratigraphy (SQS), approved by the International Commission on Stratigraphy (ICS) and formally ratified by the Executive Committee of the International Union of Geological Sciences (IUGS) (Walker et al. 2018, 2019; Helama and Oinonen 2019). Despite various attempts to investigate the spatio-temporal climate patterns and causes of abrupt climatic events, the underlying mechanisms causing such perturbations remain uncertain.

Unlike glacial–interglacial cycles, when the boundary conditions are drastically perturbed, the Meghalayan Stage (~4.2 ka–present) seldom showed major climatic variability. However, the onset of the Meghalayan Stage, identified as the 4.2 ka event, has been recognised based on the severe dry climate well registered in the speleothem of Mawmluh Cave, Meghalaya, NE India (Berkelhammer et al. 2015). The continuous climate records during the Meghalayan Stage and the response of the Indian monsoon towards globally identified climatic events remain uncertain. The global annual surface temperature increased by 0.85 °C from 1880 to 2012 (IPCC 2013), demonstrating the prevalence of global warming since the post-industrial era (1750

U. S. Banerji (✉)
National Centre for Earth Science Studies (NCESS), Thiruvananthapuram, Kerala, India
e-mail: upasana.s.banerji@gmail.com

N. Khare (ed.), *Changing Climate over India*,
https://doi.org/10.1007/978-981-92-1098-5_13

AD), resulting in severe impacts on various earth's processes such as global hydrological cycle (Alfieri et al. 2017; Herring et al. 2018) along with spatial and temporal changes in the precipitation pattern (IPCC 2007). Considering climate change and global warming being the pressing issues in the ongoing climate debates, a plethora of studies have invoked the necessity to document the last two millennia of climate reconstruction. This is because a more accurate comparison of the pre-industrial scenarios can be reconstructed to distinguish between natural and anthropogenically induced climatic variability (Jansen et al. 2007). Based on the temperature and precipitation patterns, the last two millennia have been broadly divided into five divisions, viz. Roman Warm Period (RWP; 2500–1450 yr BP), Dark Ages Cold Period (DACP; 1450–1050 yr BP), Medieval Climate Anomaly (MCA; 1050–650 yr BP), Little Ice Age (LIA; 650–100 yr BP) and Modern Warming (MW; 100 yr BP–present) (Lamb 1965; Friis-Christensen and Lassen 1991; Norgaard-Pedersen and Mikkelsen 2009; Oppo et al. 2009; Banerji et al. 2020).

The climate of the Indian subcontinent is closely intertwined with the rainfall intensities caused by the Indian summer monsoon (ISM; June–September) and the Indian winter monsoon [IWM; north-east monsoon (NEM: October–December) and western disturbances (WD: December–February)]. The ISM is a crucial part of the global climate system, responsible for agricultural productivity and socio-economic well-being in our country, as it accounts for over 80% of rainfall in the Indian subcontinent and its surrounding regions. The intensities of both ISM and IWM are exhibited by the lateral shifts in the Intertropical Convergence Zone (ITCZ), wherein the ITCZ is essentially a narrow latitudinal wind convergence zone over the tropics and subtropics that governs the rainy season over the tropical region and, thus, influences the Indian climate (Gadgil 2003; Fleitmann et al. 2007). The stark patterns in both monsoons, with sudden floods in some parts of the country, whilst severe droughts prevailing in other parts, perhaps has become the most compelling reason to understand and forecast monsoon trends. However, the limited temporal data for instrumental rainfall, which seldom extends beyond the last two centuries, has prompted climate scientists to peer into the past climate through natural archives and their associated proxies.

The western state, Gujarat, is geomorphologically divided into three distinct regions, viz. Gujarat mainland, Saurashtra peninsula and Kachchh peninsula, whilst it falls under a tropical climate with climatic zones varying from sub-humid, semi-arid to arid regions (Ray et al. 2009). There have been reasonable studies pertaining to Holocene climate variability in Gujarat (Fig. 1). The multiproxy study on the Nal Sarovar Lake sediment (Central Gujarat) demonstrated a wet climate during 4.8 – 3 ka, followed by an arid climate that persisted till ~2 ka (Prasad et al. 1997). A coupling of both summer and winter monsoon has been inferred during 3400 – 3000 cal yr BP, followed by an arid climate from 2850 cal yr BP onwards from the Mahi Estuary (mainland Gujarat) using palynological proxies (Prasad et al. 2007). Further, the slack water deposits from the Mahi River basin also demonstrate four enhanced monsoon events, viz. >5 ka, 4.6 ka and during 4.6 to 1.7 ka (Sridhar 2007) along with a mega flood event recorded during the MCA (Sridhar 2009). A nearly similar observation of sub-humid conditions during 3 ka has been indicated by the lower Narmada valley, mainland Gujarat, followed by two drier events, viz. ~2.1 ka and ~1.3 ka (Laskar et al. 2013). Another investigation from Narmada Valley sediments elucidated wet phases prior to 2185 cal yr BP and after 1809 cal yr BP, interrupted by an arid phase (Sridhar et al. 2015). A multiproxy investigation from Wadhwan Lake sediment, mainland Gujarat, revealed a dry climate during 5560 – 4250 cal yr BP and 3238 – 2709 cal yr BP with intermittent strengthening of ISM after 3500 cal yr BP (Prasad et al. 2014).

As most of the studies conducted were predominantly from mainland Gujarat, the Saurashtra region remained largely unexplored for its paleoclimate and paleomonsoon significance. The southern Saurashtra coast comprises extensive mudflats that provide a unique opportunity to decode the paleomonsoon reconstruction. Thus, in the last decade, the relict mudflats of Vasoj village and active mudflats of Diu Island, Rohisa and Jaffrabad were investigated using multiproxy approaches, and climate vis-à-vis monsoon reconstruction has been attempted (Banerji et al. 2015, 2017, 2019, 2021, 2024). The present chapter attempts to overview and compare the paleoclimate studies conducted on the mudflats of the southern Saurashtra coast

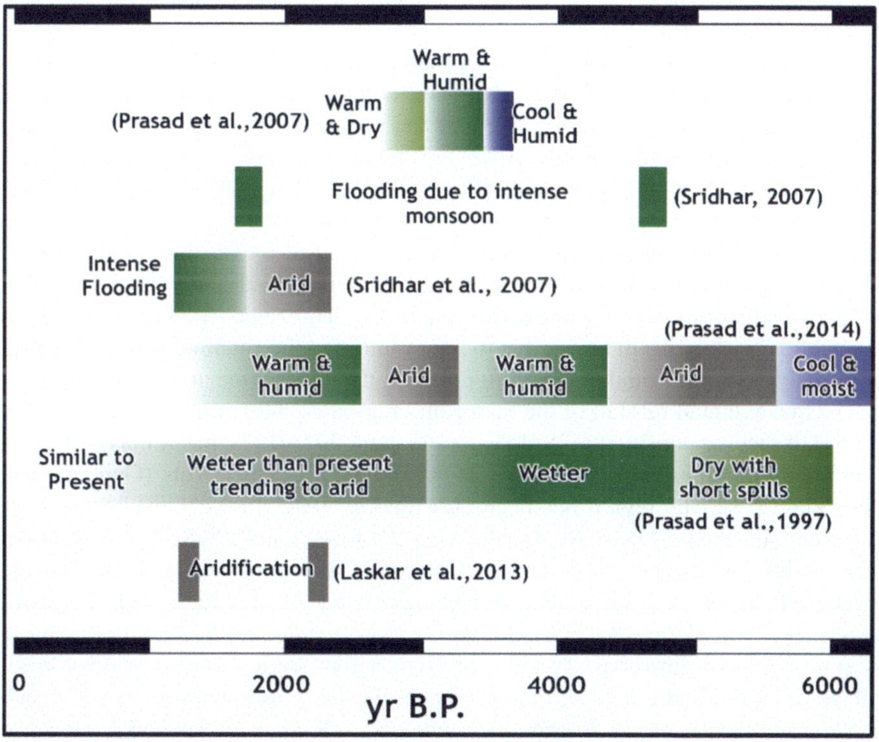

Fig. 1 Compilation and comparison of the palaeoclimatic studies conducted in Gujarat, western India, during the last 6 ka. The comparison suggests a significant data gap during the Meghalaya stage (4.2 ka to Present)

to (a) address continuous records of climate reconstruction during the Meghalayan Stage; (b) understand the response of western India towards the globally recognised abrupt climatic events; and (c) delineate the global climate teleconnection and its association with the Indian monsoon system.

2 Study Area

The coastline of Saurashtra is unique as its various segments provide considerable diversity with respect to their geologic and geomorphic evolutions and are characterised by varied assemblages of erosional and depositional landforms (Ganapathi et al. 1982). Geologically, the majority of the Saurashtra peninsula comprises basalts and their derivatives of the Deccan Trap Formation under the upper Cretaceous period (Bhonde and Bhatt 2009). The coastal plains are fringed by Quaternary sediments of both marine and continental origin, including miliolite limestone, oyster beds, alluvial valley fills and aeolianites (Fig. 2). The miliolitic limestone is exposed in the southern, eastern and western coasts of Saurashtra from Porbandar to south-east Veraval (Marathe 1981).

The southern Saurashtra coast is irregular and marked by 40–50 m vertical cliffs of miliolite limestone. A cliffed southern coast along Diu Island is observed, whilst beyond Diu, the coast is characterised by the presence of a rocky foreshore with occasional beaches and miliolite cliffs. Furthermore, the north-east of Diu Island consists of extensive tidal flats, and towards the east, along the coast of Saurashtra, locally extensive mudflats are observed near Rohisa and Jaffrabad (Merh 1995). The region is devoid of any perennial rivers, but it is drained by various small seasonal streams such as Dhantravardi, Raidi, Rupen, Malan, Rawal and Machhundri that get activated during the monsoon (Verma 1982). The rivers flowing in southern Saurashtra are sluggish in nature due to poor rainfall in the catchment area, and they are usually lost in the tidal flats before reaching the sea (Verma and Mathur 1979).

The mean annual rainfall of the Saurashtra region is ~600 mm, with the majority of rainfall received during the ISM (Farooqui et al. 2013). The mean maximum and minimum temperatures vary between 34 °C and 19 °C, respectively (Gundalia and Dholakia 2013). The region has been classified as 5A/C-1a (very dry teak forest) following sub-type 5/DS1 (dry deciduous scrub forest) and 5/DS1 (dry savannah) type of the forest type classification (Champion and Seth 1968). Presently, the vegetation such as Avicennia alba, Avicennia marina, A. officinalis with fringes of *Aegiceras corniculatum*, *Acanthus ilicifolius* and *Suaeda nudiflora* are dominant in the southern Saurashtra coastal region, whilst the palynological investigation of Vasoj village relict mudflat elucidated the presence of other core mangrove species such as *Rhizophora mucronata*, *Bruguiera gymnorrhiza*, S*onneratia sp.* and *Excoecaria agallocha* (Banerji et al. 2015).

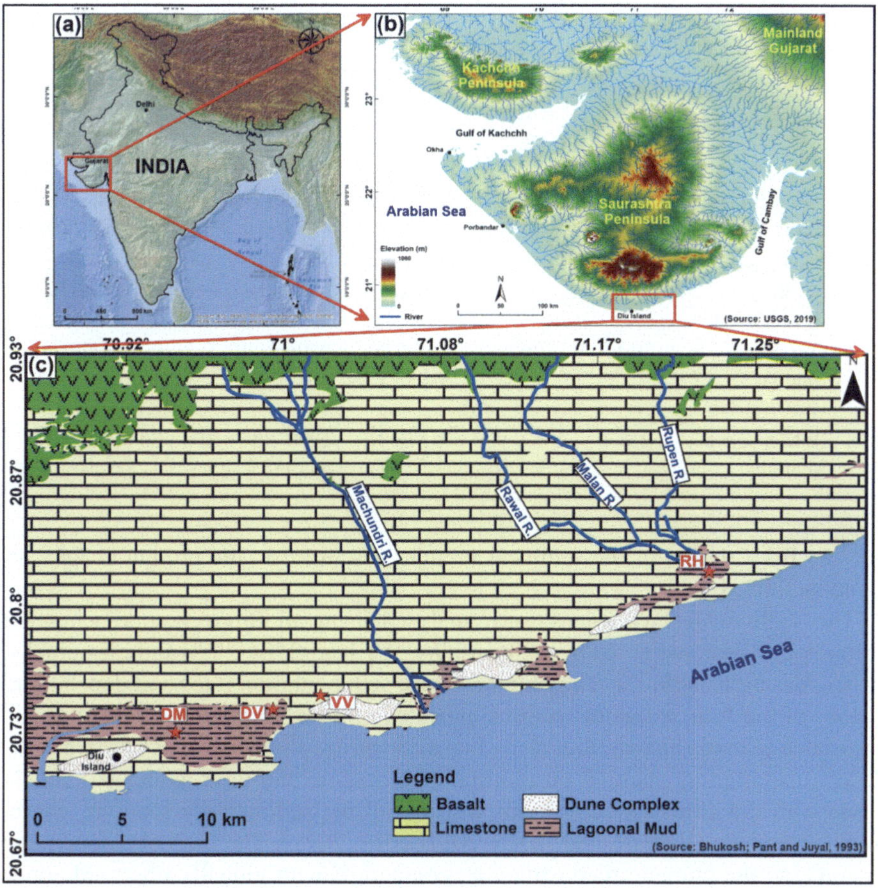

Fig. 2 Map of India demarcating Gujarat (**a**); The digital elevation model (DEM; U.S. Geological Survey 2019) of Gujarat depicting the study area (**b**); And the geological map of the southern Saurashtra coast [Pant and Juyal 1993; Geological Survey of India (GSI) 2025] consisting of miliolite limestone and deccan basalts with mudflats at selected locations of Diu and Rohisa; The core sites [VV (Banerji et al. 2015); DV (Banerji et al. 2017); DM (Banerji et al. 2021); and RH (Banerji et al. 2019)] are marked with the red star (**c**) and discussed in the present chapter

3 Approach

The present chapter aims to provide an overview of the studies conducted on the mudflats of the southern Saurashtra coast to demonstrate a continuous reconstruction of the monsoon vis-à-vis the climate system and to delineate the global climate linkages. Here, a relict mudflat section (VV) and three sediment cores from the active mudflats (DV, DM and RH) are discussed, which have provided a climatic history extending back to the last ~5 ka (representing the Meghalayan Stage; last

Table 1 Details of the relict sediment section (VV) and sediment core (DV, DM and RH) investigated in previous studies and discussed in this chapter are as follows

Sr. no	Site details	Section ID	Section length (cm)	Latitude (°N); longitude (°E)	Depositional history (yr BP/CE)	Reference
1.	Vasoj village Relict Mudflat	VV	100	20.750°N; 71.004°E	4700–1500 yr BP (−2750 to 450 CE)	(Banerji et al. 2015)
2.	Diu active Mudflat (landward side)	DV	70	20.748°N; 71.000°E	4100–70 yr BP (−2150 to 1942 CE)	(Banerji et al. 2017)
3.	Diu active Mudflat (seaward side)	DM	70	20.737°N; 70.933°E	2006 yr BP–Present (−156 to 2012 CE)	(Banerji et al. 2021)
4.	Rohisa active Mudflat	RH	62	20.816°N; 71.225°E	1970–80 yr BP (−20 to 1932 CE)	(Banerji et al. 2019)

4.2 ka). Brief details of the sediment sections discussed in the present chapter are given in Table 1.

The sedimentary sequences were analysed using multiproxy approaches and chronologically constrained by radiocarbon (^{14}C), lead-210 (^{210}Pb) and caesium-137 (^{137}Cs) dating techniques. The obtained ages, sedimentation rates and depositional history of the four sediment sections indicated an overlap in the time span and its associated climatic history, thus providing a cue on the climatic events observed. The details on sampling, sample processing, data quality, and their associated details, along with the climatic reconstruction using all the sedimentary sections, can be found elsewhere (Banerji et al. 2015, 2017, 2019, 2021). The present chapter aims to combine the climatic reconstruction, compare it with globally observed abrupt climatic events and decipher the climate teleconnection.

4 Results and Discussions

Generally, the geochemical signatures in the sediment and sedimentary rocks have been essentially used to decipher the paleo-hydrology, paleo-weathering, paleo-tectonics, paleo-redox environment and sedimentary sources (Sun et al. 2012a; Claes et al. 2019; Joshi et al. 2021; Banerji et al. 2022a; Slabunov et al. 2024). The Al_2O_3 and TiO_2, being the prime components of the rock-forming minerals, act as conservative elements by retaining their properties under weathering and remaining resistant to redox reactions (Nesbitt and Markovics 1997; Banerji et al. 2022b). Thus, they act as a crucial terrestrial proxy (Agnihotri et al. 2003; Kemp and Izumi 2014) whilst their statistical analysis (Srivastava et al. 2024) also provides clues to infer hydrological changes (Banerji et al. 2022b). The titanium (Ti) in the form of rutile (TiO_2), sphene ($CaTiSiO_5$) and ilmenite ($FeTiO_3$) can be transported with silt, fine sand and coarser quartz grains (Dolcater et al. 1970; Spears and Kanaris-Sotiriou 1976; Calvert and Pedersen 2007). The Al_2O_3 is extensively recognised as a lithogenous reference element and is predominantly used for elemental normalisation purposes, such as the Ti/Al or TiO_2/Al_2O_3 ratio provides cues to grain size as the heavy minerals and/or coarser sediment fraction are the prime Ti carrier in a variety of sediments (Correns 1969; Spears and Kanaris-Sotiriou 1976). Furthermore, the terrestrial contribution can also be deciphered through the ratio of total organic carbon (TOC) to total nitrogen (TN). Terrestrial vegetation with predominantly C3 plants is composed of lignin and cellulose with poor N content, resulting in TOC/TN > 12 (Tyson 1995), whilst the C4 terrestrial grasses represent TOC/TN >30 (Meyers 1994). In an aquatic environment, the phytoplankton and bacteria are mainly composed of proteins and nucleic acids, thereby showing typically low TOC/TN values of <10 and 4–6, respectively (Blackburn 1983; Meyers 1994; Tyson 1995).

The marine carbon cycle is driven by primary productivity in the ocean that regulates the global climate system. The past productivity variations in the form of export productivity have been reconstructed based on paleontological and geochemical variability (Sanchez 2019). The paleoproductivity indicators, such as the opal, organic carbon and calcium carbonate content, have played a major role in the CO_2 partitioning between the deep-sea environment and atmosphere (Ragueneau et al. 2000). However, they are prone to variable degrees of recycling within the water column, causing diminished signals of the respective proxies (Dymond et al. 1992). Thus, the geochemical proxies such as $CaCO_3$ and CaO/Al_2O_3 (Bhushan et al. 2001; Wei et al. 2004; Achyuthan et al. 2014), TOC and Al_2O_3-normalised Ba, Cu, Zn and Ni have been significantly used to decode the paleoproductivity (Tribovillard et al. 2006).

In the previous investigations (Banerji et al. 2015, 2017, 2019, 2021) on the mudflats of southern Saurashtra coast, various geochemical proxies were used to decipher the paleoclimate reconstruction. However, in the present chapter, selected proxies, such as Al_2O_3, TiO_2, TiO_2/Al_2O_3, TOC, TOC/TN, Ni/Al_2O_3, Cu/Al_2O_3 and Ba/Al_2O_3 have been presented and discussed for mutual comparison and demonstrating the global climate dynamics (Figs. 3, 4, 5 and 6). Other supporting evidence and proxies may be seen in the previous publications (Banerji et al. 2015, 2017, 2019, 2021).

4.1 Paleoclimate Proxies as a Function of Monsoon Variations

4.1.1 Early Meghalayan Stage (4.7–2.0 ka)

The pervasive prevalence of the '4.2 ka event' demarcating the onset of the Meghalayan Stage has been frequently described as an abrupt climate excursion that occurred ~4200 years ago. There have been various attempts to demonstrate the ISM variability during the 4.2 ka event (Misra et al. 2018; Banerji et al.

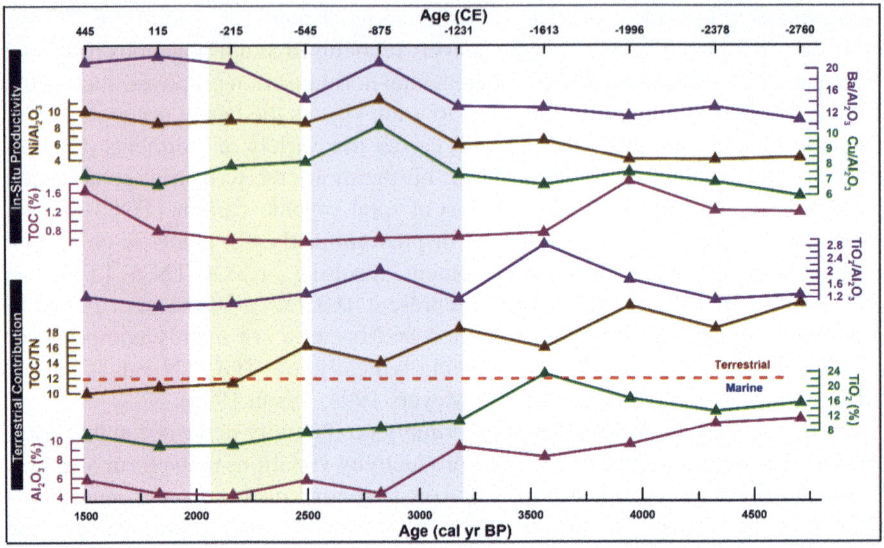

Fig. 3 Geochemical proxies for the terrestrial contribution and in-situ productivity for the VV relict mudflat section [purple: MWP; yellow: post-MWP; pink: RWP] demonstrating enhanced monsoon during ~4700 cal yr BP with a gradual declining ISM patterns

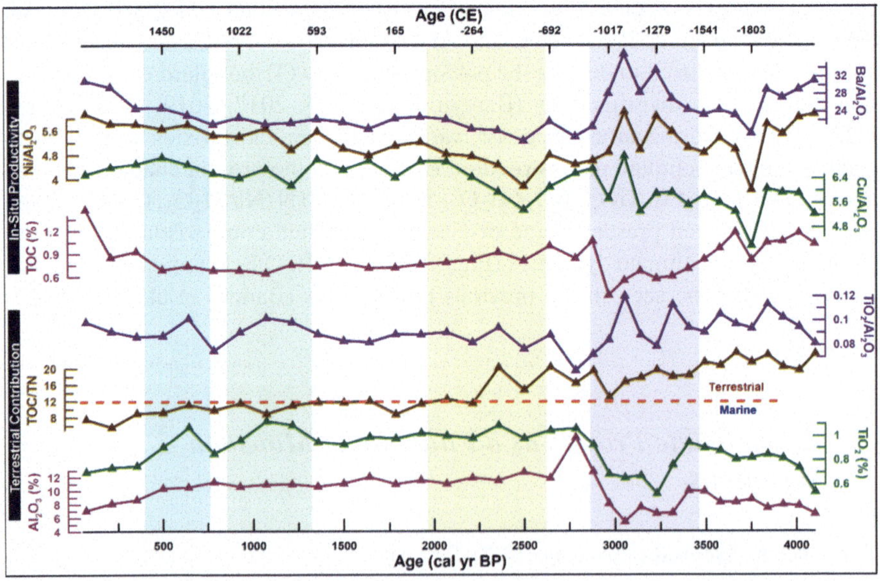

Fig. 4 Variations in the geochemical proxies depicting the terrestrial contribution and in-situ productivity for the DV active mudflat section [purple: MWP; yellow: post-MWP; green: MCA; blue: LIA] showing intermittent monsoon enhancement during 3230–2790 cal yr BP

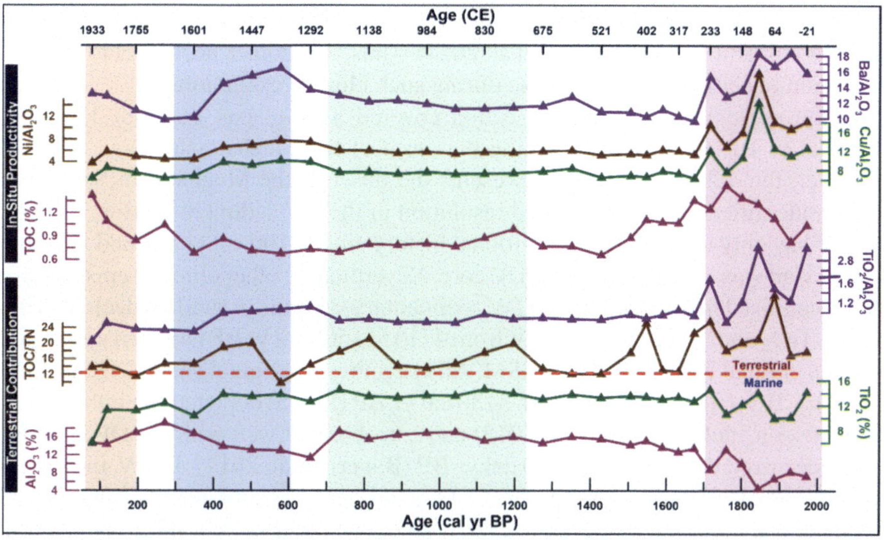

Fig. 5 Geochemical variability in the RH mudflat sediment core shows terrestrial contribution and in-situ productivity [pink: RWP; green: MCA; blue: LIA; peach: MW]. A gradual increase in Al_2O_3 and reduced in-situ productivity is observed during LIA

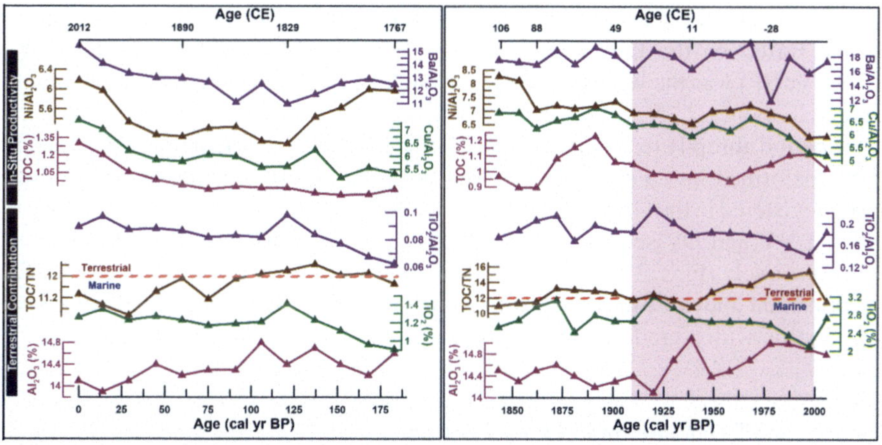

Fig. 6 Geochemical proxies (terrestrial contribution and in-situ productivity) of DM active mudflat core for the time span ranging from 2006 to 1825 cal yr BP, depicting RWP, and during the last 200 cal yr BP showing MW time span

2020; Banerji 2025). However, the coherence of the event across regional and larger scales is still ambiguous (McKay et al. 2024). Thus, it becomes necessary to identify the region-specific climate response during such climate excursions.

In the present study, the VV section covered a time span between 4710 and 1500 cal yr BP, thereby representing the onset of the Meghalayan Stage (Fig. 3). However, the 4.2 ka event that represents the onset of the Meghalayan Stage was seldom identified due to the limited resolution of the VV sediment section. Further, the DV has only covered the depositional history since 4100 cal yr BP, and thus, the 4.2 ka event was also missed in the DV core. Nevertheless, other climatic episodes are well recognised from both VV and DV sedimentary sections. A gradual decline in the Al_2O_3, TiO_2 and TOC/TN in the VV from 4710 to 1500 cal yr BP has been associated with the gradual weakening of the ISM. An intermittent marginal enhancement in the TiO_2 and TiO_2/Al_2O_3 during 3560–3180 cal yr BP (Fig. 3) depicted the influence of the monsoon spell (Banerji et al. 2015) and corroborated well with the DV (Fig. 4), which occurred during 3230–2790 cal yr BP (Banerji et al. 2017). In DV, the simultaneous increase in the Cu/Al_2O_3, Ni/Al_2O_3 and Ba/Al_2O_3 demonstrated an increase in the in-situ productivity during 3230–2790 cal yr BP. The simultaneous increase in terrestrial contribution and in-situ productivity suggested an improved monsoonal event (Fig. 4).

The time span corresponds to the lesser-explored climatic event known as the Minoan Warm Period [MWP; 3400–3250 BP; (Lécuyer et al. 2018)], which was specifically identified when the Minoan culture flourished in the Mediterranean region. Based on the Greenland ice core, it has been suggested that the MWP (~3300 yr BP) was the warmest period during the last 7 ka, wherein the temperature increased about ~2.75 °C above present temperatures in Greenland. However, the warm period abruptly terminated with a decline in ~2.0 °C temperatures in Greenland and transformed into a cooler period (Easterbrook 2016). The cooler period post-MWP registered in the Greenland ice core seems to be reflected as a dry climate (weak ISM) in VV and DV sections during 2825–1835 cal yr BP and 2640–1930 cal yr BP, respectively (Figs. 3 and 4). However, in both sections VV and DV, the TOC/TN has remained above 12 until 2 ka with a declining trend, depicting a high terrestrial contribution with a gradual reduction towards 2 ka as a result of ISM weakening.

4.1.2 Last Two Millennia (2 ka—Present)

The lens of the available studies has been on recent climate change and its future trends, with an implicit assumption of the anthropogenic influence. However, the gap in understanding the independent impacts of natural and human-induced climate change emerged as a major consequence of limitations in future climate predictions. Thus, the climate of the last two millennia and its teleconnection with global climate variables are considered the most relevant discourses of present-day global climate change (Huang et al. 2023). Surprisingly, all the sedimentary sections in the present study have recorded some part of the last two millennia in the form of depositional history, thereby providing insight into climate and monsoon variability. The Al_2O_3

and TiO_2 in the VV showed a marginal increment during 2000–1500 cal yr BP, whilst a similar enhancement has been recorded during 1969–1546 cal yr BP and 2006–1800 cal yr BP in RH and DM (Figs. 5 and 6), respectively (Banerji et al. 2019, 2021).

On the contrary, such enhanced terrestrial flux was seldom recorded in the DV sediment core. The climatic period of monsoon strengthening aligns with the RWP. Further, the RH has also shown an increase in the terrestrial contribution during 1546 – 503 cal yr BP (Fig. 5), which corroborated well with the DV and DM during 1200–1070 cal yr BP (Fig. 4) and 1300–800 cal yr BP, respectively. The time span is well recognised as MCA, and its evidence has been found in various global records (Trouet et al. 2009; Moreno et al. 2012; Calò et al. 2013; Schmitt et al. 2020). The RWP and MCA are assumed to be two warm climatic anomalies analogous to the MW and that can thus, to some extent, serve as a prologue to the ongoing global warming of the twenty-first century (Neukom et al. 2019; Jiang et al. 2023). Both the warming period, viz. the RWP and MCA, were interrupted by the DACP event, which is predominantly characterised by cool climate and glacier advance during 950 – 1450 yr BP (Bianchi and McCave 1999; Cristea et al. 2014; Helama et al. 2017). Amongst all four sedimentary sections (VV, DV, RH and DM), the RH has shown signatures of the DACP event (Fig. 5) in the form of marginal weakening of the ISM during 1590 and 1506 cal yr BP (Banerji et al. 2019). Nevertheless, it required further support from other sedimentary sections and needs to be addressed properly. In the RH, a gradual increase in Al_2O_3 during 500 – 200 cal yr BP indicated an improved terrestrial contribution due to the activation of seasonal rivers caused by monsoonal rainfall. However, the concomitant reduction of weathering (TiO_2/Al_2O_3) and in-situ productivity (Ni/Al_2O_3, Cu/Al_2O_3 and Ba/Al_2O_3) during 500 – 200 cal yr BP signify the absence of climate warming. Thus, a cool and wet climate caused by winter precipitation instead of a stronger ISM has been suggested (Banerji et al. 2019). The time span corroborates with the globally recognised LIA, which is widely known for its long-term cooling period with the latest glacier expansion in the high latitudes and altitudes during the recent phase of the Holocene Epoch (650 – 100 yr BP; 1300 to 1850 CE) (Matthews and Briffa 2005; Wanner et al. 2022).

With the continuous climate warming and anthropogenic emissions of greenhouse gases (GHGs), there has been a profound need to improve the climate projections. Although the climate models have offered unprecedented results and projections, the large uncertainties continue to impact the reliability of the projections (Tierney et al. 2020). Further, understanding the complex interaction of various climatic processes and their spatial and temporal variations on the decadal- to-centennial scale highlights the need for extensive datasets beyond observational records that seldom exceed the last two centuries (Gergis and Henley 2017; Armstrong et al. 2020).

Comparison and calibration of paleoclimate proxies with observational records pave the way for the model evaluation, detangling climate interactions and constraining their uncertainties. Amongst the sedimentary cores discussed in the present chapter, the DM recorded a sediment deposition history for the last 180 yr, nearly similar to instrumental data that goes up to ~150 yr. Nevertheless, the DV and RH have also indicated the climate of the last two centuries, but the resolution is relatively poor compared to the DM core. Thus, the climatic scenarios with respect to instrumental records are compared with the DM core. During the Meghalayan Stage, although the ISM strength remained subdued over the past 200 yr BP, nevertheless, within this two centuries span, a marginal strengthening of ISM during 200–120 cal yr BP (1750–1890 CE) followed by the poor ISM with a warmer climate during the last 120 cal yr BP (1890–2012 CE) has been elucidated (Banerji et al. 2021). The ISM strengthening depicted by the terrestrial proxies of DM is in-line with the observational rainfall dataset from NW India (Fig. 7), suggesting a relatively wet climate during 1870–1895 CE intermittently occurred between the dry climates during 1848–1869 CE and 1896–1941 CE (Sontakke et al. 2008). The climate warming of the last 120 years BP in the DM core was substantiated by the rise in the annual sea surface temperature (SST), thereby confirming the severity of ongoing global warming and climate change.

The mudflats of the southern Saurashtra coast exhibited an overlapping depositional history, which in turn provided clues to decipher climate variability as a function of ISM perturbations (Fig. 7). Additionally, the geochemical proxies also varied in tandem with instrumental records. Broadly, it can be suggested that the ISM has been variable during the last 4.7 ka to present, with simultaneous strengthening and weakening of the ISM as a function of globally observed climatic events. Furthermore, it is also worth noting that the mudflats of the southern Saurashtra coast responded well to the globally recognised climatic events (Fig. 8).

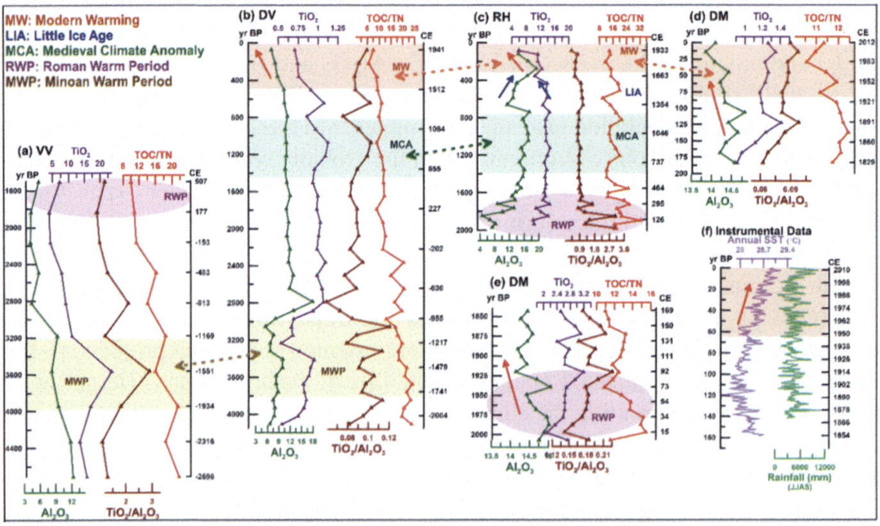

Fig. 7 Comparison of geochemical proxies of all the sedimentary sections (VV, DV, RH and DM) for the observed climatic events such as MWP, RWP, MCA, LIA and MW in addition the MW event is 5 subsequently compared with the available instrumental records representing annual sea surface temperature (SST) and rainfall (June, July, August and September: JJAS) data

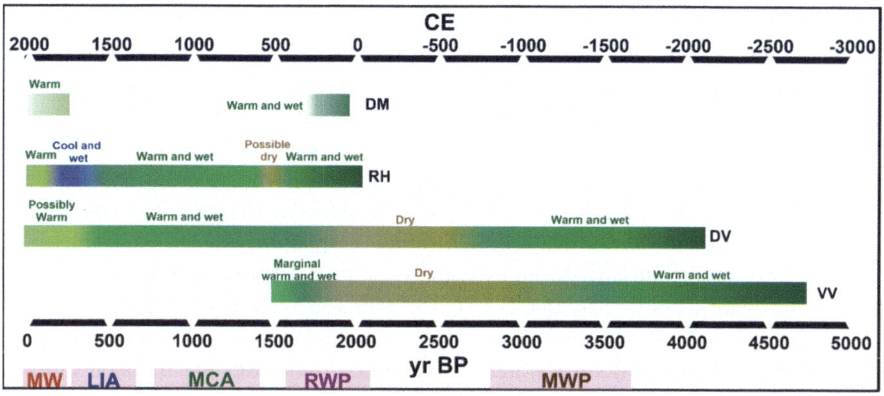

Fig. 8 The schematic representation and comparison of the reconstructed climate of the Meghalayan Stage for the sedimentary sections VV, DV, RH and DM with green, brown and blue coloured bar representing wet and warm climate, dry and cool and wet climate in the southern Saurashtra region

4.2 Possible Natural Forcing Mechanism Causing Climate Perturbations

Climate models are indispensable tools for making future climate projections and proposing appropriate adaptation and mitigation strategies. The certainty of such models and projections relies on their ability to reproduce the variability in the historical climate (Fyfe et al. 2021). The primary inputs in climate models are the solar and volcanic forcings under natural mechanisms, whilst aerosol and GHG emissions under anthropogenic forcings are typically adopted for reliable simulation of past and future climate perturbations. However, a trivial change in the forcing parameters included in the models may significantly impact the results of climate simulations (Fyfe et al. 2021). Thus, to reduce the forcing uncertainties incorporated in the models, properly understanding the association of the forcing factors and their feedback mechanisms on the climate system is essential to disentangle the complexities involved in monsoon–climate dynamics. In the present section (Fig. 9a-h), the natural forcing mechanisms such as volcanic [volcanic sulphate for Greenland and Antarctica ice cores (Zielinski and Mershon 1997; Castellano et al. 2004)], solar [Total solar irradiance (TSI) (Steinhilber et al. 2009); solar insolation (30°N) (Berger and Loutre 1991) and sunspot activity (Solanki et al. 2004)] and GHGs [CO_2 and CH_4 (Blunier et al. 1995; Bereiter et al. 2015)] forcings (Fig. 9 a–c) along with the climate variables (Fig. 9d–h) such as ITCZ (Haug et al. 2001; Sun et al. 2012b), Siberian high, Icelandic low (Mayewski et al. 1997), Atlantic meridional overturning circulation (AMOC), North Atlantic Oscillation (NAO) Index (McManus et al. 2004; Olsen et al. 2012), El Nino Southern Oscillation (ENSO) (Moy et al. 2002; Yamamoto 2009) and Global temperature reconstruction (Alley 2004; Bintanja et al. 2005) for the Meghalayan Stage have been discussed to infer the possible mechanism and its signatures in the mudflats of the southern Saurashtra coast. The volcanic eruption ejects sulphur gases in the form of aerosols into the stratosphere, which reside for one to three years in the stratosphere, causing sunlight scattering and leading to negative radiative forcing, thereby cooling the surface of the Earth (Robock 2000; Kremser et al. 2016), and thus acts as a prime driver of climate variability (Hegerl et al. 2003; Sigl et al. 2015). The profound impact of volcanic eruption on the climate has been discerned from the historical events of Mount Tambora in 1815 CE, the subsequent year was known as the 'Year Without a Summer' (Oppenheimer 2003; Raible et al. 2016), whilst the eruption of Mount Pinatubo in 1991 CE resulted in global-mean lower-tropospheric cooling by 0.4–0.5 °C for over a year (McCormick et al. 1995).

Considering the geochemical signatures in the mudflats of Saurashtra, the RWP and the LIA have been demonstrated to be linked with volcanic-induced climate perturbations. The volcanic sulphate of the Antarctic ice core (Castellano et al. 2004) seldom showed any prominent volcanic event during the RWP; on the contrary, the volcanic episode of ~1950 yr BP in the Greenland ice core (Zielinski and Mershon 1997) corroborated with the marginal weakening of ISM during 1950–1970 cal yr BP in DM core, suggesting an intermittent climate cooling within the RWP phase

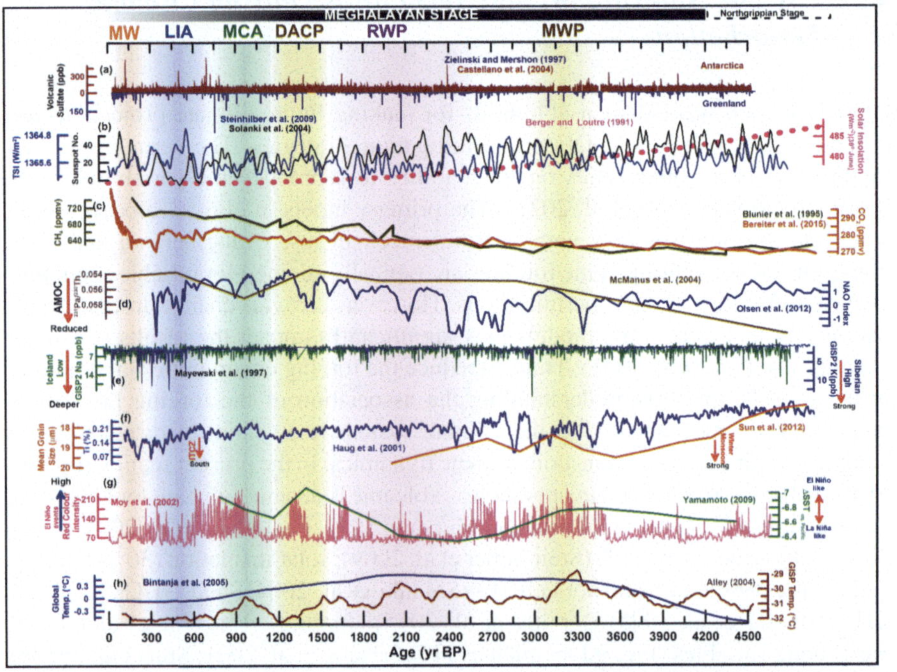

Fig. 9 Parameters depicting natural forcings (**a–c**) such as volcanic, solar and greenhouse gases, along with the climate variables (**d–h**) such as NAO, AMOC, ENSO, ITCZ migration and global temperature curve, are represented to decipher the possible mechanism that caused the abrupt climatic events of the Meghalayan Stage and its possible influence along southern Saurashtra, western India

(Banerji et al. 2021). In the case of the LIA, the higher volcanic sulphate content in both the ice cores (Greenland and Antarctica) during 800 – 660 yr BP demonstrated volcanic events just preceding the onset of the LIA. It could be possible that a major volcanic event occurred between MCA and LIA that was recorded by both the ice cores, which plausibly kick-started the LIA. The volcanic episode in the Greenland ice core is considered to be the largest eruption in 7000 yr and has been identified as the Mt Samalas eruption (1258 AD) (Langway et al. 1988). The climate cooling during the LIA in RH was also linked to the lowering of solar forcing as depicted by the reduced TSI (Banerji et al. 2019). The Maunder Minimum identified in the TSI records during the mid-1600s to early 1700 CE appeared to be concurrent with the lowest surface temperatures in the northern hemisphere during the last millennium, which also corroborated the LIA (Shindell et al. 2001; Steinhilber et al. 2009). A gradual decline in solar insolation has been observed from the early Meghalayan Stage towards the present (Berger and Loutre 1991). Thus, the LIA prevailed due to both volcanic and solar forcings. Moreover, volcanic eruption may also play a key role in altering other modes of climate variables such as the AMOC, NAO (Driscoll et al. 2012; Ortega et al. 2012), ENSO (Li et al. 2013; McGregor et al. 2020), ITCZ migration (Mukherjee et al. 1987; Pausata and Camargo 2019; Erez and Adam 2021) and the tropical monsoons (Anchukaitis et al. 2010; Haywood et al. 2013; Banerji et al. 2020). The geochemical proxies (terrestrial contribution increases and in-situ productivity decreases) in the RH core suggested a cool and wet climate during the LIA caused by the strengthening of winter precipitation (Banerji et al. 2019). The winter precipitation in the region was instigated by the southward migration of the ITCZ [as depicted by the decline in the Titanium concentration in the Cariaco basin sediment (Haug et al. 2001)] caused by the volcanic eruption (Banerji et al. 2019). The weakening of the ISM also prevailed due to the frequent occurrence of El Niño-like events (Moy et al. 2002), led by the southward migration of the ITCZ. Further, the negative (-ve) phase of NAO during the LIA resulted in the enhanced contribution of the WD, thereby causing winter precipitation in the region (Olsen et al. 2012).

The MCA and RWP are predominantly associated with the ISM strengthening, which in turn resulted from stronger solar forcings as depicted by the increased TSI. The stronger ISM was further supported by the absence of conspicuously major volcanic events, positive (+ve) NAO phase, weak Siberian high and northward persistence of the ITCZ. However, the stronger ISM during the RWP than the MCA, as depicted from the RH and DM sediment cores, resulted from the frequent occurrence of El Niño-like events during the MCA than the RWP. Nevertheless, no major variation in the ISM has been demonstrated from the DV core plausibly due to sample resolution. In addition, the higher global temperatures (Bintanja et al. 2005) were observed during the RWP than MCA during the Meghalayan Stage, which might have triggered the stronger ISM during RWP. The intermittent climate of MCA and RWP, i.e. DACP, has been identified as a cool climate associated with North Atlantic events and low solar forcings (Helama et al. 2017). However, the RH core has provided some limited signals of DACP event in the form of ISM weakening, which needs to be supported by additional evidence that can further be interpreted for its global linkages (Banerji et al. 2019).

The VV and DV demonstrated an intermittent monsoon spell during 4000 – 3000 yr BP, corroborating the lesser-known MWP. The volcanic episodes during the MWP were seldom prominent, but a shift of the ITCZ towards the south and a stronger winter monsoon have been depicted from the Ti concentration in Cariaco basin sediment and grain size variation in the north-western Chinese Loess Plateau (Haug et al. 2001; Sun et al. 2012b) with marginally low TSI (Steinhilber et al. 2009). Subsequently, the southward migration of the ITCZ must have led to increased episodes of El Niño-like events (Moy et al. 2002; Yamamoto 2009). Thus, it can be suggested that a simultaneous interaction of ISM and possibly the winter rainfall prevailed during the MWP. Nevertheless, additional supporting records are needed to attest to the prevalence of winter precipitation and the ISM in the region.

The annual mean temperature reconstruction from the Northern Hemisphere revealed that the last decade of the twentieth century was probably the warmest interval of the preceding thousand years (Jones and Mann 2004; Osborn and Briffa 2006). The rise in atmospheric GHG concentrations during the last two centuries have played a key role in the ongoing warming (Hegerl et al. 2003). Even though water vapour is the most important GHG, its residence time in the troposphere is ~10 days, and thus, it can lead to changes in the feedback mechanism rather than a direct impact on the climate. The concentration of other GHGs, such as CO_2, CH_4 and N_2O, in the past atmosphere has been estimated through the gas bubbles trapped in ice cores (Ahn et al. 2004; Masson-Delmotte et al. 2013; Ryu et al. 2020). The variations of these gases have been remarkably stable prior to the industrial era, during the Holocene. In general, the CO_2 and CH_4 have impacted the global temperature proportionally to their forcing, and exclusively for the Holocene, the global annual mean changes would have been within the range of ~0 to 0.5 °C (Schmidt et al. 2004). The terrestrial and in-situ productivity proxies for the DV, DM and RH depicted a warm climate with reduced monsoon spells in the last two centuries. Furthermore, the decline in solar insolation possibly led to the weakening of ISM; however, the subsequent increase in GHGs such as CO_2 and CH_4 likely contributed to climate warming. Despite the realisation of the significance of GHGs over the last two centuries, the appropriate portion of the anthropogenic forcing is yet to be deciphered. Nevertheless, it can be underscored that during most of the Meghalayan Stage, the climate vis-à-vis monsoon has been predominantly influenced by the natural forcing mechanisms and climate variables, whilst the last two centuries seem to be impacted by both natural forcings as well as anthropogenically induced GHGs.

5 Conclusions

The testimony of natural archives, obtained through multiproxy techniques, offers a unique opportunity to comprehend the past and present climate and evaluate climate models for reliable future projections. However, there remain uncertainties in Holocene climatic reconstructions due to the differences between climate models and proxy-based climate reconstructions, as well as the limitations in observational and reconstructed climate datasets on spatial and temporal scales. Thus, the region-specific climate reconstruction and its linkage with the global climate system require careful attention. In view of this, the present chapter aims to provide an overview of the climate variations during the Meghalayan Stage as reflected in the mudflats of the southern Saurashtra coast, Gujarat, western India. The study emphasised the critical role of the ISM in shaping the climatic conditions of western India and its linkage with global climatic events, thereby exhibiting the teleconnection of regional and global climate systems during the Meghalayan Stage. Despite advancements in understanding past climates, significant uncertainties persist regarding the mechanisms driving abrupt climatic events and their global correlations. Thus, the present chapter invokes further investigation into paleoclimate records to enhance our understanding of historical climate variability and its implications for current climate change discussions.

Acknowledgements The author is grateful to the Director, NCESS, for his encouragement and support. The author is indebted to Prof. Ravi Bhushan, Dr. Navin Juyal, Dr. D Padmalal and Dr. Maya K. for their guidance and support. The author extends special thanks to the editor, Dr. Neloy Khare, for providing the opportunity to contribute a chapter to the edited book volume. The author is also thankful to Mr. Rakesh Kumar Panda for helping in the preparation of the geological map.

References

Achyuthan H, Nagasundaram M, Gourlan AT, et al (2014) Mid-Holocene Indian Summer Monsoon variability off the Andaman Islands, Bay of Bengal. Quat Int 349:232–244. https://doi.org/10.1016/j.quaint.2014.07.041

Agnihotri R, Sarin MM, Somayajulu BLK, et al (2003) Late-Quaternary biogenic productivity and organic carbon deposition in the eastern Arabian Sea. Palaeogeogr Palaeoclimatol Palaeoecol 197:43–60. https://doi.org/10.1016/S0031-0182(03)00385-7

Ahn J, Wahlen M, Deck BL, et al (2004) A record of atmospheric CO_2 during the last 40,000 years from the Siple Dome, Antarctica ice core. J Geophys Res D Atmos 109:1–8. https://doi.org/10.1029/2003JD004415

Alfieri L, Bisselink B, Dottori F, et al (2017) Global projections of river flood risk in a warmer world. Earth's Futur 5:171–182. https://doi.org/10.1002/2016EF000485

Alley RB (2004) Abrupt climate changes: Oceans, ice, and Us. Oceanography 17:194–206. https://doi.org/10.5670/oceanog.2004.20

Anchukaitis KJ, Buckley BM, Cook ER, et al (2010) Influence of volcanic eruptions on the climate of the Asian monsoon region. Geophys Res Lett 37:1–5. https://doi.org/10.1029/2010GL044843

Armstrong MS, Kiem AS, Vance TR (2020) Comparing instrumental, palaeoclimate, and projected rainfall data: Implications for water resources management and hydrological modelling. J Hydrol Reg Stud 31:100728. https://doi.org/10.1016/j.ejrh.2020.100728

Banerji US, Pandey S, Bhushan R, Juyal N (2015) Mid-Holocene climate and land-sea interaction along the southern coast of Saurashtra, western India. J Asian Earth Sci 111:428–439. https://doi.org/10.1016/j.jseaes.2015.06.021

Banerji US, Bhushan R, Jull AJT (2017) Mid−late Holocene monsoonal records from the partially active mudflat of Diu Island, southern Saurashtra, Gujarat, western India. Quat Int 443:200–210. https://doi.org/10.1016/j.quaint.2016.09.060

Banerji US, Bhushan R, Jull AJT (2019) Signatures of global climatic events and forcing factors for the last two millennia on the active mudflats of Rohisa, southern Saurashtra, Gujarat, western India. Quat Int 507:172–187. https://doi.org/10.1016/j.quaint.2019.02.015

Banerji US, Arulbalaji P, Padmalal D (2020) Holocene climate variability and Indian Summer Monsoon: An overview. Holocene 30:744–773. https://doi.org/10.1177/0959683619895577

Banerji U, Bhushan R, Joshi KB, et al (2021) Hydroclimate variability during the last two millennia from the mudflats of Diu Island, Western India. Geol J 3584–3604. https://doi.org/10.1002/gj.4116

Banerji US, Dubey CP, Goswami V, Joshi KB (2022a) Geochemical indicators in provenance estimation. In: John A-A, Kailasa P, Verma SK (eds) Geochemical Treasures and Petrogenetic Processes. Springer Nature, pp 95–121

Banerji US, Joshi KB, Pandey L, Dubey CP (2022b) An outline of geochemical proxies used on marine sediments deposited during the Quaternary Period. pp 1–35

Banerji US, Bhushan R, Joshi KB, et al (2024) Geochemical records of mudflat sediments from southern Saurashtra, Western India: Implications for Holocene climate and global teleconnection. Holocene 34:1700–1720. https://doi.org/10.1177/09596836241266398

Banerji US (2025) Decoding the conundrum of indian monsoon variability and its global linkage during the Meghalayan Stage (4.2 ka to present). In: Khare N (Ed) Climate dynamics over the Indian subcontinent-geological perspectives. Cambridge Scholars Publishing, London, UK, p 308

Bereiter B, Eggleston S, Schmitt J, et al (2015) Revision of the EPICA Dome C CO2 record from 800 to 600-kyr before present. Geophys Res Lett 42:542–549. https://doi.org/10.1002/2014GL061957

Berger A, Loutre MF (1991) Insolation values for the climate of the last 10 million years. Quat Sci Rev 10:297–317. https://doi.org/10.1016/0277-3791(91)90033-Q

Berkelhammer M, Sinha A, Stott L, et al (2015) An Abrupt Shift in the Indian Monsoon 4000 Years Ago. In: Climates, Landscapes, and Civilizations. pp 75–87

Bhonde U, Bhatt N (2009) Joints as Fingerprints of Stress in the Quaternary Carbonate Deposits along Coastal Saurashtra, Western India. J Geol Soc India 74:703–710

Bhushan R, Dutta K, Somayajulu BLK (2001) Concentrations and burial fluxes of organic and inorganic carbon on the eastern margins of the Arabian Sea. Mar Geol 178:95–113. https://doi.org/10.1016/S0025-3227(01)00179-7

Bianchi GG, McCave IN (1999) Holocene periodicity in North Atlantic climate and deep-ocean flow south of Iceland. Nature 397:515–517. https://doi.org/10.1038/17362

Bintanja R, Van De Wal RSW, Oerlemans J (2005) Modelled atmospheric temperatures and global sea levels over the past million years. Nature 437:125–128. https://doi.org/10.1038/nature03975

Blackburn TH (1983) The microbial nitrogen cycle. In: Krumbein WE (ed) Microbial Geochemistry. Blackwell Scientific, Oxford, pp 63–89

Blunier T, Chappellaz J, Schwander J, et al (1995) Variations in atmospheric methane concentration during the Holocene epoch. Nature 374:46–48

Calò C, Henne PD, Eugster P, et al (2013) 1200 years of decadal-scale variability of Mediterranean vegetation and climate at Pantelleria Island, Italy. Holocene 23:1477–1486. https://doi.org/10.1177/0959683613493935

Calvert SE, Pedersen TF (2007) Elemental Proxies for Palaeoclimatic and Palaeoceanographic Variability in Marine Sediments: Interpretation and Application. Dev Mar Geol 1:567–644. https://doi.org/10.1016/S1572-5480(07)01019-6

Castellano E, Becagli S, Jouzel J, et al (2004) Volcanic eruption frequency over the last 45 ky as recorded in Epica-Dome C ice core (East Antarctica) and its relationship with climatic changes. Glob Planet Change 42:195–205. https://doi.org/10.1016/j.gloplacha.2003.11.007

Champion HG, Seth SK (1968) A Revised Survey of the Forest Types of India.

Claes H, Huysmans M, Soete J, et al (2019) Elemental geochemistry to complement stable isotope data of fossil travertine: Importance of digestion method and statistics. Sediment Geol 386:118–131

Correns CW (1969) Titanium. In: Wedepohl KH, Correns CW, Shaw DM, et al. (eds) Handbook of geochemistry. Springer Berlin, Berlin, p 22

Cristea G, Cuna SM, Fărcaş S, et al (2014) Carbon isotope composition as indicator for climatic changes during the middle and late Holocene in a peat bog from Maramureş Mountains (Romania). Holocene 24:15–23. https://doi.org/10.1177/0959683613512166

Dolcater DL, Syers JK, Jackson ML (1970) Titanium as free oxide and substituted forms in kaolinites and other soil minerals. Clays Clay Miner 18:71–79. https://doi.org/10.1346/CCMN.1970.0180202

Driscoll S, Bozzo A, Gray LJ, et al (2012) Coupled Model Intercomparison Project 5 (CMIP5) simulations of climate following volcanic eruptions. J Geophys Res Atmos 117:. https://doi.org/10.1029/2012JD017607

Dymond J, Suess E, Lyle M (1992) Barium in Deep-Sea Sediment: A Geochemical Proxy for Paleoproductivity. Paleoceanography 7:163–181. https://doi.org/10.1029/92pa00181

Easterbrook DJ (2016) Temperature fluctuations in Greenland and the Arctic. Evidence-Based Clim Sci Data Opposing CO2 Emiss as Prim Source Glob Warm Second Ed 137–160. https://doi.org/10.1016/B978-0-12-804588-6.00008-2

Erez M, Adam O (2021) Energetic Constraints on the Time-Dependent Response of the ITCZ to Volcanic Eruptions. J Clim 34:9989–10006. https://doi.org/10.1175/JCLI-D-21-0146.1

Farooqui A, Gaur AS, Prasad V, Paula D (2013) Climate, vegetation and ecology during Harappan period: Excavations at Kanjetar and Kaj, mid-Saurashtra coast, Gujarat. J Archaeol Sci 40:2631–2647. https://doi.org/10.1016/j.jas.2013.02.005

Fleitmann D, Burns SJ, Mangini A, et al (2007) Holocene ITCZ and Indian monsoon dynamics recorded in stalagmites from Oman and Yemen (Socotra). Quat Sci Rev 26:170–188. https://doi.org/10.1016/j.quascirev.2006.04.012

Friis-Christensen E, Lassen K (1991) Length of the solar cycle: an indicator of solar activity closely associated with climate. Science 254:698–700

Fyfe JC, Kharin V V., Santer BD, et al (2021) Significant impact of forcing uncertainty in a large ensemble of climate model simulations. Proc Natl Acad Sci U S A 118:1–6. https://doi.org/10.1073/pnas.2016549118

Gadgil S (2003) The Indian Monsoon and Its Variability. Annu Rev Earth Planet Sci 31:429–467. https://doi.org/10.1146/annurev.earth.31.100901.141251

Ganapathi S, Patel MP, Merh SS (1982) Evolution of the coastline between okha-madhi and Kodinar, western Saurashtra with special reference to Quaternary morphology and sedimentation. In: Merh SS (ed) Geology and geomorphology of quaternary terrains of gujarat with special reference to coastline evolution. M.S. University of Baroda, Vadodara, pp 95–122

Geological Survey of India (GSI) (2025) Bhukosh. https://bhukosh.gsi.gov.in/Bhukosh/MapViewer.aspx. Accessed 30 Nov 2025

Gergis J, Henley BJ (2017) Southern Hemisphere rainfall variability over the past 200 years. Clim Dyn 48:2087–2105. https://doi.org/10.1007/s00382-016-3191-7

Gundalia MJ, Dholakia MB (2013) Estimation of pan evaporation using mean air temperature and radiation for monsoon season in Junagadh region. Int J Eng Res Appl 3:64–70

Haug GH, Hughen KA, Sigman DM, et al (2001) Southward Migration of the Intertropical Convergence Zone through the Holocene. Science 293:1304–1308. https://doi.org/10.1126/science.105972

Haywood JM, Jones A, Bellouin N, Stephenson D (2013) Asymmetric forcing from stratospheric aerosols impacts Sahelian rainfall. Nat Clim Chang 3:660–665. https://doi.org/10.1038/nclimate1857

Hegerl GC, Crowley TJ, Baum SK, et al (2003) Detection of volcanic, solar and greenhouse gas signals in paleo-reconstructions of Northern Hemispheric temperature. Geophys Res Lett 30:94–97. https://doi.org/10.1029/2002gl016635

Helama S, Oinonen M (2019) Exact dating of the Meghalayan lower boundary based on high-latitude tree-ring isotope chronology. Quat Sci Rev 214:178–184. https://doi.org/10.1016/j.quascirev.2019.04.013

Helama S, Jones PD, Briffa KR (2017) Dark Ages Cold Period: A literature review and directions for future research. Holocene 27:1600–1606. https://doi.org/10.1177/0959683617693898

Herring SC, Christidis N, Hoell A, et al (2018) Explaining Extreme Events of 2016- From a climate perspective. Bull Am Meteorol Soc 99:S1–S157

Huang Y, Xiao J, Wen R, et al (2023) High- and low- latitude forcing on hydroclimate variability in southern China over the last two millennia. Palaeogeogr Palaeoclimatol Palaeoecol 610:111352. https://doi.org/10.1016/j.palaeo.2022.111352

IPCC (2007) Climate change 2007: the physical science basis. Intergov Panel Clim Chang. https://doi.org/10.1038/446727a

IPCC (2013) Working Group I Contribution To The Fifth Assessment Report Of The Intergovernmental Panel On Climate Change

Jansen E, Overpeck J, Briffa KR, et al (2007) Palaeoclimate. In: Solomon S, D. Qin, M. Manning, et al. (eds) Climate Change 2007: The Physical Science Basis. Contribution of Working Group I to the Fourth Assessment Report of the Intergovernmental Panel on Climate Change. Cambridge University Press, Cambridge, United Kingdom and New York, NY, USA, pp 433–497

Jiang L, Yu K, Tao S, et al (2023) Modulation of East Asian monsoon strength by ENSO during the warm periods of the late Holocene: Evidence from Porites corals in the northern South China Sea. Glob Planet Change 225:104136. https://doi.org/10.1016/j.gloplacha.2023.104136

Jones PD, Mann ME (2004) Climate over past millennia. Rev Geophys 42:1–42. https://doi.org/10.1029/2003RG000143

Joshi KB, Banerji US, Dubey CP, Oliveira EP (2021) Heavy minerals in provenance studies: an overview. Arab J Geosci 14:1330. https://doi.org/10.1007/s12517-021-07687-y

Kemp DB, Izumi K (2014) Multiproxy geochemical analysis of a Panthalassic margin record of the early Toarcian oceanic anoxic event (Toyora area, Japan). Palaeogeogr Palaeoclimatol Palaeoecol 414:332–341. https://doi.org/10.1016/j.palaeo.2014.09.019

Kremser S, Thomason LW, von Hobe M, et al (2016) Stratospheric aerosol—Observations, processes, and impact on climate. Rev Geophys 54:278–335. https://doi.org/10.1002/2015RG000511.Received

Lamb HH (1965) The early medieval warm epoch and its sequel. Palaeogeogr Palaeoclimatol Palaeoecol 1:13–37

Langway CC, Clausen HB, Hammer CU (1988) An inter-hemispheric volcanic time-marker in ice cores from Greenland and Antarctica. Ann Glaciol 10:102–108

Laskar AH, Yadava MG, Sharma N, Ramesh R (2013) Late-Holocene climate in the Lower Narmada valley, Gujarat, western India, inferred using sedimentary carbon and oxygen isotope ratios. Holocene 23:1115–1122. https://doi.org/10.1177/0959683613483621

Lécuyer C, Atrops F, Amiot R, et al (2018) Tsunami sedimentary deposits of Crete records climate during the 'Minoan Warming Period' ($\approx$3350 yr BP). Holocene 28:914–929. https://doi.org/10.1177/0959683617752840

Li J, Xie SP, Cook ER, et al (2013) El Niño modulations over the past seven centuries. Nat Clim Chang 3:822–826. https://doi.org/10.1038/nclimate1936

Marathe A. (1981) Geoarcrheology of the Hiran valley, Saurashrra, India. Deccan college Post Graduate and Research Institute, Pune

Masson-Delmotte V, Schulz M, Abe-Ouchi A, et al (2013) Information from paleoclimate archives. In: Stocker TF, Qin D, Plattner G-K, et al. (eds) Climate Change 2013 the Physical Science Basis: Working Group I Contribution to the Fifth Assessment Report of the Intergovernmental Panel on Climate Change. Cambridge University Press, Cambridge, United Kingdom and New York, NY, USA., pp 383–464

Matthews JA, Briffa KR (2005) The "Little Ice Age": Re-evaluation of an evolving concept. Geogr Ann Ser A Phys Geogr 87:17–36. https://doi.org/10.1111/j.0435-3676.2005.00242.x

Mayewski PA, Rohling EE, Stager JC, et al (2004) Holocene climate variability. Quat Res 62:243–255. https://doi.org/10.1016/j.yqres.2004.07.001

Mayewski PA, Meeker LD, Twickler MS, et al (1997) Major features and forcing of high latitude northern hemisphere atmospheric circulation using a 110,000 year long glaciochemical series. J Geophys Res 102:26,345–26,366

McCormick MP, Thomason LW, Trepte CR (1995) Atmospheric effects of the Mt Pinatubo eruption. Nature 373:399–404

McGregor S, Khodri M, Maher N, et al (2020) The Effect of Strong Volcanic Eruptions on ENSO. Geophys Monogr Ser 253:267–287. https://doi.org/10.1002/9781119548164.ch12

McKay NP, Kaufman DS, Arcusa SH, et al (2024) The 4.2 ka event is not remarkable in the context of Holocene climate variability. Nat Commun 15:1–12. https://doi.org/10.1038/s41467-024-50886-w

McManus JF, Francois R, Gherardl JM, et al (2004) Collapse and rapid resumption of Atlantic meridional circulation linked to deglacial climate changes. Nature 428:834–837. https://doi.org/10.1038/nature02494

Merh SS (1995) Geology of Gujarat. Geological Society of India, Bangalore

Meyers PA (1994) Preservation of elemental and isotopic source identification of sedimentary organic matter. Chem Geol 114:289–302. https://doi.org/10.1016/0009-2541(94)90059-0

Misra P, Tandon SK, Sinha R (2018) Holocene climate records from lake sediments in India: assessment of coherence across climate zones. Earth-Sci Rev 190:370–397. https://doi.org/10.1016/j.earscirev.2018.12.017

Moreno A, Pérez A, Frigola J, et al (2012) The Medieval Climate Anomaly in the Iberian Peninsula reconstructed from marine and lake records. Quat Sci Rev 43:16–32. https://doi.org/10.1016/j.quascirev.2012.04.007

Moy CM, Seltze GO, Rodbell DT, Anderson DM (2002) Variability of El Nino/ Oscillation activity at millennial timescales during the Holocene epoch. Nature 420:162–165. https://doi.org/10.1038/nature01163.1.

Mukherjee BK, Indira K, Dani KK (1987) Low-latitude volcanic Eruptions and their effects on Sri Lankan rainfall during the north-east monsoon. J Climatol 7:145–155

Nesbitt HW, Markovics G (1997) Weathering of granodioritic crust, long-term storage of elements in weathering profiles, and petrogenesis of siliciclastic sediments. Geochim Cosmochim Acta 61:1653–1670

Neukom R, Steiger N, Gómez-Navarro JJ, et al (2019) No evidence for globally coherent warm and cold periods over the preindustrial Common Era. Nature 571:550–554. https://doi.org/10.1038/s41586-019-1401-2

Norgaard-Pedersen N, Mikkelsen N (2009) 8000 year marine record of climate variability and fjord dynamics from Southern Greenland. Mar Geol 264:177–189

Olsen J, Anderson NJ, Knudsen MF (2012) Variability of the North Atlantic Oscillation over the past 5,200 years. Nat Geosci 5:808–812. https://doi.org/10.1038/ngeo1589

Oppenheimer C (2003) Climatic, environmental and human con sequences of the largest known historic eruption: Tambora volcano (Indonesia) 1815. Prog Phys Geogr 27:230–259. https://doi.org/10.1191/0309133303pp379ra

Oppo DW, Rosenthal Y, Linsley BK (2009) 2,000-Year-long temperature and hydrology reconstructions from the Indo-Pacific warm pool. Nature. https://doi.org/10.1038/nature08233

Ortega P, Montoya M, González-Rouco F, et al (2012) Variability of the Atlantic meridional overturning circulation in the last millennium and two IPCC scenarios. Clim Dyn 38:1925–1947. https://doi.org/10.1007/s00382-011-1081-6

Osborn TJ, Briffa KR (2006) Erratum: The spatial extent of 20th-century warmth in the context of the past 1200 years. Science 312:697. https://doi.org/10.1126/science.312.5774.697

Pant RK, Juyal N (1993) Late quaternary coastal instability and sea level changes: new evidences from Saurashtra Coast, western India. Zeitschrift fir Geo-morphologie NF 37:29–40

Pausata FSR, Camargo SJ (2019) Tropical cyclone activity affected by volcanically induced ITCZ shifts. Proc Natl Acad Sci U S A 116:7732–7737. https://doi.org/10.1073/pnas.1900777116

Prasad S, Kusumgar S, Gupta SK (1997) A mid to late Holocene record of palaeoclimatic changes from Nal Sarovar: a palaeodesert margin lake in western India. J Quat Sci 12:153–159. https://doi.org/10.1002/(sici)1099-1417(199703/04)12:2<153::aid-jqs300>3.0.co;2-x

Prasad V, Phartiyal B, Sharma A (2007) Evidence of enhanced winter precipitation and the prevalence of a cool and dry climate during the mid to late Holocene in mainland Gujarat, India. The Holocene 17:889–896. https://doi.org/10.1177/0959683607082403

Prasad V, Farooqui A, Sharma A, et al (2014) Mid-late Holocene monsoonal variations from mainland Gujarat, India: A multi-proxy study for evaluating climate culture relationship. Palaeogeogr Palaeoclimatol Palaeoecol 397:38–51. https://doi.org/10.1016/j.palaeo.2013.05.025

Ragueneau O, Tréguer P, Leynaert A, et al (2000) A review of the Si cycle in the modern ocean: Recent progress and missing gaps in the application of biogenic opal as a paleoproductivity proxy. Glob Planet Change 26:317–365. https://doi.org/10.1016/S0921-8181(00)00052-7

Raible CC, Brönnimann S, Auchmann R, et al (2016) Tambora 1815 as a test case for high impact volcanic eruptions: Earth system effects. Wiley Interdiscip Rev Clim Chang 7:569–589. https://doi.org/10.1002/wcc.407

Ray K, Mohanty M, Chincholikar JR (2009) Climate variability over Gujarat, India. In: Workshop Proceedings: Impact of Climate Change on Agriculture. pp 38–43

Robock A (2000) Volcanic eruptions and climate. Rev Geophys 38:191–219

Ryu Y, Ahn J, Yang JW, et al (2020) Atmospheric Nitrous Oxide Variations on Centennial Time Scales During the Past Two Millennia. Global Biogeochem Cycles 34:. https://doi.org/10.1029/2020GB006568

Sanchez MR (2019) Global assessment of marine paleoproductivity proxies. Universitat Autònoma de Barcelona

Schmidt GA, Shindell DT, Miller RL, et al (2004) General circulation modelling of Holocene climate variability. Quat Sci Rev 23:2167–2181. https://doi.org/10.1016/j.quascirev.2004.08.005

Schmitt D, Gischler E, Anselmetti FS, Vogel H (2020) Caribbean cyclone activity: an annually-resolved Common Era record. Sci Rep 10:1–18. https://doi.org/10.1038/s41598-020-68633-8

Shindell DT, Schmidt GA, Mann ME, et al (2001) Solar forcing of regional climate change during the Maunder Minimum. Science 294:2149–2152. https://doi.org/10.1126/science.1064363

Sigl M, Winstrup M, McConnell JR, et al (2015) Timing and climate forcing of volcanic eruptions for the past 2,500 years. Nature 523:543–549. https://doi.org/10.1038/nature14565

Slabunov A, Joshi KB, Singh SK, Rai VK (2024) Depositional age and formation conditions of Archean banded iron formations, Bundelkhand Craton, Central India: Geochemistry, neodymium isotopes and U-Pb zircon geochronology. Precambrian Res 401:107254

Solanki SK, Usoskin IG, Kromer B, et al (2004) Unusual activity of the Sun during recent decades compared to the previous 11,000 years. Nature 431:1084–1087. https://doi.org/10.1038/nature02995

Sontakke N a. A, Singh N, Singh HNN (2008) Instrumental period rainfall series of the Indian region (AD 1813–2005): Revised reconstruction, update and analysis. Holocene 18:1055–1066. https://doi.org/10.1177/0959683608095576

Spears DA, Kanaris-Sotiriou R (1976) Titanium in some Carboniferous sediments from Great Britain. Geochim Cosmochim Acta 40:345–351

Sridhar A (2009) Evidence of a late-medieval mega flood event in the upper reaches of the Mahi River basin, Gujarat. Curr. Sci. 96:1517–1520

Sridhar A, Laskar A, Prasad V, et al (2015) Late Holocene flooding history of a tropical river in western India in response to southwest monsoon fluctuations: A multi proxy study from lower Narmada valley. Quat Int 371:181–190. https://doi.org/10.1016/j.quaint.2014.10.052

Sridhar A (2007) A mid-late Holocene flood record from the alluvial reach of the Mahi River, Western India. Catena 70:330–339. https://doi.org/10.1016/j.catena.2006.10.012

Srivastava D, Dubey CP, Banerji US, Joshi KB (2024) Geochemical trends in sedimentary environments using PCA approach. J Earth Syst Sci 133:. https://doi.org/10.1007/s12040-024-02306-2

Steinhilber F, Beer J, Fröhlich C (2009) Total solar irradiance during the Holocene. Geophys Res Lett 36:L19704. https://doi.org/10.1029/2009gl040142

Sun L, Gui H, Chen S (2012a) Geochemistry of sandstones from the Neoproterozoic Shijia Formation, northern Anhui Province, China: implications for provenance, weathering and tectonic setting. Geochemistry 72:253–260

Sun Y, Clemens SC, Morrill C, et al (2012b) Influence of Atlantic meridional overturning circulation on the East Asian winter monsoon. Nat Geosci 5:46–49. https://doi.org/10.1038/ngeo1326

Tierney JE, Poulsen CJ, Montañez IP, et al (2020) Past climates inform our future. Science (80-) 370:. https://doi.org/10.1126/science.aay3701

Tribovillard N, Algeo TJ, Lyons T, Riboulleau A (2006) Trace metals as paleoredox and paleoproductivity proxies: An update. Chem Geol 232:12–32. https://doi.org/10.1016/j.chemgeo.2006.02.012

Trouet V, Esper J, Graham NE, et al (2009) Medieval Climate Anomaly. Science (80-) 78:1256–1260. https://doi.org/10.1126/science.1166349

Tyson R V (1995) Sedimentary Organic Matter: Organic Facies and Palynofacies. Chapman and Hall, London

U.S. Geological Survey (2019) 3D Elevation Program 1-Meter Resolution Digital Elevation Model, accessed October 23, 2025. URL https://www.usgs.gov/the-national-map-data-delivery

Verma KK, Mathur UB (1979) Report on the studies on miliolite limestones and associated tertiary and quaternary deposits of Delvada coast,Junagarh district, Gujarat. Geological survey of India Western Region

Verma KK (1982) Geology of southern Saurashtra coastline with special reference to the miliolite problem. In: Merh SS (ed) Geology and geomorphology of quaternary terrains of Gujarat with special reference to coastline evolution. Department of Geology, M.S. University of Baroda, Vadodara

Walker M, Gibbard P, Head MJ, et al (2019) Formal Subdivision of the Holocene Series/Epoch: A Summary. J Geol Soc India 93:135–141. https://doi.org/10.1007/s12594-019-1141-9

Walker M, Head MJ, Berkelhammer M, et al (2018) Formal ratification of the subdivision of the Holocene Series/ Epoch (Quaternary System/Period): two new Global Boundary Stratotype Sections and Points (GSSPs) and three new stages/ subseries. Episodes 41:213–223. https://doi.org/10.18814/epiiugs/2018/018016

Wanner H, Beer J, Bütikofer J, et al (2008) Mid- to Late Holocene climate change: an overview. Quat Sci Rev 27:1791–1828. https://doi.org/10.1016/j.quascirev.2008.06.013

Wanner H, Pfister C, Neukom R (2022) The variable European Little Ice Age. Quat Sci Rev 287:. https://doi.org/10.1016/j.quascirev.2022.107531

Wei G, Liu Y, Li XH, et al (2004) Major and trace element variations of the sediments at ODP Site 1144, South China Sea, during the last 230 ka and their paleoclimate implications. Palaeogeogr Palaeoclimatol Palaeoecol 212:331–342. https://doi.org/10.1016/j.palaeo.2004.06.011

Yamamoto M (2009) Response of mid-latitude North Pacific surface temperatures to orbital forcing and linkage to the East Asian summer monsoon and tropical ocean-atmosphere interactions. J Quat Sci 24:836–847. https://doi.org/10.1002/jqs.1255

Zielinski GA, Mershon GR (1997) Paleoenvironmental implications of the insoluble microparticle record in the GISP2 (Greenland) ice core during the rapidly changing climate of the Pleistocene-Holocene transition. Geol Soc Am Bull 109:547–559. https://doi.org/10.1130/0016-7606(1997)109<0547:PIOTIM>2.3.CO;2

Chapter 14
Geothermal Record of Climate Change in India: A Review

Sukanta Roy and Vyasulu V. Akkiraju

1 Introduction

Studying past climates can help position the 20th-century global warming into a broader context of past centuries or millennia, better understand the climate system as a whole, and improve projections of climate change in the future (see, for example, National Research Council 2006 and references therein). Models developed and tested for future warming scenarios critically depend on how well they predict the past, taking into account the internal climate system variability, and using the known natural climate forcing such as solar variability, dust, and aerosols in the atmosphere from volcanic eruptions, for example. However, surface air temperature (SAT), which are the most definitive records of climate change, are most reliably known only for the period from 1850 to present for which there is a reasonable coverage of meteorological stations with systematic temperature measurements and on a global scale (Hansen et al. 2010, 2025). Extensive measurements started only around ~1900 A.D. in most parts of the world including India. A reliable SAT record can be reconstructed for North America up to 1880 only (Hansen and Lebedeff 1987). The average annual SAT during the period 1880–2024, compiled from measurements over land and sea reveal that temperature has increased by more than 1.2 °C relative to 1880–1920 mean SAT, with much of the warming occurring since 1975 A.D. On smaller, more recent timescales, observations reveal that the 1970–2010 average warming rate of 0.18 °C/decade has almost doubled during 2010–2023, and the most recent 10 years are the warmest on record (Hansen et al. 2025; www.climate.nasa.gov). The anomalous warming during the post-1850 period has been attributed primarily to anthropogenic forcing associated with anomalous increase in greenhouse gas concentrations in the atmosphere. However, the relatively short instrumental record of about 150 years is inadequate to derive a reliable long-term mean dating back to the pre-industrial era

S. Roy (✉) · V. V. Akkiraju
Ministry of Earth Sciences, Borehole Geophysics Research Laboratory, Karad, India
e-mail: sukanta.roy@nic.in

N. Khare (ed.), *Changing Climate over India*,
https://doi.org/10.1007/978-981-92-1098-5_14

over which the warming/cooling departures may be unambiguously interpreted and projected into the future. The short record is also inadequate to capture the internal, long-term climate system variability. Given other evidences for both cooler and warmer periods even during the past millennia, a much longer past climate record is essential to train climate models and predict future climate scenarios. Moreover, the globally averaged SAT record also suffers from varying observational procedures, changes in station location, increasing urban heat island effects (see, for example, Ellsaesser et al. 1986; Hansen and Lebedeff 1987; Karl and Quayle 1988). The warming signal we are looking for is small compared with the variations in temperature taking place from day to day and season to season. Individual meteorological stations and groups of stations exhibit large inter-annual variability as well as decadal or longer periods of both warming and cooling. Thus, the available SAT records are limited, both by duration and quality.

To study how the climatic conditions varied prior to the time of the Industrial Revolution (c.a. 1800 A.D.), a number of proxy evidences such as tree rings, corals, ocean and lake sediments, cave deposits, fossils, ice cores, borehole temperatures, and glacier length records have been used to reconstruct past temperatures with varying degrees of success. Succinct reviews of the different methods are given by Trotter et al. (2002), Cole (2003), Haug et al. (2003), Zolitschka (2003), Lauritzen (2003), Alley and Cuffey (2001), Lachenbruch and Marshall (1986), Oerlemans (2005). Knowledge of the physical, chemical, biological, and/or ecological processes is used to guide the sampling, analysis, and conversion of natural proxy data into surface temperature reconstructions. Most proxies rely largely on statistical techniques for calibration, i.e. to define the relationship between the proxy measurements and the concurrent instrumental temperature record, which in turn is used to reconstruct past temperature variations from the remaining proxy data. Because the individual proxy records possess both inherent advantages and shortcomings when compared to other proxies in terms of length of record, temporal resolution, and geographical coverage for example, studies of global temperature trends frequently consider multi-proxy reconstructions. Despite significant uncertainties, these methods provide robust estimates of past surface temperature changes and facilitate temperature reconstructions dating back to the past 1000 years (Moberg et al. 2005; Zorita and Storch 2005; von Storch 2004; Chapman and Davis 2010).

The present chapter deals with the geothermal signature of past climate change, specifically the record preserved in borehole temperature measurements, as a tool for studying the temperature changes during the past few centuries. We review the (i) method of sampling, measurements and data analysis, (ii) strengths and weaknesses, (iii) comparison with other proxy reconstructions, (iv) results of analyses using an extensive geothermal and meteorological database for peninsular India (8–28°N), a region that covers the low-latitude gap in global studies, and discuss the results in the larger context of future surface temperature projections.

2 Thermal Memory of the Earth

Surface air temperatures are highly sensitive to immediate weather conditions, such as solar radiation, cloud cover, and wind, which can cause rapid, sub-hourly changes at a measurement station. On the other hand, the ground's thermal properties, including thermal conductivity (or diffusivity) and heat capacity, dampen these fluctuations, resulting in less variable subsurface temperatures with increase in depth.Therefore, temperatures beneath the ground surface represent a systematic running mean of the recent temperature history at the surface (Lachenbruch and Marshall 1986). The Earth serves as a low-pass filter, progressively filtering out the high frequency surface noise, as the temperature change at the surface propagates downward with time by diffusion. Accordingly, the deeper the temperature measurement, the farther back in time is the interval of surface temperature history that is captured in the geothermal record. Thus, the ground 'remembers' the major events in its surface temperature history (Birch 1948), and careful temperature measurements made today to depths of a few hundred metres can potentially provide information on the local surface temperature history during the past few decades to a few centuries. The thermal memory of the Earth for typical surface temperature events is explained and illustrated succinctly in Fig. 1 of the seminal work of Lachenbruch and Marshall (1986).

The depth to which equilibrium temperatures are perturbed in a given time is governed by the thermal diffusivity (α) of the Earth medium. A simple rule of thumb is obtained from the solution of heat diffusion equation and is given by:

$$z = \sqrt{4\alpha t} \tag{1}$$

where z is the depth and t is the time. For typical rocks, with an average thermal diffusivity $\alpha = 32\ \mathrm{m^2 y^{-1}}$, a thermal front (e.g., a 5% change) propagates to about 15 m in one year, 160 m in 100 years, and 500 m in 1000 years. Therefore, the surface ground temperature (SGT) history of the Earth over the last millennium is captured in the uppermost 500 m of the crust. The Earth thus acts as a recorder of past surface temperature variations and these records can be effectively used for constraining temperature trends prior to the occurrence of instrumental meteorological records and in areas where there is a paucity of instrumentally recorded data.

3 Borehole Temperatures and Climate Change: Method of Analysis

Under ideal conditions, the subsurface temperature regime below 15–20 m depth (where the influence of diurnal and seasonal variations in ground temperature decay to negligible levels) is expected to exhibit a systematic linear increase in temperature with depth in accordance with Fourier's Law. The slope of this increase, referred to as the geothermal gradient, is controlled by upward heat flow from the Earth's

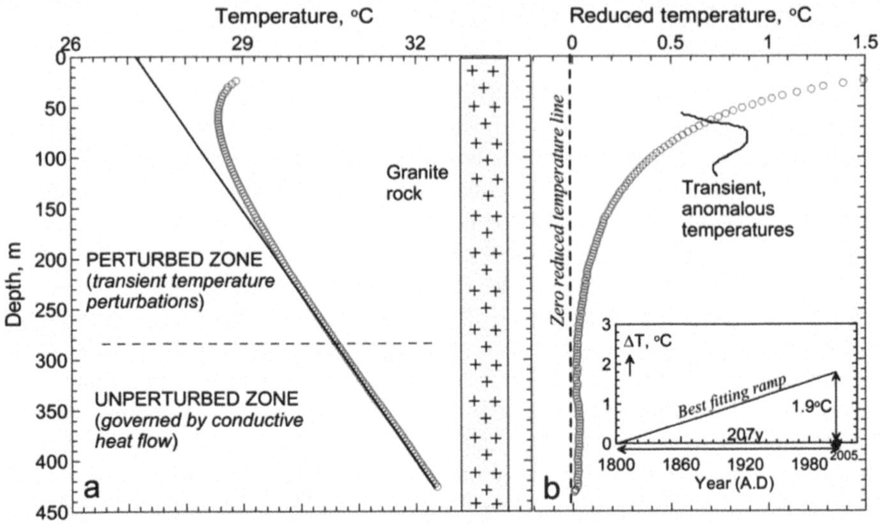

Fig. 1 Method of analysis. **a** Temperature-depth (t-z) profile (open circles) measured in a 430 m deep borehole passing through granitic rock in Pavagada, Karnataka in the year 2005 A.D (*Source of data*: Roy et al. 2008). Background thermal regime is shown by black solid line. The uppermost part of the profile, within ~20 m from the surface, is affected by the diurnal, annual, and seasonal surface temperature variations. The middle portion, down to ~280 m, represents the diffusion of decadal to century scale surface temperature variations. The deepest portion is least affected by the recent surface temperature variations and is used to determine a background thermal gradient and a long-term mean surface temperature ($T_o = 27.2\,°\mathrm{C}$). This portion is unaffected by transient perturbations, reflecting the steady-state heat flow from the interior of the Earth by conduction in accordance with Fourier's Law. b Reduced temperature profile (open circles) calculated by subtracting the background thermal regime from the measured temperatures. The transient anomalies are inverted to obtain the magnitude of temperature change (ΔT) and the duration time (t^*), as described in Sect. 3 of the text. The vertical dashed line corresponds to zero reduced temperature. The best-fitting ramp model ($\Delta T = 1.9\,°\mathrm{C}$; $t^* = 207$ years) is shown in the inset panel

interior and the thermal conductivity of the Earth materials through which heat is conducted. The intercept of the linear portion of the temperature-depth (t-z) profile corresponds to the present-day mean annual ground temperature at the surface. In reality, however, past changes in mean surface air temperature have caused corresponding changes in surface ground temperature. These temperature perturbations diffuse downward below the surface over time, which result in systematic departures from linearity in the observed t-z profile, generally extending to depths of about 150–300 m. The depth extent of the perturbations is controlled primarily by thermal diffusivity of the medium. The process is illustrated with a t-z profile measured in a 430 m deep borehole passing through granitic rock (Fig. 1a). The pioneering study in the detection of this phenomenon is that of Lachenbruch and Marshall (1986) in Alaska, where permafrost conditions and absence of ground water movement provided an ideal natural condition to detect this diffusion process. The study demonstrated that present-day borehole t-z measurements at carefully selected sites can preserve information on past surface ground temperature variations spanning the last several decades to a few centuries. Such geothermal records of past surface temperature changes are particularly valuable for constraining temperature trends prior to the advent of instrumental meteorological observations (that started only in the latter part of the nineteenth century and more extensively since ~1880 A.D.) and for regions where instrumental climate data are sparse or unavailable.

The method of analysis is described by previous researchers (e.g., Chisholm and Chapman 1992; Roy et al. 2002, Roy and Chapman 2012; Akkiraju et al. 2018a) and summarized here for completeness. The temperature distribution in the subsurface is governed by the outward conductive heat flow in accordance with Fourier's law and the temperature boundary condition at the surface. Changes in surface temperature with time diffuse downward below the surface, causing transient temperature perturbations to the t-z profile, $\Delta T(z, t)$, that can be represented by the one-dimensional heat diffusion equation (Carslaw and Jaeger 1959):

$$\frac{\partial^2 \Delta T(z,t)}{\partial z^2} = \frac{1}{\alpha}\frac{\partial T}{\partial t}, \tag{2}$$

where z represents the depth and α represents the thermal diffusivity of the earth medium. The transient temperature perturbations $\Delta T(z, t)$, which manifest as a curvature in the upper part of the temperature profile, can be interpreted in terms of temperature changes at the surface with time by solving Eq. (1) with appropriate initial and boundary conditions.

While carrying out the analysis of a t-z profile, the background thermal field is removed from the climatically perturbed field and the residual transient temperatures (also referred to as reduced temperatures), $\Delta T_R(z)$, are computed at each depth, using the relation:

$$\Delta T_R(z) = T(z) - \left[T_o + gz\right], \tag{3}$$

where T_o is the surface temperature intercept on the temperature-axis and g is the temperature gradient estimated through a least square regression of the t-z observations. The method of analysis is illustrated using a representative t-z profile obtained from borehole measurements in Pavagada area of Karnataka (Fig. 1a), with the corresponding reduced temperature profile shown in Fig. 1b. Inversion of the reduced temperature profile using a 'ramp' model of surface temperature change provides estimates of past changes in surface ground temperature. The method can be easily adapted for other models of surface temperature change, e.g. step and linear, where there are evidences (e.g., in land-use records) justifying such changes, by making small changes to the Eq. (3) as given in Lachenbruch and Marshall (1986). While the step model is used to represent scenarios such as rapid deforestation, the ramp model is generally preferred in most regions where *apriori* information is unavailable. This approach captures the first-order variations in the reduced temperature profiles and provides robust estimates of recent change in surface ground temperature.

The ramp model is parameterized in terms of a temperature change at a depth z, $\Delta T(z)$, and a duration time t^*, and is expressed following Lachenbruch and Marshall (1986):

$$\Delta T(z) = 4\Delta T i^2 erfc\left(\frac{z}{\sqrt{4\alpha t^*}}\right) \tag{4}$$

where i^2erfc is the second integral of the complimentary error function. The thermal diffusivity of the Earth medium, α, is determined using the relationship $\alpha = k/\rho c$, where k is thermal conductivity (Wm^{-1} K^{-1}), ρ is density (kg m^{-3}), and c is specific heat capacity (Jkg^{-1} K^{-1}). For consistency of the analysis, a uniform thermal diffusivity (α) of 10^{-6} m^2 s^{-1} (or 31.6 m^2/y) for rocks is assumed throughout the study.

It is expressly clarified that the curvature in the upper part of a t-z profile can be interpreted unambiguously in terms of past climate change only when other non-climatic perturbations such as hydrologic (moving groundwater), geologic (systematic thermal conductivity variations with depth), and sudden or rapid changes to surface conditions due to human interventions are minimal. Site selection criteria for heat flow measurements usually account for such effects (Roy and Rao 2000; Akkiraju and Roy 2011a).

4 Advantages and Shortcomings

The geothermal reconstruction of surface ground temperature history has advantages and shortcomings as a means of studying contemporary climate change (see, for example, Lachenbruch and Marshall 1986; Chisholm and Chapman 1992; Roy et al. 2002). Advantages include the following:

(i) The long-term signal of major events is preserved, whereas the high frequency noise is not. Unlike time series of surface air temperature measurements, which depend on a thermal environment that changes quickly and therefore contain large variations on all time scales, and sometimes suffer from the effects of changing measurement conditions, the Earth integrates the surface ground temperature continuously through day and night and through the seasons of the year, filtering out the high frequency noise and maintains only a running record of long-term mean (low frequency signal) and departures from it.

(ii) The measured anomaly is a direct thermophysical consequence of the thermal conditions at the ground surface, not a proxy. It does not rely on empirical relationships, for example between tree ring growth and temperature or isotopic fractionation and temperature.

(iii) The information on surface temperature history can be recovered where and when it is desired, without real-time monitoring, by drilling an observation borehole. The geothermal record extends farther back in time when compared to meteorological records that are available only for the past ~150 years.

The shortcomings of the method include:

(i) Observations require drilling of boreholes, ideally with continuous core for thermal conductivity measurements.
(ii) Reconstructions of surface temperature history suffer from the unavoidable smearing of the climatic signal caused by thermal diffusion into the Earth, which results in loss of temporal resolution with increase in depth. The magnitude of the surface temperature change (ΔT) is much better resolved compared to the onset time (t^*).
(iii) The relation between changes in surface ground temperature and the routinely measured climatological parameters (surface air temperature, humidity, rainfall, solar radiation, wind speed, and wind direction) under different surface and subsurface conditions is not fully understood.

In spite of the ambiguity of historical details, the geothermal data are robust indicators of the general time scale, the total heat absorbed by the Earth, and the total temperature increase or decrease associated with the surface temperature changes. The geothermal anomalies are sensitive to small departures from heat balance on the ground surface that cannot be detected otherwise such as by direct measurement of solar and other fluxes (Lachenbruch and Marshall 1986; Lachenbruch et al. 1988).

5 Case study—Peninsular India

Since the early 1960s, borehole temperature measurements have been conducted by the CSIR–National Geophysical Research Institute (NGRI) for determining heat flow and crustal thermal structure (see, for example, reviews by Gupta et al. 1995; Roy and Rao 2000; Rao et al. 2003; Roy 2008 and references therein), and more recently, by the Ministry of Earth Sciences—Borehole Geophysics Research Laboratory (MoES-BGRL), as part of the scientific drilling investigations in Koyna region of Maharashtra. The measurements were conducted primarily in boreholes of opportunity, drilled for exploration of minerals or groundwater, and sometimes for other specific purposes such as potential sites for waste repositories and other infrastructure projects, at several hundred sites in peninsular India. The sites are distributed over the major climatic provinces in peninsular India, which also offers a good opportunity to analyse the t-z profiles for studying potential records of past climate change captured in them. Moreover, climate change in peninsular India has been extensively studied by the India Meteorological Department (IMD). The region is covered by detailed meteorological SAT and other observations acquired at several tens of sites since the year 1900 (see, for example, Pramanik and Jagannathan 1954; Jagannathan and Parthasarathy 1972; Hingane et al. 1985; Sarker and Thapliyal 1988; Thapliyal 1990; Thapliyal and Kulshrestha 1991; Srivastava et al. 1992; Rupa Kumar et al. 1994). The region has been divided into six major climatic provinces, viz., interior peninsula (IP), east coast (EC), west coast (WC), north-central (NC), north-east (NE), and north-west (NW) (Hingane et al. 1986). The availability of extensive meteorological

SAT records in the region allows us to test the consistency of results obtained from geothermal records and make inter-comparisons between the two data sets over the common period of record. Accordingly, peninsular India is chosen as a test region to demonstrate the efficacy of the geothermal signature of climate change.

Temperature-depth profiles from 146 sites, distributed over the six major climatic provinces, have been analysed so far to infer the past climate change in peninsular India. Also, the geothermal records have been combined with the SAT records from 49 meteorological stations located in the vicinity of the borehole sites to obtain an estimate of pre-observational mean surface temperature prior to the onset of instrumental observations in the region. The spatial distribution of borehole sites and meteorological stations is shown in Fig. 2. A detailed account of the datasets and analysis is provided in Roy et al. (2002), Roy and Chapman (2012), Akkiraju et al. (2018a, b), Akkiraju (2024). The salient results obtained from those studies are summarized below.

T-z Measurements

Temperature measurements were conducted in boreholes under equilibrium conditions, i.e. after allowing sufficient time for the temperatures in the water column to reach equilibrium with the formation temperatures in the hole. Temperature data were collected at 3 m depth intervals by manually lowering a calibrated, 1 KΩ thermistor probe and measuring the transducer resistance to a precision of 0.1 Ω using a Wheatstone bridge. Since 2008, measurements have been carried out using a Hart Scientific CHUB E4 data logger. The temperature precision is typically ±3 mK, but comparisons of multiple and repeated temperature-depth logs in boreholes show that temperatures are generally reproducible to better than 20 mK (Roy and Rao 2000).

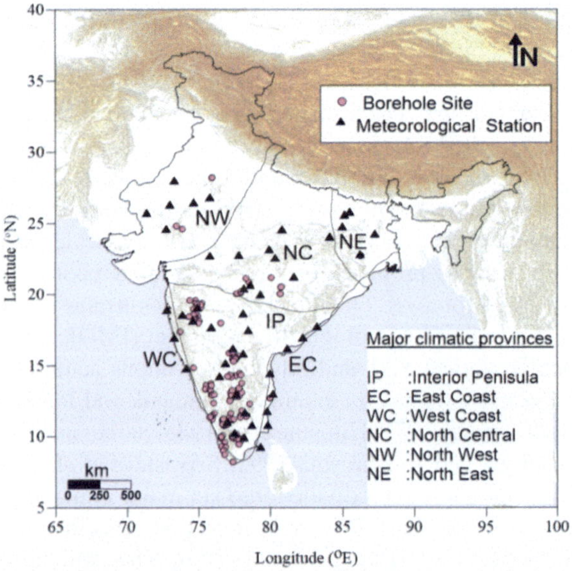

Fig. 2 Map of India showing 146 geothermal and 49 meteorological stations distributed among major climatic provinces in peninsular India. Geothermal stations refer to the borehole sites from where t-z profiles have been considered for analysis in the present work

Selection Criteria for t-z Profiles

As indicated in Sect. 3, the interpretation of past climate change from analysis of the curvature in t-z profiles depends on a careful selection of t-z profiles so that non-climatic sources of curvature are excluded. The following criteria were adopted to screen the borehole temperature database of peninsular India and select t-z profiles from 146 sites for our analyses (Roy et al. 2002): (1) boreholes pass through rock types that do not show systematic increase of decrease in thermal conductivity with depth; (2) boreholes are located in crystalline terrains away from mountains and valleys to minimize terrain effects and fluid flow related perturbations; and (3) t-z profiles do not show signatures of fracture-controlled groundwater flow in the boreholes. An additional requirement specific to geothermal climate change studies is the ability to distinguish the steady-state background thermal regime from the component affected by climatic perturbations. Accordingly, t-z logs extending to depths of at least 150 m were considered. The selected t-z profiles in the present study span a range of depths from 150 to 1522 m. A representative subset of t-z profiles used in the study is presented in Fig. 3. Among the six climatic provinces, the interior peninsula (IP) and west coast (WC) regions are more densely sampled than the others. Nevertheless, the full set of 146 temperature profiles constitutes a robust dataset covering the low-latitude range of 8°–28°N, a region that is otherwise sparsely represented in geothermal climate change investigations.

Analysis and Results

The selected t-z profiles were analysed following the methodology outlined in Sect. 3. Reduced temperature profiles were calculated by removing the background thermal regime from the observed temperature-depth datasets. The resulting transient anomalies were then inverted by considering a ramp model for surface temperature change and a duration time. As discussed earlier, this approach highlights first-order variations in the reduced temperature profiles and yields robust estimates of the most recent changes in surface ground temperature, typically the last events in a given area. Reduced temperature profiles thus computed for the 146 borehole sites are plotted together in Fig. 4. The plot resembling a 'wine glass' reveals certain important features. Most borehole sites in peninsular India reveal positive reduced temperatures (implying recent warming) that extend to depths of about 200 m. At a few sites, the anomalies extend deeper to about 300 m. In contrast, some borehole sites reveal negative reduced temperatures confined to the upper 100 m of the t-z profiles, implying recent cooling of surface ground temperature in certain localities. Following a simple rule of thumb ($t = z^2/4\alpha$), typical thermal diffusion time (t) associated with 200 m depth (z) is ~300 years. Overall, 128 boreholes (~88%) show positive reduced temperatures, indicating widespread ground surface warming across peninsular India over the past three centuries. Reduced temperature profiles from the remaining 18 sites, largely clustered in a few localities, reveal negative anomalies that imply local ground surface temperature cooling during the decade to a few decades preceding the year of t-z measurements. For sites exhibiting warming, the mean temperature change (ΔT) is 1.3° ± 0.1 °C at the 95% confidence level. The inferred duration

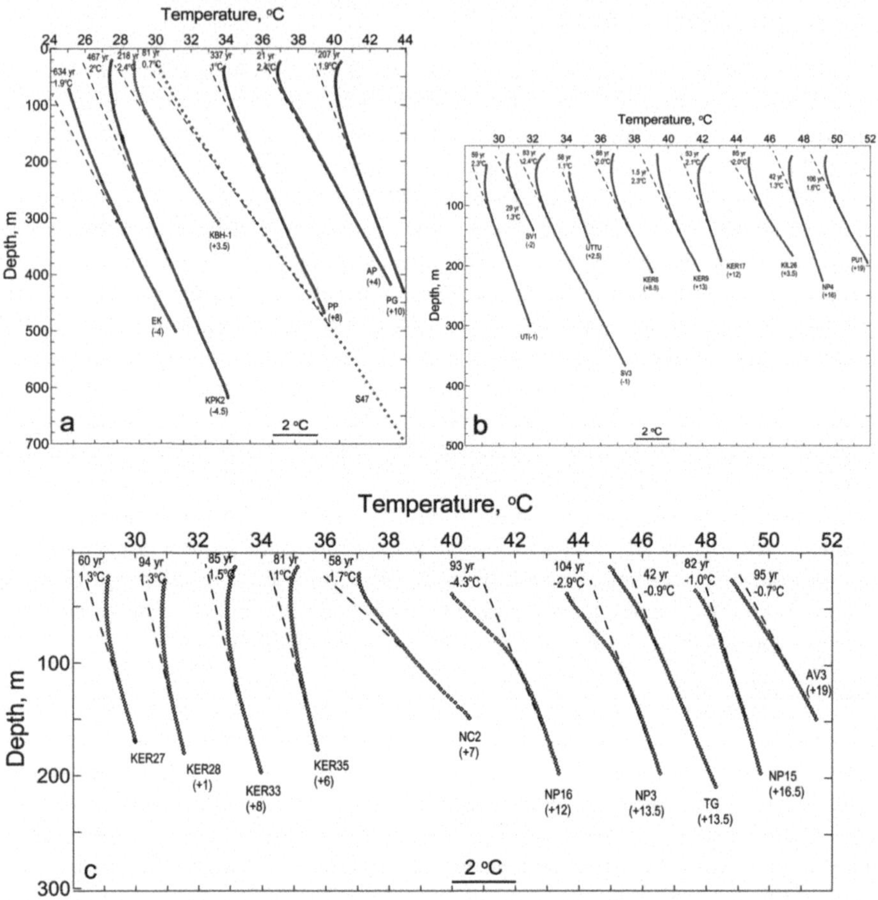

Fig. 3 **a–c** Representative t-z profiles from the Dharwar craton, Southern Granulite Terrain, Aravalli and Deccan Trap provinces in peninsular India showing the magnitude of temperature change (ΔT) and duration time for the change prior to the t-z data acquisition (t^*), as inferred from analysis in terms of a ramp model for the surface temperature change (*Source of data* Roy et al. 2002; Roy and Chapman 2012; Akkiraju 2024). Borehole identification numbers are indicated at the bottom of each profile, together with horizontal temperature offsets (in °C) to avoid overlap

of warming varies considerably among sites, ranging from a few decades to a few centuries, and shows greater variability than the amplitude estimates. On the other hand, the remaining 18 sites (reduced t-z profiles) show local cooling, with a mean temperature change of –1.2 ± 0.6 °C for duration of 84 ± 33 years at 95% level of confidence. The significant cooling during the recent past at many of these sites are likely associated with local site-specific reasons such as irrigation and land-use changes (Akkiraju and Roy 2011a; Roy and Chapman 2012). Overall, the inter-site variability is substantial, with temperature changes ranging between 3.6 °C to –4.3 °C, possibly reflecting differences in physiographic conditions and local microclimatic

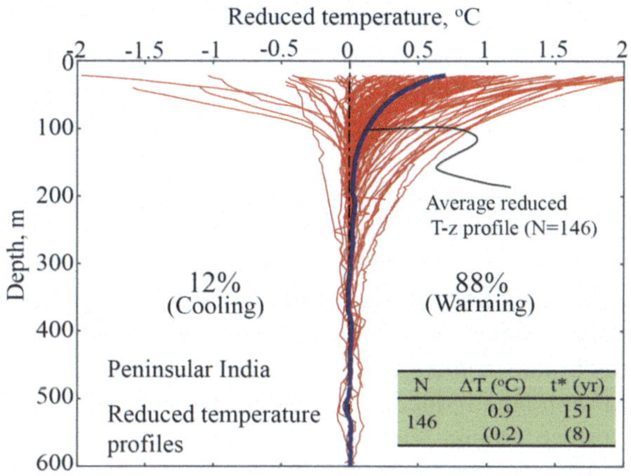

Fig. 4 Plot showing the reduced t-z profiles (red lines) at 146 sites in peninsular India. The mean reduced t-z profile computed by averaging the 146 profiles is also shown (thick blue line). Inversion of the mean reduced temperature profile reveals a total warming of 0.9 ± 0.2 °C that started 151 ± 8 years before the time of the t-z logs, i.e. c.a. 1850 A.D. Averaging the individual ΔT *and* t^* values for 146 individual sites yield a mean warming of 1.0 ± 0.2 °C and mean duration of 129 ± 18 years, not much different from the inversion of the mean reduced temperature profile

conditions. On average, the results for peninsular India yield a mean amplitude of surface temperature change 1.0 ± 0.2 °C and a mean duration time of 129 ± 18 years (95% confidence level), corresponding to onset of warming in the 1860s. Collectively, these results provide clear and compelling evidence of widespread ground surface warming across peninsular India over the past two centuries.

Discussion

Temperature-depth profiles measured in boreholes (depth range 150–1522 m) at 146 sites in peninsular India (latitude band 8–28°N) show a clear-cut curvature in the uppermost 100–200 m, in a few cases extending to ~300 m below the surface. In majority of the sites, the curvature is indicative of surface ground warming in the past one to two centuries. Only a limited number of sites display curvature towards negative temperature anomalies, reflecting localized cooling that likely began a decade, or at most a few decades, prior to the time of temperature log. The variability in reduced temperature profiles capture both natural climate variability and site-specific microclimate influences. Overall, the analyses of geothermal records covering the major climatic provinces in peninsular India reveal an average surface ground warming of about 1.0 ± 0.2 °C that started 129 ± 18 years prior to the time of the borehole temperature measurements. When all the 146 reduced temperature profiles are first averaged and the mean reduced temperature profile is inverted, the inferred surface ground temperature history corresponds to 0.9 ± 0.2 °C of warming over the past 151 ± 8 years, a result that is consistent with the previous estimate obtained by inverting individual t-z profiles and averaging the results. Therefore, the geothermal record of climate change is consistent with about 1 °C of surface ground warming that started c.a. 1850 A.D. across a large region in peninsular India. For comparison, the mean ground warming in the Northern Hemisphere mid-latitude region (30°–60°N), obtained by analysing 439 temperature profiles using a ramp model, was reported to be 0.8 ± 0.1 °C with onset c.a. 1800 A.D. (Harris and Chapman 2001). A subsequent analyses of 687 borehole t-z profiles covering substantial parts of the Northern Hemisphere, including some from India, yielded a similar estimate of ~0.8 °C warming that started c.a. 1840 A.D. (Harris and Chapman 2005). In this context, the Indian geothermal dataset, drawn from the low-latitude band of 8°–28° N, provides valuable complementary evidence from a region that is otherwise sparsely represented in borehole-based climate studies.

The analysis of borehole reduced temperature profiles provides an independent estimate of the surface ground temperature history with reference to the year of borehole t-z log. The t-z profiles in India were acquired from different climatic provinces and importantly, over a relatively long, 48-year period, from 1965 to 2013. The profiles acquired in the earlier part of this period may contain slightly different surface ground temperature history than those acquired in the later decades. While most studies have ignored this aspect, the study by Harris and Chapman (2005) addressed this issue in a limited way by forward continuing the profiles in time using a Laplace transform, assuming a constant SGT between the year of the individual T-z logs and the year of the last T-z log to obtain a consistent (albeit conservative) reduced temperature anomaly prior to the inversion. Forward continued t-z profiles using a linear trend derived from SAT records between the year of temperature logging and the last year of t-z log generally yields slightly larger warming as a result of the thermal diffusion process. Such corrections have not been applied to the analyses from India presented here.

The availability of meteorological SAT records at 49 stations located in the six climatic provinces of peninsular India where borehole sites are located allows us to compare the rate of surface temperature change obtained from the two data sets over the common period of record. The SAT records exhibit substantial inter-site variability, comparable to the borehole temperature dataset (Roy and Chapman 2012; Akkiraju 2024). The mean annual SAT anomalies computed relative to the 1961–1990 mean and averaged over the 49 meteorological stations yield a linear trend of 0.5 ± 0.2 °C/100 years (95% confidence limit for the mean). Despite large spatial variability, the average surface temperature trends derived independently from geothermal and SAT data are generally consistent, recognizing the inherent uncertainty in the onset time (t^*) compared to the more robust magnitude of temperature change (ΔT) obtained from the inversion of borehole temperatures. Importantly, the geothermal records extend farther back in time than the instrumental SAT observations. The slightly larger magnitude of recent warming inferred from geothermal data likely reflects the Earth's integration of surface temperature variations over a longer interval than covered by instrumental records. Overall, these results highlight the value of geothermal data in providing independent evidence for surface warming over the past few centuries, extending well beyond the start of systematic meteorological observations.

The close agreement between surface ground warming inferred from borehole data and surface air warming derived from SAT records in peninsular India suggests that ground temperatures effectively track changes in surface air temperature, consistent with earlier findings from North America (Chisholm and Chapman 1992; Harris and Chapman 1997). Under such conditions, geothermal and meteorological records can be combined to estimate a long-term baseline temperature predating instrumental observations. This approach was first demonstrated for regions such as Alaska and Utah (Lachenbruch and Marshall 1986; Chisholm and Chapman 1992; Harris and Chapman 1997) and later applied to peninsular India (Roy et al. 2002; Roy and Chapman 2012; Akkiraju et al. 2018b). In view of the different frequency and temporal contents of the two datasets, the SAT time series for 1900–2006 constructed by averaging records from the 49 IMD stations was used as a surface forcing function to generate a synthetic temperature-depth profile, which was then compared with the mean borehole temperature-depth profile derived from the 146 sites. This method requires an initial surface temperature prior to the onset of the SAT record. For a surface temperature history represented by n discrete step changes, each with amplitude ΔT_i occurring at time τ_i* prior to the borehole temperature measurement, the resulting transient temperature profile, $T_{RSAT(z)}$, at the time of observation is expressed following Carslaw and Jaeger (1959) (Fig. 5):

$$T_{RSAT} = \sum_{i=1}^{N} \Delta T_i erfc\left(\frac{z}{\sqrt{4\alpha\tau_i}}\right) \tag{5}$$

This formulation requires specification of an initial surface temperature preceding the start of the SAT time series. Harris and Chapman (1997) referred to this initial value as the pre-observational mean (POM) temperature. The POM represents a weighted average of surface temperatures prior to the onset of the instrumental SAT record at τ_1, the onset time corresponding to the first annual mean. Accordingly, the first step change in temperature, ΔT_1, can be written as:

$$\Delta T_1 = T_1 - \text{POM} \tag{6}$$

In practice, the POM temperature step is determined by finding the value that minimizes the misfit between the observed reduced temperature profile and the synthetic transient temperature profile derived from the SAT record (Harris and Chapman 1997). For peninsular India, the mean reduced temperature profile from boreholes, extending to a depth of 600 m, is compared with synthetic transient profiles generated using the average SAT time series and a range of POM values tested at 0.01 °C intervals in the range −5 to + 5 °C. The optimum fit corresponds to a POM that is 0.7 ± 0.1 °C lower than the 1961–1990 mean SAT, as illustrated in Fig. 5a,b. To demonstrate the sensitivity of the estimate, two additional POM–SAT models offset by ±0.2 °C from the best-fitting value (−0.7 °C) are also shown. The synthetic temperature profiles associated with POM values of −0.5 and −0.9 °C provide visibly poorer fits to the observations than the −0.7 °C model. RMS misfits for all POM-SAT models

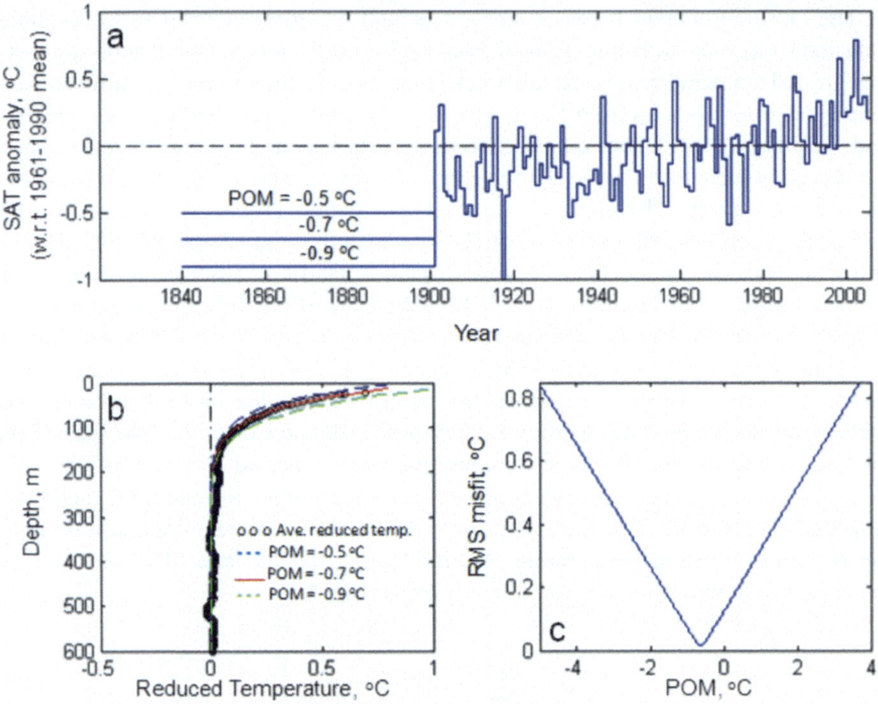

Fig. 5 Combined analysis of geothermal and meteorological SAT data for peninsular India. **a** Mean SAT anomalies from 49 weather stations data in peninsular India (staircase plot) relative to the 1961–1990 mean (dashed line). Horizontal lines illustrate three possible choices of pre-observational mean (POM) –0.5, 0.7, and 0.9 °C lower than the 1961–1990 mean SAT. **b** Synthetic (computed) t-z profiles for the three POM-SAT hybrid model scenarios shown in (**a**), along with the t-z profile obtained by averaging the set of 146 borehole reduced temperature profiles in peninsular India. The best-fitting POM-SAT model corresponds to a POM 0.7 °C lower than the 1961–1990 mean SAT. **c** The sharp trough in RMS misfit as a function of the POM suggests that the inferred POM from combined analyses of geothermal and meteorological SAT records is quite robust (Akkiraju et al. 2018b) (Reproduced with permission from Journal of Geological Society of India)

across the full POM range reveal a pronounced minimum, indicating that the POM estimate is robust (Fig. 5c). This robustness arises because only a single free parameter, the POM, is involved in the analysis. Moreover, if the observed temperature perturbations are not driven by surface temperature changes, the correlation between the SAT time series and reduced temperature profiles remains weak regardless of the chosen POM value.

Our findings are in close agreement with the estimate of about –0.7 °C for the Northern Hemisphere mid-latitudes reported in comparable studies. In addition, when referenced to the 1961–1990 mean temperature, annual SAT averages for peninsular India during the period 1995–2024 indicate substantial additional warming of

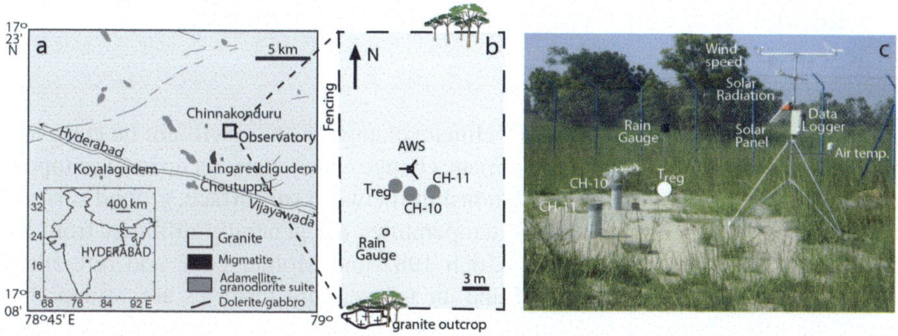

Fig. 6 Geothermal climate change observatory in Choutuppal area near Hyderabad, India. **a** Location map; **b**, **c** layout and photograph of the observatory showing the different components. AWS: automatic weather station with various sensors for air temperature, incident shortwave solar radiation, rain gauge, wind speed and direction, powered by a solar panel. T_{reg}: 1.2 m long, six sensor ground temperature probe inserted into the granite regolith. The datalogger of AWS records both meteorological parameters as well as ground temperatures to 1.19 m depth. CH-10 and CH-11 are 210 and 21 m deep boreholes, respectively (Akkiraju et al. 2024) (Reproduced with permission from Elsevier)

up to 0.5 °C. Consequently, the hybrid approach that integrates geothermal and meteorological records suggests a total surface warming of approximately 1.2 °C in peninsular India from the mid-1800s to the present. Despite the potentially large uncertainties associated with borehole temperature analyses, meteorological datasets, and their combined interpretation, the evidence clearly indicates that peninsular India has warmed by more than 1 °C relative to a baseline broadly representative of 19th-century mean temperatures. Such a robust conclusion could not be established using meteorological records alone, given the limited length of the instrumental SAT series beginning around 1900 AD.

The primary question about global warming concerns reliable surface temperature reconstructions over the past few centuries and the projections for global temperature change for the next century. A study by Chapman and Davis (2010) provides valuable insights to address the questions by integrating three different datasets: (a) 1000-year reconstructions of past climate change based on tree rings, corals, glacier lengths, and borehole temperatures, (b) the instrumental SAT record, and (c) Intergovernmental Panel on Climate Change (IPCC) projections for temperature change from 2000 to 2100 (four scenarios) from a consideration of multiple parameters. All of the scenarios reveal substantially higher (1–4 °C) global warming estimates in the twenty-first century compared to the warming observed in the past millennium. More recent IPCC assessments further reinforce these conclusions, underscoring the seriousness of projected warming and the emerging challenges for climate mitigation and adaptation in the coming decades.

6 Linkage Between Geothermal and Meteorological Records

A review of the geothermal record of climate change in India will not be complete without a mention of the geothermal climate change observatory set up in Choutuppal near Hyderabad to study the inter-relationships between subsurface, ground surface, and surface air temperature. Ground temperatures are generally different from air temperatures (see, for example, Goodrich 1982 and Molnar 2022 and references therein). Reconstructions of ground and air temperature histories are comparable only if the offset between SGT and SAT remains stable over the timescales relevant to climate change studies, i.e. several years to decades or longer (Mann and Schmidt 2003). To examine ground-air temperature coupling, several investigations have focussed on direct measurements of ground temperatures at sites co-located with meteorological stations (Putnam and Chapman 1996; Cermak et al. 2000, 2016; Beltrami 2001; Schmidt et al. 2001; Akkiraju and Roy 2011a, b). Such sites are referred to as geothermal climate change observatories in this chapter. One such observatory was set up at a relatively remote site located in CSIR-NGRI campus in Choutuppal area near Hyderabad (17.29°N, 78.92°E), southern India in July 2009. This is the only observatory of its kind in the low-latitude region of peninsular India, where the climate is influenced by factors such as solar forcing, monsoon rainfall, northern Indian Ocean sea surface temperature and moisture, air temperature and zonal winds that operate over a range of time scales. A comprehensive description of the instrumentation, datasets from the first five years of operation, and data analyses is provided in Akkiraju et al. (2024). The key observations and results are summarized below.

The observatory consists of (i) automatic weather station that records surface air temperature (SAT), solar radiation, rainfall, humidity, wind speed, and wind direction, (ii) two boreholes CH-10 and CH-11 drilled into granite to depths of 21 and 210 m, and (iii) 1.2 m long ground temperature probe with temperature sensors embedded at 1.5, 9 19, 49, 99, and 119 cm depth, all located at the same site and within 3 m distance of one another. Location of the Choutuppal geothermal observatory along with layout and photograph of the site showing different components is given in Fig. 6.

The data sets for SAT, SGT, SGT-SAT, rainfall, and solar radiation acquired during the first five years of operation (2009–2014) are plotted in Fig. 7a–d. The data are relatively continuous, with only minor data gaps. Observations reveal that air temperature variations diffuse downward into the granitic subsurface with associated attenuation of high frequency components and phase lag characteristic of conductive heat transfer. Ground temperatures effectively track daily and seasonal SAT variations; diurnal, and the annual/seasonal changes of surface temperature decay to insignificant levels beyond 1 and 15 m depths, respectively. Overall, the ground is systematically warmer than the air, with the mean ground-air temperature difference ($\Delta T_{SGT-SAT}$) averaging 3.4 ± 1.8 °C but varying seasonally from -2.1 °C to $+7.6$ °C. The $\Delta T_{SGT-SAT}$ offset is relatively constant from winter to summer but decreases and

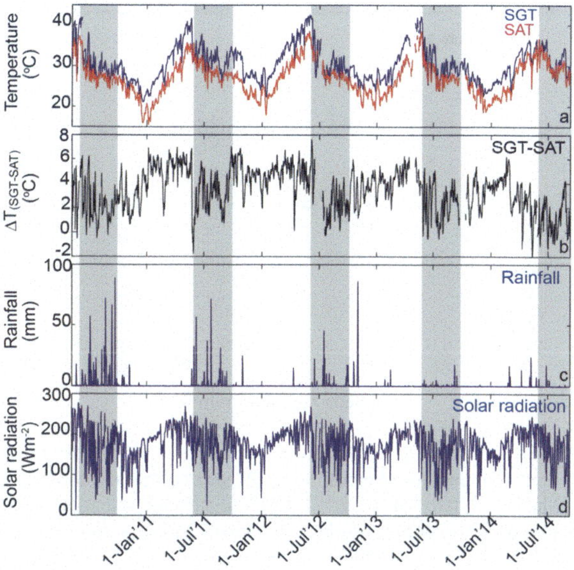

Fig. 7 Plots of surface meteorological data recorded at the geothermal climate change observatory in Choutuppal during a ~5 years period (26 July 2009 to 10 September 2014): **a** Surface air temperature at 1.25 m height above the ground surface, **b** relative humidity, **c** incoming shortwave solar radiation, (d) rainfall, **e** wind speed, and **f** wind direction (degrees from North). Data gaps are minimal and shown as breaks in the time series. Each panel (except rainfall) contains daily averages calculated from 30 min averages based on 2-s samples of each parameter. Rainfall displayed in panel (**d**) represents the total for each day. The monsoon months (June through September) are shaded grey (Akkiraju et al. 2024) (Reproduced with permission from Elsevier)

shows larger variability during monsoon season in each year. Thus, observations reveal that the SAT and SGT appear to track each other over annual to multi-annual cycles but with significant time varying offsets.

The data further reveal substantial inter-day variability in solar radiation, largely driven by daily weather phenomenon. The mean annual solar radiation at incident at the site varies from 172 to 196 Wm^{-2}. A linear correlation between weekly ground-air temperature difference $\Delta T_{SGT\text{-}SAT}$ and incident solar radiation during the period 2009–2014 indicates that on average, the ground warms by 2.5 °C for every 100 Wm^{-2} incident solar radiation at the Choutuppal observatory site. This relationship highlights the critical role of local surface energy balance in governing ground-air temperature tracking (Fig. 8). Consequently, much of the inter-annual variability in ground-air temperature difference at the Choutuppal site can be attributed to variations in incident solar radiation.

The study underway in Choutuppal geothermal climate change observatory is complementary to the ground-air temperature studies carried out at a number of sites in the mid- to high-latitude regions: northwest Utah, USA (Putnam and Chapman 1996; Bartlett et al. 2006; Davis et al. 2010), Prague, Czech Republic (Cermak et al. 2014, 2016), Fargo, North Dakota, USA (Schmidt et al. 2001; Smerdon et al. 2004);

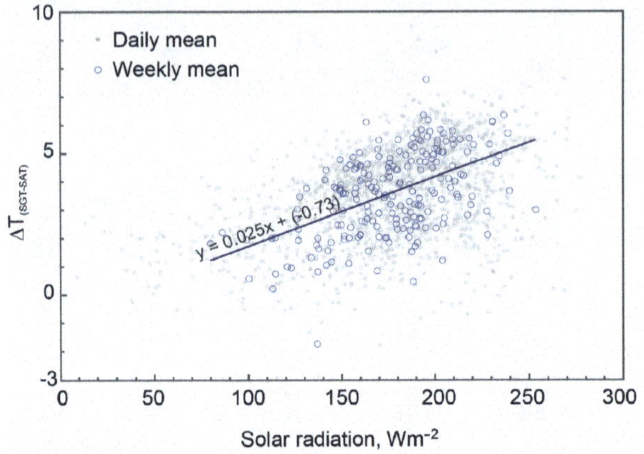

Fig. 8 Plot showing the relationship between the ground-air temperature difference $\Delta T_{(\mathrm{SGT\text{-}SAT})}$ and incident shortwave solar radiation measured at the geothermal climate change observatory in Choutuppal during May 5, 2010 to Sep 10, 2014. Daily mean solar radiation (open grey circles) and weekly mean solar radiation (open blue circles) are shown. A linear fit through the weekly means (dark blue line) suggests that the ground becomes, on average, 2.5 °C warmer than air for every 100 Wm^{-2} increase in solar radiation (Modified after Akkiraju et al. 2024)

Pomquet, Nova Scotia, Canada (Beltrami 2001) that are located in quite different climatic settings (arid, desert, snow cover, permafrost) and ground surface conditions (granite, soil, regolith, asphalt). A comparative study of these stations is given in Akkiraju et al. (2024). When compared to similar observatories in mid- to high-latitude regions, the observations at the Choutuppal site extend the global observational coverage to conditions characterized by the highest mean annual surface air temperatures and strong monsoon influence. Also, the Choutuppal site in peninsular India avoids complications such as snow cover and variable insolation that are common in higher latitude regions in the Northern Hemisphere. Therefore, the observations over the five year period (2009–2014) at Choutuppal in peninsular India support the reliability of using borehole ground temperature profiles to reconstruct the surface ground temperature history, while also emphasizing the need to account for local surface factors and seasonal variability. Extension of similar studies to other climatic provinces in the low latitude region would provide valuable datasets to strengthen the geothermics of climate change.

7 Conclusions

(a) Analysis of borehole temperature-depth profiles at 146 sites distributed over six climatic provinces in peninsular India, namely, Interior Peninsula, West Coast, East Coast, North Central, North West, and North East, provide unequivocal evidence of surface ground temperature (SGT) warming in a large region of peninsular India during the past two centuries. On average, the magnitude of warming over the region is 1.0 ± 0.2 °C corresponding to onset *ca* 1850 AD, considering a ramp model for surface temperature change. For comparison, analysis of the mean reduced temperature-depth profile, obtained by averaging all the 146 reduced temperature profiles, yields a SGT history of 0.9 ± 0.2 °C during the past 151 ± 8 years, which is consistent with the former estimate obtained by inverting individual t-z profiles.

(b) SGT reconstructions from geothermal records have decadal- to century-scale resolution and thus constrain the long-term temperature trends, but not the annual or sub-annual scale variations.

(c) The rate of warming is generally consistent with the linear trend of 0.5 ± 0.2 °C/ 100 y (at 95% confidence level for the mean) obtained from the mean annual SAT data for the period 1901–2006 at 49 stations of the India Meteorological Department in peninsular India.

(d) A combined analysis of the geothermal and meteorological SAT datasets offers valuable insights into the 19th-century baseline to the 20th-century surface air temperature change in peninsular India. The study reveals that the long-term mean temperature for the nineteenth century was ~0.7 °C lower than the 1961–1990 mean temperature. This is a robust result that compares favourably with the estimate of −0.7 °C for the Northern Hemisphere mid-latitude region from similar studies but using much larger datasets. Furthermore, relative to the 1961–1990 mean temperature, SAT annual averages for peninsular India from 1995 to 2024 imply significant additional warming of up to 0.5 °C relative to the 1961–1990 mean. Thus, the hybrid model constrained by geothermal and meteorological records suggests that the total surface warming in peninsular India from the mid-1800s to the present time is about 1.2 °C. Despite the potential uncertainties inherent in trends derived from geothermal datasets, meteorological SAT records, and their combined analysis, the evidence clearly indicates that peninsular India has undergone warming exceeding 1 °C relative to a baseline broadly representative of 19th-century mean temperatures.

(e) The efficacy of the geothermal method to infer past climate change hinges upon the extent to which the air and ground temperatures track each other. Co-located ground and air temperature datasets recorded during 2009–2014 at the Geothermal Climate Change Observatory at Choutuppal (near Hyderabad, India) reveal that the ground temperatures track the diurnal and seasonal changes in SAT. Over annual to multi-annual cycles, SAT and SGT appear to track each other but with important time varying offsets ranging from −2.1 to +7.6 °C with a mean value of 3.4 ± 1.8 (s.d.) °C, predominantly governed by incident solar radiation, during the period of study.

(f) The analyses of geothermal and meteorological records, together with the results of ground-air temperature tracking experiment in peninsular India, demonstrate the potential of the geothermal method to constrain climate change in the past few centuries relevant to post-industrial revolution anomalous greenhouse warming, and complement the information available from proxy records of climate change such as tree rings, sediment cores, ice cores, and corals towards surface temperature reconstructions for the past millennium and projections for temperature change in the twenty-first century.

Acknowledgements We are grateful to Secretary, Ministry of Earth Sciences for continued support and encouragement to pursue the study at BGRL. The review is largely based on the work carried out during our tenure at CSIR-National Geophysical Research Institute (NGRI), which was continued at BGRL. The borehole temperature data was collected and curated by members of the Heat Flow Studies group at CSIR-NGRI during the period 1962–2014. We acknowledge the financial support from the Department of Science and Technology, Government of India which was crucial for setting up of the Geothermal Climate Change Observatory in CSIR-NGRI campus at Choutuppal in 2009. The surface air temperature data used in the study was obtained from the India Meteorological Department. SR acknowledges the detailed interactions at various stages of the work on Geothermics of Climate Change during 1995-2024 with R.U.M. Rao, David S. Chapman (deceased), Harsh K. Gupta, V.P. Dimri, Vladimir Cermak, Henry Pollack, Robert N. Harris, Derrick Hasterok, and Mike Davis among others. We thank Neloy Khare Editor for the invitation to write this chapter, constructive suggestions, and patient handling of the manuscript.

References

Akkiraju, V.V., 2024. Geothermal and Meteorological records of Climate Change in Peninsular India. *Ph.D. Thesis (unpublished)*, Indian Institute of Technology (Indian School of Mines), Dhanbad, India, 216 p.

Akkiraju, V.V., Roy, S., 2011a. Geothermal Climate Change Observatory in South India 1: Borehole temperatures and inferred surface temperature histories. *Physics and Chemistry of the Earth*, 36, 1419-1427. https://doi.org/10.1016/j.pce.2011.01.004

Akkiraju, V.V., Roy, S., 2011b. Geothermal Climate Change Observatory in South India 2: Setup and First Results. *Physics and Chemistry of the Earth*, 36, 1428–1436. https://doi.org/10.1016/j.pce.2011.01.011

Akkiraju, V.V., Roy, S., Shalivahan, 2018a. Ground surface warming in peninsular India: evidence from geothermal records. *Current Science*, 115, No. 8, pp 1567–1571. https://doi.org/10.18520/cs/v115/i8/1567-1571

Akkiraju, V.V., Roy, S., Tiwari. V.M., Shalivahan, 2018b. A 19th Century baseline temperature estimate for peninsular India from combined analysis of geothermal and meteorological records. *Journal of Geological Society of India*, 92, Issue 5, pp. 542–547. https://doi.org/10.1007/s12594-018-1064-x

Akkiraju, V.V., Roy, S., Chapman, D.S., 2024. Ground-air temperature tracking from a geothermal climate-change observatory in South India. *Tectonophysics*, 870, https://https://doi.org/10.1016/j.tecto.2023.230154

Alley, R.B, Cuffey, K.M., 2001. Oxygen- and hydrogen-isotopic ratios of water in precipitation: beyond paleothermometry. *Reviews in Mineralogy and Geochemistry* 43, 527–553.

Bartlett, M.G., Chapman, D.S., Harris R.N., 2006. A decade of ground-air temperature tracking at emigrant pass observatory, Utah, *J. Climate*, 19, 3722–3731.

Beltrami, H., 2001. On the relationship between ground temperature histories and meteorological records: a report on the Pomquet station, *Global Planet. Change*, 29, 327–348.

Birch, F., 1948. The effects of Pleistocene climatic variations upon geothermal gradients, *Am. J. Sci.*, 246, 729–760.

Carslaw, H.S., Jaeger, J.C., 1959. Conduction of Heat in Solids, second ed. *Oxford University Press*, New York. 386 p.

Cermak, V., Safanda, J., Kresl, M., Dedecek, P., Bodri, L., 2000. Recent climate warming: surface air temperature series and geothermal evidence. *Studia Geophysica et Geodaetica* 44, 430–441.

Cermak, V., Dedecek, P., Safanda, J., Kresl, M., Dedecek, P., 2014. Ground-air temperature tracking and multi-year cycles in the subsurface temperature time series at geothermal climate-change observatory. *Studia Geophysica et Geodaetica*, 58, 403–424.

Cermak, V., Bodri, L., Kresl, M., Dedecek, P., Safanda, J., 2016. Eleven years of ground-air temperature tracking over different land cover types, Int. J. Climatology, 37, 1084–1099, https://doi.org/10.1002/joc.4764.

Chapman, D.S., Davis, M.G., 2010. Climate Change: Past, Present, and Future. *EOS*, 91(37), 325–326.

Chisholm, T.J., Chapman, D.S., 1992. Climate change inferred from analysis of borehole temperatures: An example from Western Utah. *J. Geophys. Res.*, 97, 14155–14176.

Cole, J.E. 2003. Holocene coral records: windows on tropical climate variability. In: *Global Change in the Holocene.* pp. 168–184 A. McKay, R. Battarbee, J. Birks, and F. Oldfield (eds.). Arnold, London.

Davis, M.G., Harris, R.N., Chapman, D.S., 2010. Repeat temperature measurements in boreholes from northwestern Utah link ground and air temperature changes at the decadal time scale. *J. Geophys. Res.*, 115, B05203, https://doi.org/10.1029/2009JB006875.

Ellsaesser, H. W., MacCracken, M. C., Walton, J. J., Grotch, S. L., 1986. Global climatic trends as revealed by the recorded data, Reviews of Geophysics, 24, 745–792.

Goodrich, L.E., 1982. The influence of snow cover on the ground thermal regime. *Can. Geotech. J.*, 19, 421–432.

Gupta, M. L., 1995. Thermal regime of the Indian shield, in Terrestrial Heat Flow and Geothermal Energy in Asia, edited by M. L. Gupta and M. Yamano, Oxford and IBH, New Delhi, 63–81.

Hansen, J., Lebedeff, S., 1987. Global trends of measured surface air temperatures, *J. Geophys. Res.*, 92, 13345–13372.

Hansen, J., Ruedy. R., Sato. M., Lo.K., 2010. Global surface temperature change. *Rev. Geophys.*, 48, RG4004, https://doi.org/10.1029/2010RG000345

Hansen, J.E., Kharecha, P., Sato, M., Tselioudis, G., Kelly, J., Bauer, S.E., Ruedy, R. Jeong, E., Jin, Q., Rignot, E., Velicogna, I., Schoeberl, M.R., von Schuckmann, K., Amponsem, J., Cao, J., Keskinen, A., Li, J., Pokela, A., 2025. Global warming has accelerated: Are the United Nations and the public well-informed? *Environ. Sci. Pol. Sustain. Devel.*, 67, no. 1, 6–44, https://doi.org/10.1080/00139157.2025.2434494.

Harris, R.N., Chapman, D.S., 1997. Borehole temperatures and a baseline for 20th-century global warming estimates. *Science*, 275, 1618–1621.

Harris, R.N., Chapman, D.S., 2001. Mid-latitude (30°–60° N) climatic warming inferred by combining borehole temperatures with surface air temperatures. *Geophys. Res. Lett.* 28, 747–750.

Harris, R.N., Chapman, D.S., 2005. Borehole temperatures and tree rings: Seasonality and estimates of extra-tropical Northern Hemispheric warming, J. Geophys. Res. 110, F04003, 10/1029/2005JF000303.

Haug, G.H., Gunther, D., Peterson, L.C., Sigman, D.M., Hughen, K.A., Aechlimann. B., 2003. Climate and the collapse of Maya civilization. *Science,* 299, 1731–1735.

Hingane, L.S., Rupakumar, K., Ramamurthy, B.V., 1985. Long term trends of surface air temperature in India, *J. Climatol.*, 5, 521–528.

Jagannathan, P., Parthasarathy, B., 1972. Fluctuations in the seasonal oscillation of temperature in India, *Indian J. Meteorol. Geophys.*, 23, 15–22.

Karl, T.R., Quayle, R.G., 1988. Climate change in fact and in theory: Are we collecting the facts? Climate change. 13, 5–17. https://doi.org/10.1007/BF00140159

Lachenbruch, A.H., Marshall, B.V., 1986. Changing climate: Geothermal evidence from permafrost in the Alaskan Arctic, *Science*, 234, 689–696.

Lachenbruch, A. H., Cladouhos, T. T., Saltus, R. W., 1988. Permafrost temperature and the changing climate, in Proc., 5th International Conference on Permafrost, *Tapir, Trondheimx*, pp. 1535–1543.

Lauritzen, S.E., 2003. Reconstructing Holocene climate records from speleothems. In: *Global Change in the Holocene*, 242–263. A. Mackay, R. Battarbee, J. Birks, and F. Oldfield (eds.). Arnold, London.

Mann, M.E., Schmidt, G.A., 2003. Ground vs. surface air temperature trends: implications for borehole surface temperature reconstructions. *Geophys. Res. Lett.*, 30(12), 1607. https://doi.org/10.1029/2003GL017170

Moberg, A., Sonechkin, D.M., Holmgren, K., Datsenko, N.M., Karlen.W., 2005b. Highly variable Northern Hemisphere temperatures reconstructed from low- and high-resolution proxy data. *Nature* 433, 613–617.

Molnar, P., 2022. Differences between soil and air temperatures: Implications for geological reconstructions of past climate, *Geosphere* 18 (2), 800–824. https://doi.org/10.1130/GES02448.1

National Research Council, 2006. Surface temperature reconstructions for the last 2000 years. In: Report of the Committee on surface temperature reconstructions for the last 2000 years, Board on Atmospheric Sciences and climate, Division on Earth and Life Studies, National Research Council of the National Academies. The National Academies Press, Washington, D.C., USA, 145 p.

Oerlemans, J., 2005a. Extracting a climate signal from 169 glacier records. Supporting online material available:http://www.sciencemag.org/cgi/content/full/1107046/DC1. *Science* 308, 675–677.

Pramanik, S. K., Jagannathan, P., 1954. Climatic changes in India, II Temperature, *Indian J. Meteorol. Geophys.,* 5, 29–47.

Putnam, S.N., Chapman, D.S., 1996. A geothermal climate change observatory: First year results from Emigrant Pass in northwest Utah, *J. Geophys. Res.* 101, 21877–21890.

Rao, R.U.M., Roy, S. Srinivasan, R., 2003. Heat-flow researches in India: results and perspectives. In: Mahadevan, T.M., Arora, B.R. and Gupta, K.R., Eds., Indian Continental Lithosphere, Emerging Research Trends. *Mem. Geol. Soc. India*, 53, 347–391.

Roy, S, 2008. Heat Flow Studies in India during the past five decades. *Geol. Soc. India Memoir*, v.68, pp. 89–112.

Roy, S., Chapman, D.S., 2012. Borehole temperatures and climate change: Ground temperature change in south India over the past two centuries. *J. Geophys. Res.* 117, https://doi.org/10.1029/2011JD017224.

Roy, S., Rao, R.U.M., 2000. Heat flow in the Indian shield. *J. Geophys. Res.* 105, 25587–25604.

Roy, S., Harris, R,N., Rao, R.U.M., Chapman, D.S., 2002. Climate change in India inferred from geothermal observations. J. Geophys. Res. 107. https://doi.org/10.1029/2001JB000536

Roy, S., Ray, L., Bhattacharya, A., Srinivasan, R., 2008. Heat flow and crustal thermal structure in the late Archean Closepet granite batholiths, south India, *Int. J. Earth Sci.*, https://doi.org/10.1007/s00531-007-0239-2.

Rupa Kumar, K., Krishna Kumar, K., Pant, G.B., 1994. Diurnal asymmetry of surface temperature trends over India, *Geophys. Res. Lett.*, 21, 677–680.

Sarker, R. P., Thapliyal, V., 1988.Climate change and variability, *Mausam*, 39, 127–138.

Schmidt, W.L., Gosnold, W.D., Enz, J.W., 2001. A decade of air-ground temperature exchange from Fargo, North Dakota, *Global Planet. Change*, 29, 311–325.

Smerdon, J.E., Pollack,H.N., Cermak, V., Enz, J.W., Kresl, M., Safanda, J., Wehmiller, J.F., 2004. Air-ground temperature coupling and subsurface propagation of annual temperature signals, *J. Geophys. Res.*, 109, https://doi.org/10.1029/2004JD005056.

Srivastava, H. N., Dewan, B. N., Dikshit, S. K., Prakash Rao, G. S., Singh, S. S., Rao, K. R., 1992. Decadal trends in climate over India, *Mausam,* 43, 7–20.

Thapliyal, V., 1990. Perspective of climate change in India, Report of the Expert Meeting on Climate Change Detection Project, Toronto, Niagara-on-the-Lake, 26–30 Nov., World Meteorol.Org., Geneva.

Thapliyal, V., Kulshrestha, S. M., 1991. Climate changes and trends over India, *Mausam,* 42, 333–338.

Trotter, R.T.I., Cobb, N.S., Whitham, T.G., 2002. Herbivory, plant resistance, and climate in the tree ring record: interactions distort climatic reconstructions. *Proceedings of the National Academy of Sciences* 99, 15, 10197–10202.

von Storch, H., Zorita, E., Jones, J., Dimitriev, Y., Gonzalez-Rouco, F., Tett. S., 2004. Reconstructing past climate from noisy data. *Science* 306, 679–682.

Zolitschka, B., 2003. Dating based on freshwater- and marine-laminated sediments. In: *Global Change in the Holocene*. pp. 92–106. A. Mackay et al. (eds.). Arnold, London.

Zorita, E., von Storch. H., 2005. Methodical aspects of reconstructing non-local historical temperatures. *Memoriedella Società Astronomica Italia* 76, 794–801.

Chapter 15
Climate Regime During Glauconite Formation in Adhaura Block, Rohtas District, Bihar: Its Search as a Potash Fertilizer

Avinash Niranjan, Sandhya Verma, Akhouri Bishwapriya, Janardan Prasad, and Ranbir Nandan

1 Introduction

In recent times, the world's population has increased, mostly in the developing countries. The consumption of fertilizers in the country has risen in recent years. To meet the growing demand, production of fertilizer needs to be enhanced. Potassium (K) is one of the key chemical elements that is used in fertilizers. Potassium is obtained from potash minerals (Sylvite, sylvinite, Langbeinite, Carnallite, Polyhalite, Glauconite, etc.) and associated marine deposits.

Glauconite is a group of dioctahedral, potassium, iron-rich, compositionally heterogeneous, phyllosilicate (clay) mineral. The general chemical composition of Glauconite is $(K, Na) (Fe, Al, Mg)_2 (Si, Al)4O_{10}(OH)_2$ (Odin and Matter 1981). Term Glauconite was named in 1828 by Christian Keferstein from the Greek 'glaukos', blue-green, in allusion to its colour; but the term 'glauconite' was used earlier, in 1823 by Alexandre Brongniart, as a name for the English term 'green sand' (Source: https://www.mindat.org/min-1710.html#autoanchor1).

A. Niranjan
Geological Survey of India, SU: Arunachal Pradesh, Itanagar, Arunachal Pradesh, India

S. Verma
Geological Survey of India, State Unit Bihar, Patna, India

A. Bishwapriya (✉)
Geological Survey of India, SU: Jharkhand, Ranchi, India
e-mail: akhouri.bishwapriya@gsi.gov.in

J. Prasad
Geological Survey of India, Kolkata, India

R. Nandan
Department (Geology), B. N. College, Patna University, Patna, India

N. Khare (ed.), *Changing Climate over India*,
https://doi.org/10.1007/978-981-92-1098-5_15

Glauconite is found commonly associated with sand/sandstone, shale, marl, and occasionally with limestone in the rock formations ranging in age from Precambrian (Bhima sediments of Karnataka and Andhra Pradesh; Vindhyans of U.P., M.P., Rajasthan, and Bihar) through Cretaceous and Eocene (Ladakh, Garhwal, Rajasthan and Assam) to Recent off-shore sediments off the Trivandrum coast, Bay of Bengal and coasts of Andaman Nicobar Island. The glauconite deposits of U.P, M.P., Rajasthan, Gujarat, A. P., and Karnataka are well developed. The Glauconite Sandstone of (Chorhat Sandstone, Semri Group) has been dated 1504–1409 Ma by Rb–Sr (Kumar et al. 2001).

In Rohtas district, Bihar, the rocks of Vindhyan Supergroup are represented by Kaimur Group and Semri Group (Bose et al. 2015; Mandal et al. 2019). Glauconite minerals are associated with the rocks of Basuhari Formation of Semri (Delamette 1989; Garzanti et al. 1989; Jimenez-Millan et al. 1998; Amorosi 2012). The objective of this study was to investigate different zones of glauconite mineralization to check its potentiality as Fertilizer minerals within glauconitic sandstone of Basuhari Formation of Semri Group. The study highlights stages of investigation, delineation of mineralized zones, laboratory-based petrographic studies, and ore beneficiation for its uses as fertilizer minerals.

a. **Study Area and Geological Setting**

The investigation for fertilizer mineral (glauconite) was carried out to the South of Adhaura area, Rohtas district, Bihar (Fig. 1) in toposheet no 63 P/14 (latitudes N24°30′50″ to N24°36′20″ and longitudes E83°56′25″ to E83°45′00″) and toposheet no 63 P/10 (latitudes N24°33′26″ to N24°30′00″ and longitudes E83°44′59″ to E 83°31′28″). The main localities are Nawara, Tipa, Nauhatta, Tiura, Chutia, Katudanr, Paruka, Parchha, Jadunathpur, Belduria, and Jaradag area.

The investigation area lies on the eastern extremity of the axe shaped Upper Proterozoic (Algonkian) Vindhyan sedimentary basin, which extends from Sasaram to Hosangabad and from Chittorgarh to Agra and Gwalior, surrounding the batholitic mass of the Bundelkhand Granite Massif. The detailed stratigraphic succession of the Son valley Vindhyan is due to Auden (1933, 1940), who took into account the observation made by Mallet (1869) and Oldham et al. (1901). However, Auden's account does not include the central and western Son valley region (Bihar area).

The Vindhyan Supergroup comprises lithologies mainly of shale, slate, carbonate, and sandstone. The Semri Group comprising Rohinian sandstone of Rohinian Formation, porcellanite, and shales of Chopan Formation, olive shale, and khaki splintery shale of Kheinjua Formation, cherty stromatolitic limestone of Bargwan Formation, quartz arenite, sandstone with glauconite pellets, fine- to medium-grained glauconitic sandstone and shale, acid volcano-clastic with welded tuff and interlayered fine-grained sandstone of Basuhari Formation, coarse- to medium-grained sandstone/quartzite of Ghaghar Formation and rocks comprising Rohtas limestone of Rohtas Formation. In Bihar-Jharkhand the Vindhyan rocks are present as easternmost continuity of the Son Valley sequence falling in three adjacent districts, Garhwa in Jharkhand and Rohtas and Kaimur in Bihar and cover an area of about 3740 km^2. The sequence is bounded by the Mahakoshal Group and Chhotanagpur Gneissic Complex

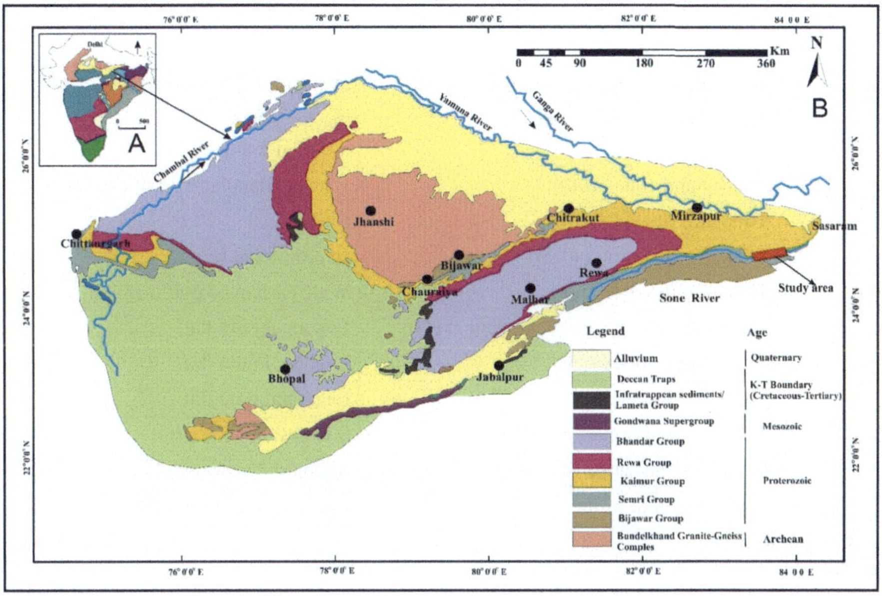

Fig. 1 Regional geological map of the Vindhyan Supergroup showing the location of the study area (Bengtson S et al. 2017)

in the south and Gangetic alluvium in the north and the east (GSI, Published in Misc. Pub. 30, 2009) and in regional stratigraphic succession (GSI, Unified Legend India, CHQ 2017). The detailed geological succession is in Table 1.

2 Geology of the Study Area

The Vindhyan Supergroup in the area is represented by Kaimur Group comprising coarse—to medium-grained sandstone/quartzite of Ghaghar Formation and rocks of Semri Group comprising Rohtas limestone of Rohtas Formation, quartz arenite, sandstone with glauconite pellets, fine- to medium-grained glauconitic sandstone and shale, acid volcano-clastic with welded tuff and interlayered fine-grained sandstone of Basuhari Formation, cherty stromatolitic limestone of Bargwan Formation; olive shale and khaki splintery shale of Kheinjua Formation, porcellanite and shales of Chopan Formation and Rohinian sandstone of Rohinian Formation. The Kaimur sandstone has a reported K–Ar age on authigenic glauconite of 910 $\pm$ 39 Ma (Vinogradov et al. 1964). Brecciated Banded Haematite Chert of Mahakoshal Group is also exposed at the contact of the rocks of Vindhayan Supergroup.

Table 1 Lithostratigraphic succession of the Vindhyan Supergroup of Bihar–Jharkhand (Singh 1998; GSI, Published in Misc. Pub. 30, 2009)

Group	Formation (thickness in metres)	Depositional Environment
Kaimur	Dhandraul Quartzite (200)	Destructive beach Alluvial fan delta
	Scarp Sandstone (100)	Barrier island and wash-over fans
	Bijaigarh Shale (+80)	Lagoon
	Ghaghar Quartzite (+100)	Intertidal flat beach
Semri	Rohtas Limestone (340)	Wide carbonate platform
	Basuhari (Glauconitic Sandstone) (130)	Lagoonal tidal flat
	Fawn Limestone (125)	Lagoonal tidal flat
	Kheinjua Shale (35)	Lagoonal tidal flat
	Porcellanite Formation (440)	Shallow water coastal
	Kajrahat Limestone (340)	Supratidal to subtidal
	Rohiniyan Sandstone (60)	Fluviotidal estuarine beach
UNCONFORMITY Mahakoshal Group Chhotanagpur Gneissic Complex		

b. **Occurrence of Glauconite in Basuhari Formation**
The Basuhari Formation has been divided into three litho units, viz., Quartz arenite, glauconitic sandstone with shale, and acid volcanoclastic with welded tuff (Tables 2, 3 and 4). The facies of Basuhari Formation exposed in Rohtas district, Bihar covering part of area of toposheet no. 63P/10 & 63P/14 and the variation in lateral thickness and vertical facies has been observed (Fig. 2). Quartz arenite is the lower most unit of this formation, exposed to the south of Paruka. Overlying this, is acid volcanoclastic with welded tuff, glauconitic sandstone with shale, and quartz arenite (reef type) at the top. The contact between quartz arenite and underlying unit Fawn limestone is gradational which is observed to the south of Paruka area.

c. **Acid Volcanoclastic with Welded Tuff (Glauconitic)**
The Lower Vindhyan sequence of Bihar contains several horizons of acidic volcanic ash and tuffaceous material within the Basuhari Formation of the Semri Group (Kumar and Mukund 2012). Grey-green acid volcanic and volcanoclastic rocks occur interbedded with glauconite-bearing sandstone in localities such as Nawara, Nawadih, Nauhatta and Tipa (Toposheet 63P/14), and at Katudanr, Paruka, Mujuraha, Madhukupia, Matiawan, Nawadih Kalan, Jadunathpur, Belduria, and Jaradag (Toposheet 63P/10) of Rohtas District.

The acid volcanoclastics and welded tuffs form compact, fine-grained rocks that commonly exhibit conchoidal fracture and distinctive spheroidal weathering. Individual spheroidal bodies range from ~1 to 1.5 m in length and 80–100 cm in width. North of Nawadih Kalan, these volcanic layers alternate rhythmically with quartz

Table 2 Lithological succession in the study area

Ghagar sandstone	Ghagar formation (+100 m)	Kaimur group	Vindhyan supergroup
Rohtas Limestone with thin black shale	Rohtas formation (340 m)	Semri group	
Fault			
Quartz arenite (reef type) Glauconitic Sandstone with shale Acid Volcanoclastics with welded tuff Quartz arenite	Basuhari formation (130 m)		
Fawn Limestone	Bargwan formation (125 m)		
Kheinjua Shale	Kheinjua formation (+4 m)		
Chopan Porcellanite	Chopan formation		
Rohinian Sandstone	Rohinian formation		
SNNF (Son Narmada North Fault)			
Brecciated Banded Haematite Chert	Agori	Mahakoshal group	

Based on the All-India Unified Legend 2017

Table 3 Different levels of maturation: nascent, slightly evolved, evolved, and highly evolved glauconite grains by Odin and Matter (1981)

Sl. No.	Glauconite types	Maturity	K_2O content	Mineralogical structure	Colour	XRD peak position
1.	Nascent	Low	<4%	Glauconite smectite ↓	Pale green	14Å ↓
2.	Slightly evolved	Moderate	4-6%		Light green	
3.	Evolved	High	6-8%		Green	
4.	Highly evolved	Very high	>8%	Glauconite mica	Dark green	10Å

Table 4 Concentration (in %) of major oxides in seven bed rock samples of glauconitic sandstone from Bandu-Chutia area used in bivariate scatter plots

S. no.	Sample no	Location	SiO_2 (%)	Al_2O_3 (%)	Fe_2O_3 (%)	MnO (%)	TiO_2 (%)	CaO (%)	MgO (%)	Na_2O (%)	K_2O (%)	P_2O_5 (%)
1	AD-61	Nawadih	76.06	11.62	3	0.01	0.45	0.14	1.24	0.32	3.67	0.08
2	AD-64	Nawadih	76.47	7.84	3.52	0.01	0.35	0.17	1.25	0.17	3.51	0.08
3	AD-66	Nawadih	78.07	8.65	3.34	0.04	0.36	0.23	0.92	0.18	3.14	0.08
4	AD-152	Chutia	72.87	11.68	5.1	0.05	0.24	0.16	1	0.18	3.63	0.06
5	AD-163	Chutia	70.93	9.67	9.88	0.28	0.11	0.29	1.03	0.2	2.11	0.06
6	AD-113B	Tiura	60.67	17.9	1.55	0.03	0.37	0.14	1.6	0.09	13.71	0.06
7	AD-113C	Tiura	60.73	17.73	1.48	0.03	0.39	0.16	1.64	0.09	13.79	0.06

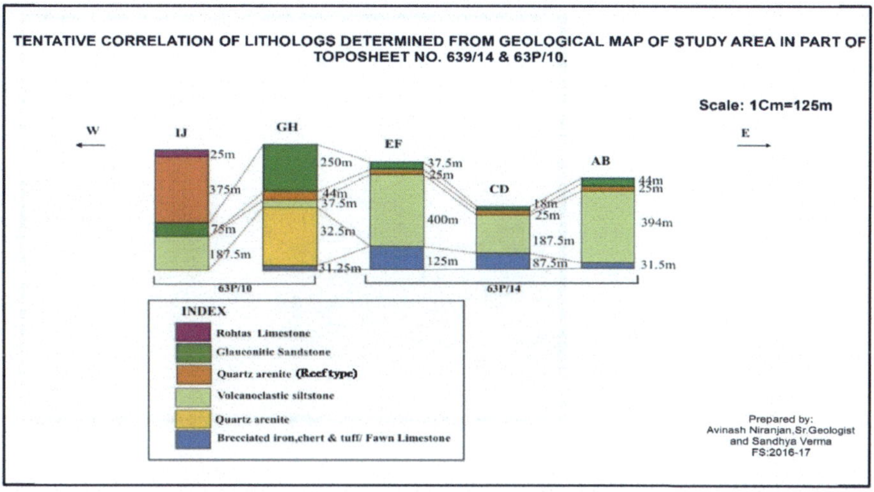

Fig. 2 Lithology of various sections of Glauconitic Sandstone (Basuhari Formation), Rohtas District, Bihar

arenite. The unit generally trends between N75°W–S75°E and E–W to N75°E–S75°W, dipping gently (10°–25°) towards the north. Chilled rims on volcanic pillows are visible near Paruka village. Several conformable lenses of acidic volcanic to volcanoclastic material—apparently unreported previously—occur within the Glauconitic Sandstone Formation of the Semri Group in this portion of the Vindhyan Basin.

Petrography

Thin-section studies show welded tuffs (ignimbrites) in shades of brown, pink, and red. They are dense, non-vesicular rocks displaying clear flow banding and a pronounced porphyritic texture, with feldspar phenocrysts up to 5 × 2 mm embedded in a microgranular to vitrophyric groundmass. The principal mineral phases are K-feldspar, quartz, plagioclase and minor biotite, with glauconite and clinochlore occurring as accessories. Glauconite content can approach 20%.

Quartz grains in these volcanoclastic rocks often exhibit markedly irregular, lobate or embayed boundaries. The ash-flow tuffs are thinly bedded rhyolitic tuffs of grey-green to brown coloration. They consist of a mixture of lithic debris and crystal fragments set within devitrified acidic volcanic glass that forms alternating fine-grained and microcrystalline bands. These bands curve to form streaks and filament-like structures. Some samples contain notable quartz and feldspar phenocrysts supported by a felsic vitrophyric groundmass. The matrix is dominated by quartz and feldspar, with minor streaks of clinochlore, glauconite and biotite.

Fig. 3 Acid volcanic (Welded Tuff) flows with spheroidal weathering observed in the Basuhari Formation of Semri group

d. **Glauconitic Sandstone**

Glauconitic sandstone is usually green, greyish-green, grey, and greenish-brown in colour, and it shows limonitic stain and presence of clay pellets. It is fine to medium grained, occasionally medium to coarse grained, and is thinly to thickly bedded. Glauconite occurs as grain form as well as pelletal form in the glauconitic sandstone unit of the Basuhari Formation (Figs. 3, 4 and 5). The glauconitic sandstone shows inter-granular filling of glauconite along with quartz and feldspar grains. Glauconite occurs as single grain as well as aggregate of grains in glauconitic sandstone unit. The glauconitic sandstone is observed in Chutia-Tiura area (in parts of toposheet 63P/14), the cumulative strike is ~4 km with width of 50–60 m (Plate-I) and in Paruka-Mujuraha area, the cumulative strike is 2 km with width of 30–500 m (discontinuous outcrop) and in Jadunathpur- Belduria-Jaradag areas the cumulative strike is 6 km and width of 80–100 m (discontinuous outcrop) (in the parts of toposheet 63P/10) (Plate-II).

East of Chutia village, medium to coarse-grained glauconitic sandstone having strike length of 500 m with width of 60–80 m trends of N70°E / S70°W and the bedding has a dip of 12° towards NW. In north of Jadunathpur village, fine- to medium-grained, locally coarse-grained glauconitic sandstone is observed, in which glauconite occurs as pellets. The bedding has an attitude of EW/20° dipping towards north. North of Jaradag and Belduria area, medium- to coarse-grained glauconitic sandstone shows trend of N80°W/ S80°E with dip of 12° towards north.

During large scale mapping on 1: 12,500 scale, glauconite mineralized zone was delineated in Paruka- Mujuraha, Nawadih Kalan, Jadunathpur, Belduria and Jaradag. Coarse- to medium-grained glauconitic sandstone unit was observed where glauconite occurs in pelletal form as well as grain form in Paruka area. Green coloured, coarse- to medium-grained glauconitic sandstone unit was observed in pelletal form as well as grain form in Paruka area (Figs. 6 and 7).

Petrographically, glauconitic sandstone is fine- to coarse-grained and composed of quartz, feldspar, biotite and glauconite pellets (Figs. 6, 8a and b, 9a and b). Model

Fig. 4 Photomicrograph of acid volcanoclastic rock showing quartz with highly irregular, lobbed/embayed grain boundaries

Fig. 5 A specimen of Glauconite-rich fine-rained sandstone, Jadunathpur village

Fig. 6 Glauconite occurring as pellet form in Glauconitic Sandstone

Fig. 7 Quartz arenite observed in Nawadih kalan area showing asymmetric ripple marks

per cent of glauconite pellets for quantifying potash of glauconite is about >5–20%. Clasts of quartz and feldspar are observed in the section, sub-angular to sub-rounded in shape. These clasts are embedded in siliceous matrix. Here glauconite occurs as pellets within matrix (Figs. 10 and 11). Under high magnification, alteration of K-feldspar and development of sericite are observed in glauconitic sandstone.

e. **Quartz Arenite**

It is composed of coarse-grained, reddish brown-coloured (due to ferruginization), hard, compact rock comprising dominantly of quartz. It overlies the acid volcaniclastic unit. Various evidences of faulting, viz., silicification, slickenside striations, dip reversal, quartz ribbon and tension gashes are observed in quartz arenite (Fig. 7). in Nawara, Nawadih, Tipa, Nauhatta, Tiura, Tenwa, Chutia areas (in parts of Toposheet no. 63P/14) and Katudanr, Paruka, Mujuraha, Nawadih Kalan, Jadunathpur, Belduria, and Jaradag areas (in parts of Toposheet no. 63P/10). A linear

Fig. 8 Photomicrograph of quartz arenite shows subangular grains of quartz, microcline, Muscovite, and few biotite

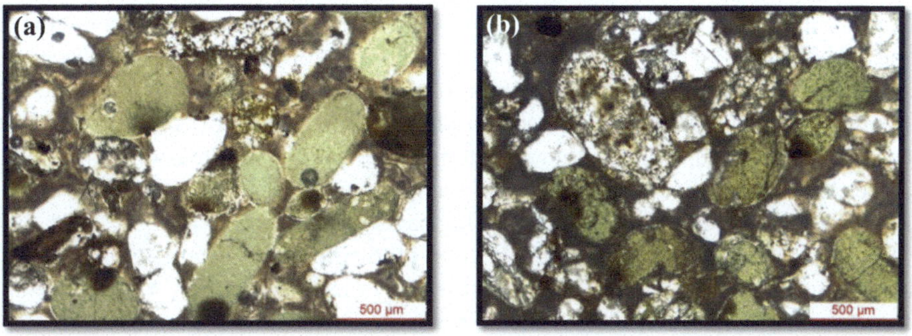

Fig. 9 **a** and **b** Photomicrographs showing glauconite pellets and rock fragments in glauconitic sandstone

Fig. 10 **a** and **b** Photomicrograph showing dark greenish to black glauconite pellets within siliceous matrix in glauconitic sandstone

Fig. 11 Banding observed in glauconitic siltstone and felsic tuff

ridge of extremely silicified quartz arenite was observed in Katudanr area having a trend of N70°E/ S70°W which corroborate with the trend of major Son-Narmada South Fault. The attitude of the bed varies from N65°E/45° towards south to N50°E/ 22° towards NW. This unit is characterized by the presence of occasional slicken side striations and vug filling of quartz often forming sigmoidal tension gashes like features. Prominent reversal of dip both in amount and direction has been observed within the quartz arenite unit while showing same NE-SW trend.

Petrographically, quartz arenite consists of medium- to coarse-grained, sub-rounded grain of quartz. The texture is primarily clastic. Matrix is <5% and both the matrix and cement are siliceous (Fig. 8).

f. **Field Photographs and Photomicrographs**

See Figs. 3, 4, 5, 6, 7 and 8.

g. **Methodology**

Large scale mapping on 1: 12,500 scale with collection of bedrock samples was carried out to delineate the Potash (Glauconite) bearing mineralized zones in the area. Samples representing all lithologies were collected comprising 347 nos. of grab samples and 16 nos. of pit samples. The pits were excavated across the strike to expose the concealed glauconite bands. Two pits (1 m × 1 m × 1 m length, width, and depth) were also made in the covered areas. The samples were then processed using iron mortar pestle and sieved through –100 mesh sieve. The last residue was not thrown, instead mixed with the sieved fraction till proper fine granulation was achieved. The sieved samples were doubly coned and quartered and 250 g of sample was submitted for chemical analysis in polythene sample bags.

Sand-grade glauconite concentrates from a variety of sediment types were prepared by sieving followed by magnetic and specific gravity separation (Jeans et al. 1982). The original samples were analysed by XRF method. Samples from fine-grained glauconitic sandstone, greyish-green siltstone, glauconitic siltstone, green shale, and fawn limestone intercalated green shale were studied through X-ray diffraction (XRD). Two number of bulk samples of 250 kg each were collected from Chutia and Jadunathpur area for ore beneficiation studies. Sample no. 50 nos. samples were collected for petrographic studies.

h. **Results and Discussion**

Glauconite in the study area occurs within fine to coarse-grained sandstone, siltstone, and shale, and sedimentary units enriched in this mineral are collectively known as greensands. The glauconitic sandstone contains quartz, glauconite, mica, microcline, pyrite, and clinochlore. Glauconite appears as pellet-shaped bodies, discrete grains, coatings on sand particles, and as material occupying pore spaces between grains. Its colour varies from green to greyish-green, grey, and greenish brown. In medium- to coarse-grained glauconitic sandstone, glauconite typically constitutes 60–70% of the concentrated fraction, with grain sizes ranging from about 80–650 μm (Fig. 8). Much of the glauconite formed by diagenetic alteration of detrital biotite in a shallow marine setting under reducing conditions, particularly within sandstone.

Glauconite forms authigenically under a relatively narrow set of environmental conditions. It develops at or just beneath the sediment–water interface in marine environments that range from oxygenated to mildly reducing, and where sediment accumulation is minimal (McRae 1972; Odin and Matter 1981; Amorosi 1997). Glauconitization is especially common in fine-grained shelf and slope deposits at depths of roughly 30–500 m (Amorosi 1997; Kelly and Webb 1999). Although typically associated with muds, it also forms within sandy siliciclastic substrates through replacement, surface coating, or pore-filling processes. Faecal pellets are the most common precursor because the passage of clay-rich sediment through the digestive tracts of organisms generates localized chemical conditions favourable for glauconite formation (Anderson et al. 1958; Pryor 1975). Beyond pellets, glauconite can replace a wide range of grains including micas, quartz, feldspar, chert, calcite, dolomite, phosphate fragments, and volcanic rock debris (McRae 1972; Pryor 1975; Odin and Matter 1981).

Within the glauconitic sandstone formation of the study area, the exposed lithological succession begins with volcanoclastic sandstone containing welded tuff and shale at the base, gradually transitioning upward into quartz arenite with pebbly layers—representing an overall coarsening-upward sequence. This stratigraphic pattern indicates deposition during a regressive phase in Meso-Proterozoic time. The slow rate of sediment supply during regression provided an extended residence time for grains close to the seafloor, creating favourable conditions for glauconite development in a shallow marine setting. A subsequent marine transgression enhanced the degree of glauconite maturation. The glauconite present may have originated from both marine and volcanic sediment inputs, though its precise source remains a topic of further investigation.

Glauconite in the glauconitic sandstone and siltstone ranges in colour from pale green to dark brownish-green and includes varieties spanning nascent to more evolved forms (Odin and Matter 1981). These authors defined four stages of glauconite maturity—nascent, slightly evolved, evolved, and highly evolved—each representing different durations of residence at or near the sediment–water interface and therefore reflecting changes in sedimentation rate.

In the study area, glauconite in sandstone is primarily nascent to slightly evolved, showing low to moderate maturity and pale to light green hues. These grains contain approximately 4% to < 6% K_2O. In contrast, glauconite from siltstone is highly evolved, with significantly greater maturity and K_2O values reaching ~13% (Figs. 9, 10, 11 and 12).

k. **Photomicrographs**

See Figs. 9, 10, 11 and 12.

Fig. 12 Glauconite occur as grain form as well as in the matrix in glauconitic siltstone

3 Geochemistry

a. **Chutia-Tiura Area**
In bed rock samples collected from glauconitic sandstone analysed by XRF, K_2O% ranges from 2.14 to 4.55%, in acid volcanoclastic rock K_2O% ranges from 0.68 to 5.12% while in welded tuff (glauconitic) analysed, K_2O% ranges from 1.38 to 5.40%. In Tiura area, three samples from the glauconitic siltstone unit from different depth of dug wells yielded K_2O value of 13.79, 13.71, and 13.44%, respectively. In the south of Tipa (Banahi), one sample of the glauconitic siltstone, near the contact of Fawn limestone yielded 13.02% K_2O, Fawn limestone intercalated with green shale K_2O% analysed K_2O% between 4.27 and 6.69%. Kheinjua Shale analysed K_2O% between 4.03 and 12.15%.

b. **Paruka-Mujuraha Area**
In bed rock samples collected from Glauconitic Sandstone analysed by XRF,K_2O% ranges from 1.79 to 4.25%, from acid volcanoclastic rock K_2O% ranges from 2.62 to 5.45% and from welded tuff (glauconitic), K_2O% ranges from 3.74 to 5.75%.One sample of Fawn limestone intercalated with green shale yielded 6.92% K_2O value.

c. **Jadunathpur-Belduria-Jaradag Areas**
Bed rock samples collected from Glauconitic Sandstone were analysed by XRF in which K_2O% ranges from 2.50% to 3.14%; acid volcanoclastic rock yielded 5.0% K_2O% and in welded tuff (glauconitic), K_2O% ranges from 2.77% to 4.99%. Check analysis of 10 bedrock samples from the mineralized zones was also carried out.

d. **Chemical Weathering of Glauconitic Sandstone**
A good measure of the degree of chemical weathering can be ascertained from the Chemical Index of Alteration (CIA, Nesbitt and Young 1982). $CIA = [Al_2O_3/(Al_2O_3 + Na_2O + K_2O)] \times 100$.

Seven samples of glauconitic sandstone showing average abundance of all major oxides were analysed by XRF method. Glauconitic sandstone shows CIA values ranging between 55 and 80. They exhibit CIA values close to 70, well within the range of typical shale. In the A-CN-K plot (Fig. 13), glauconitic sandstone falls on a line parallel to the AK side of the A-CN-K plot, indicating its shift due to the presence of glauconite mineral. The CIA values between 70 and 80 indicate low to moderate extent of chemical weathering under humid climate.

e. **Bivariate Analysis**

Bivariate analysis is one of the simplest forms of quantitative (statistical) analysis. It involves the analysis of two variables, for determining the empirical relationship between them (Table 5 and Figs. 14, 15, 16, 17, 18, 19, 20, 21, 22, 23 and 24).

Bivariate scatter plot showing relation between K_2O and major oxides in bed rock samples of glauconitic sandstone from Bandu-Chutia area.

Bivariate scatter plot showing relation between K_2O and P_2O_5 in bed rock samples of Jaradag-Katudanr Area.

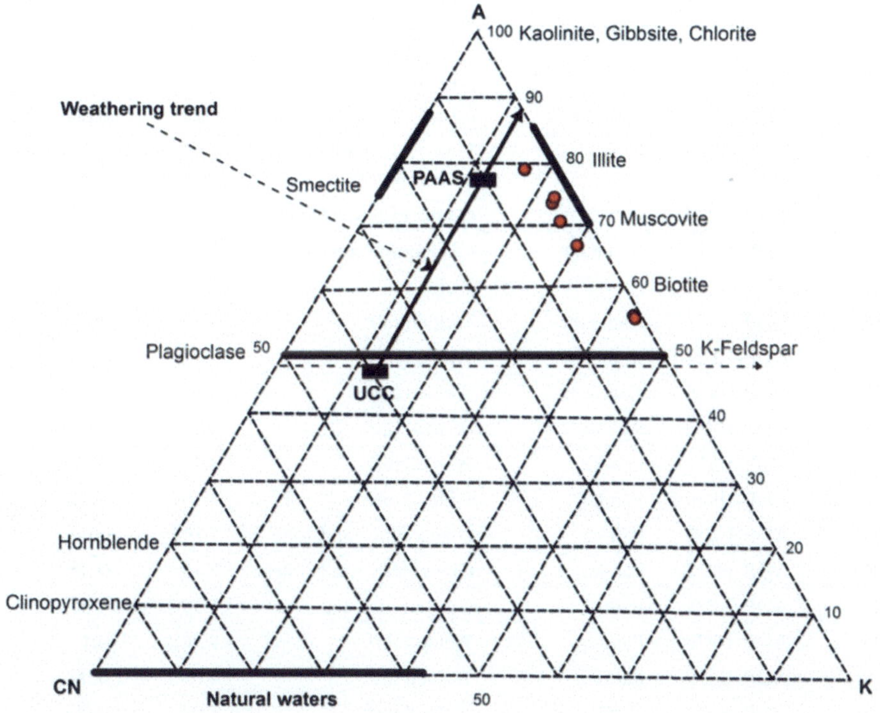

Fig. 13 CIA plot of glauconitic sandstone

Table 5 Concentration (in %) of K_2O and P_2O_5in 50 bed rock samples of glauconitic sandstone and 54 sample of welded tuff & acid volcanoclastic sandstone from Jaradag-Katudanr area used in bivariate scatter plots

S. no.	Sample no	Location	Lithology	K_2O (%)	P_2O_5 (%)
1	ADP-10/01	Belduria	Shale	5.25	0.07
2	ADP-10/03	Belduria	Welded tuff	3.72	0.1
3	ADP-10/05	Jadunathpur	Welded tuff	4.17	0.09
4	ADP-10/07	Paruka	Glauconitic sandstone	2.99	0.05
5	ADP-10/08	Paruka	Glauconitic sandstone	3.52	0.07
6	ADP-10/09	Paruka	Glauconitic sandstone	3.56	0.12
7	ADP-10/10	Paruka	Glauconitic sandstone	2.38	0.04
8	ADP-10/11	Paruka	Glauconitic sandstone	4.25	0.05
9	ADP-10/12	Paruka	Glauconitic sandstone	2.4	0.03
10	ADP-10/15	Mujuraha	Fine-grained volcanoclastic sandstone	3.99	0.1
11	ADP-10/16	Katudanr	Fine-grained volcanoclastic sandstone with Shale	3.68	0.09
12	ADP-10/17	Katudanr	Fine-grained volcanoclastic sandstone with Shale	4.78	0.08
13	ADP-10/18	Katudanr	Glauconitic sandstone	2.55	0.06
14	ADP-10/19	Paruka	Fine-grained glauconitic sandstone	2.99	0.06
15	ADP-10/20	Paruka	Glauconitic sandstone	3.66	0.07
16	ADP-10/21	Katudanr	Glauconitic sandstone	2.19	0.05
17	ADP-10/22	Paruka	Welded tuff	4.95	0.08
18	ADP-10/23	Katudanr	Welded tuff	3.36	0.12
19	ADP-10/24	Katudanr	Welded tuff	4.92	0.05
20	ADP-10/26	Katudanr	Welded tuff	4.15	0.09
21	ADP-10/28	Paruka	Welded tuff	4.77	0.08
22	ADP-10/29	Paruka	Welded tuff	3.74	0.09
23	ADP-10/30	Paruka	Welded tuff	3.79	0.09
24	ADP-10/31	Paruka	Glauconitic sandstone	1.89	0.04
25	ADP-10/32	Paruka	Glauconitic sandstone	2.15	0.05
26	ADP-10/33	Paruka	Glauconitic sandstone	1.79	0.04
27	ADP-10/38	Paruka	Welded tuff	4.56	0.1
28	ADP-10/39	Paruka	Welded tuff	4.09	0.09
29	ADP-10/40	Paruka	Welded tuff	3.93	0.09
30	ADP-10/41	Paruka	Welded tuff	4.75	0.11
31	ADP-10/42	Paruka	Welded tuff	4.8	0.1

(continued)

Table 5 (continued)

S. no.	Sample no	Location	Lithology	K_2O (%)	P_2O_5 (%)
32	ADP-10/44	Paruka	Welded tuff	4.86	0.08
33	ADP-10/45	Mujuraha	Fine-grained volcanoclastic sandstone with Shale	4.15	0.12
34	ADP-10/46	Mujuraha	Fine-grained volcanoclastic sandstone with Shale	3.95	0.09
35	ADP-10/47	Mujuraha	Fine-grained volcanoclastic sandstone with Shale	5.45	0.07
36	ADP-10/48	Mujuraha	Welded tuff	5.75	0.09
37	ADP-10/49	Mujuraha	Welded tuff	4.82	0.09
38	ADP-10/50	Mujuraha	Welded tuff	4.18	0.09
39	ADP-10/51	Mujuraha	Fine-grained volcanoclastic sandstone with Shale	4.2	0.08
40	ADP-10/53	Tiara Kalan	Welded tuff	3.82	0.09
41	ADP-10/56	Madhukupia	Welded tuff	4.91	0.09
42	ADP-10/57	Madhukupia	Welded tuff	5.11	0.11
43	ADP-10/61	Amahuan	Welded tuff	4.9	0.11
44	ADP-10/63	Nawadih Kalan	Welded tuff	3.99	0.07
45	ADP-10/64	Nawadih Kalan	Welded tuff	3.58	0.11
46	ADP-10/65	Nawadih Kalan	Glauconitic Sandstone	3.11	0.08
47	ADP-10/66	Gurur	Fine-grained volcanoclastic sandstone with Shale	2.92	0.08
48	ADP-10/67	Gurur	Fine-grained volcanoclastic sandstone with Shale	4.38	0.1
49	ADP-10/68	Gurur	Glauconitic Sandstone	2.36	0.08
50	ADP-10/69	Gurur	Glauconitic Sandstone	3.36	0.06
51	ADP-10/70	Mujuraha	Fine-grained sandstone with Shale	2.62	0.07
52	ADP-10/72	Mujuraha	Glauconitic Sandstone	3.48	0.07
53	ADP-10/73	Mujuraha	Glauconitic Sandstone	3.39	0.07
54	ADP-10/74	Tiara Kalan	Welded tuff	3.34	0.09
55	ADP-10/76	Parchha	Fine-grained volcanoclastic sandstone with Shale	4.87	0.06

(continued)

Table 5 (continued)

S. no.	Sample no	Location	Lithology	K_2O (%)	P_2O_5 (%)
56	ADP-10/77	Parchha	Fine-grained volcanoclastic sandstone with Shale	3.69	0.09
57	ADP-10/82	Jadunathpur	Welded tuff	3.42	0.11
58	ADP-10/83	Jadunathpur	Welded tuff	3.56	0.08
59	ADP-10/84	Jadunathpur	Welded tuff	3.77	0.09
60	ADP-10/85	Jadunathpur	Welded tuff	4.73	0.08
61	ADP-10/86	Jadunathpur	Welded tuff	4.99	0.11
62	ADP-10/87	Jadunathpur	Fine-grained volcanoclastic sandstone with Shale	5	0.08
63	ADP-10/88	Jadunathpur	Fine-grained volcanoclastic sandstone with Shale	3.65	0.09
64	ADP-10/89	Jadunathpur	Fine-grained volcanoclastic sandstone with Shale	2.77	0.06
65	ADP-10/90	Jadunathpur	Glauconitic Sandstone	2.63	0.06
66	ADP-10/91	Jadunathpur	Glauconitic Sandstone	2.7	0.06
67	ADP-10/93	Jadunathpur	Glauconitic Sandstone	3.14	0.07
68	ADP-10/94	Jadunathpur	Glauconitic Sandstone	3.11	0.06
69	ADP-10/95	Jadunathpur	Glauconitic Sandstone	2.5	0.04
70	ADP-10/97	Jaradag	Welded tuff	4.29	0.09
71	ADP-10/100	Belduria	Glauconitic Sandstone	3.11	0.04
72	ADP-10/101	Jaradag	Glauconitic Sandstone	2.48	0.07
73	ADP-10/102	Jaradag	Fine-grained volcanoclastic sandstone with Shale	5.01	0.06
74	ADP-10/103	Matiawan	Welded tuff	3.54	0.09
75	ADP-10/104	Nawadih Kalan	Welded tuff	4.4	0.07
76	ADP-10/105	Nawadih Kalan	Glauconitic Siltstone	2.46	0.06
77	ADP-10/106	Jaradag	Glauconitic Sandstone	3.69	0.06
78	ADP-10/107	Jaradag	Glauconitic Sandstone	1.46	0.07
79	ADP-10/108	Belduria	Welded tuff	4.26	0.08
80	ADP-10/109	Belduria	Glauconitic Sandstone	2.8	0.07
81	ADP-10/110	Belduria	Glauconitic Sandstone	2.85	0.05
82	ADP-10/111	Belduria	Glauconitic Sandstone	2.74	0.05

(continued)

Table 5 (continued)

S. no.	Sample no	Location	Lithology	K_2O (%)	P_2O_5 (%)
83	ADP-10/112	Belduria	Glauconitic Sandstone with shale	1.91	0.08
84	ADP-10/113	Belduria	Welded tuff	4.13	0.09
85	ADP-10/114	Belduria	Glauconitic Sandstone	3.06	0.05
86	ADP-10/115	Belduria	Glauconitic Sandstone	2.94	0.08
87	ADP-10/116	Belduria	Glauconitic Sandstone	2.36	0.04
88	ADP-10/117	Belduria	Glauconitic Sandstone	2.8	0.05
89	ADP-10/118	Belduria	Glauconitic Sandstone	2.45	0.04
90	ADP-10/119	Belduria	Glauconitic Sandstone	1.96	0.06
91	ADP-10/120	Belduria	Glauconitic Sandstone	1.78	0.07
92	ADP-10/121	Belduria	Glauconitic Sandstone	2.95	0.06
93	ADP-10/122	Belduria	Welded tuff	4.32	0.08
94	ADP-10/123	Belduria	Fine-grained volcanoclastic sandstone with Shale	4.73	0.06
95	ADP-10/124	Belduria	Glauconitic Sandstone	3.09	0.06
96	ADP-10/125	Belduria	Glauconitic Sandstone	3.27	0.07
97	ADP-10/126	Belduria	Glauconitic Sandstone	3.24	0.07
98	ADP-10/127	Belduria	Glauconitic Sandstone	3.57	0.05
99	ADP-10/128	Belduria	Glauconitic Sandstone	2.44	0.04
100	ADP-10/129	Belduria	Glauconitic Sandstone	3.45	0.07
101	ADP-10/130	Belduria	Glauconitic Sandstone	3.44	0.06
102	ADP-10/131	Nawadih Kalan	Glauconitic Sandstone	3.43	0.05
103	ADP-10/132	Jadunathpur	Glauconitic Sandstone	3.31	0.09
104	ADP-10/133	Jadunathpur	Glauconitic Sandstone	3.29	0.13

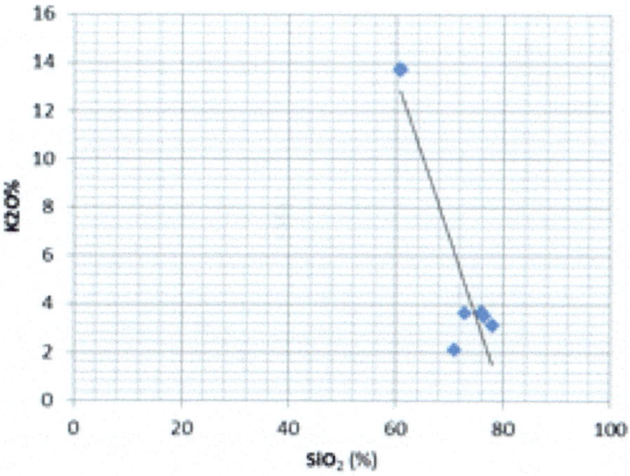

Fig. 14 The image showing negative correlation of K_2O with SiO_2 in bed rock samples of glauconitic sandstone

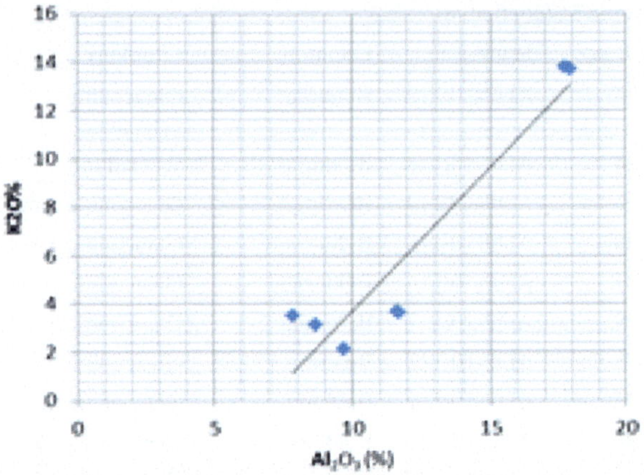

Fig. 15 Positive correlation of K_2O with Al_2O_3 in bed rock samples of glauconitic sandstone

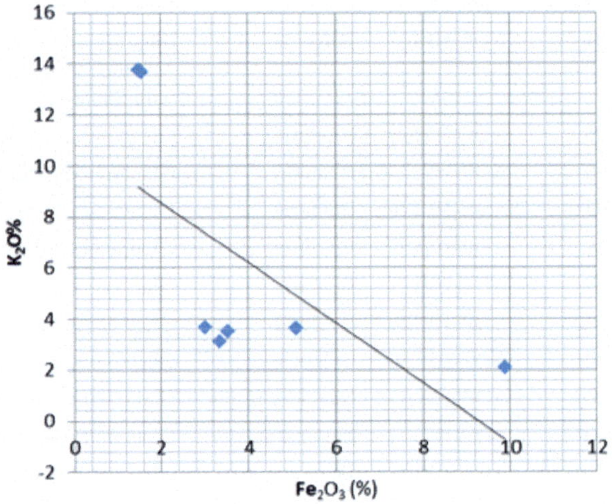

Fig. 16 Negative correlation of K_2O with Fe_2O_3 in bed rock samples of glauconitic sandstone

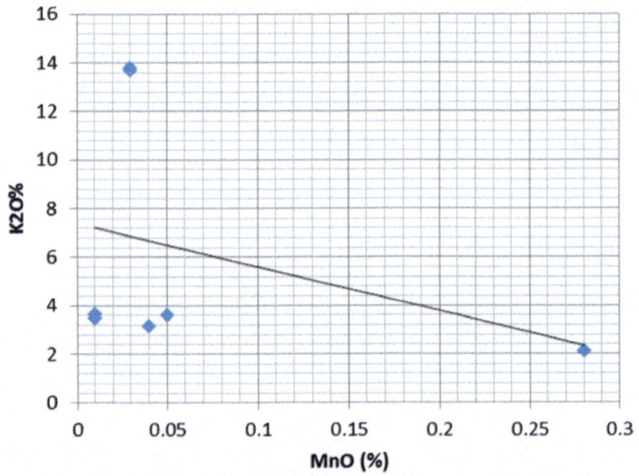

Fig. 17 Negative correlation of K_2O with MnO in bed rock samples of glauconitic sandstone

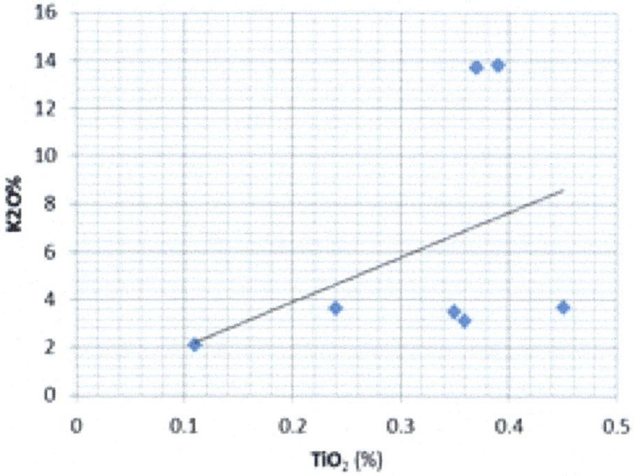

Fig. 18 Positive correlation of K_2O with TiO_2 in bed rock samples of glauconitic sandstone

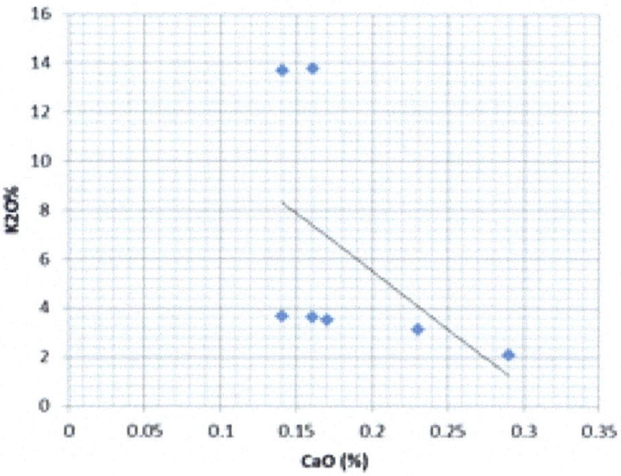

Fig. 19 Negative correlation of K_2O with CaO in bed rock samples of glauconitic sandstone

4 Correlation Matrix

a. Correlation Matrix of Glauconitic Sandstone

K_2O indicates strong positive correlation with Al_2O_3, MgO, and moderate positive correlation with TiO_2. K_2O value has negative correlation with SiO_2, Fe_2O_3, MnO, CaO, Na_2O, and P_2O_5 (Table 6). Al_2O_3 shows strong positive correlation with K_2O ($r = 0.944$). This co-variation indicates that K-bearing minerals have significant influence on Al distribution and suggests that the bulk of Al and K is primarily contributed by glauconite (Table 7).

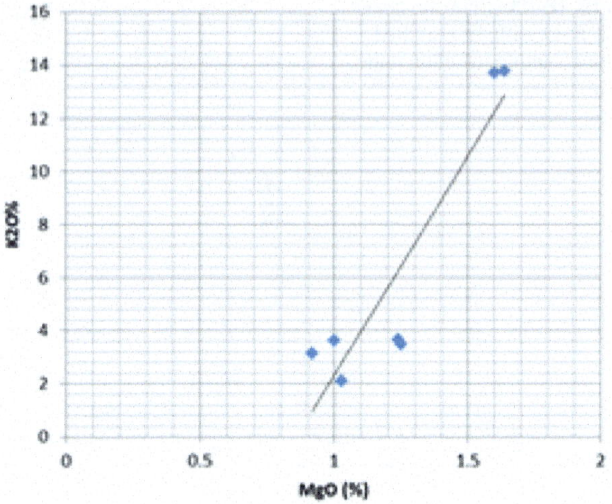

Fig. 20 Positive correlation of K_2O with MgO in bed rock samples of glauconitic

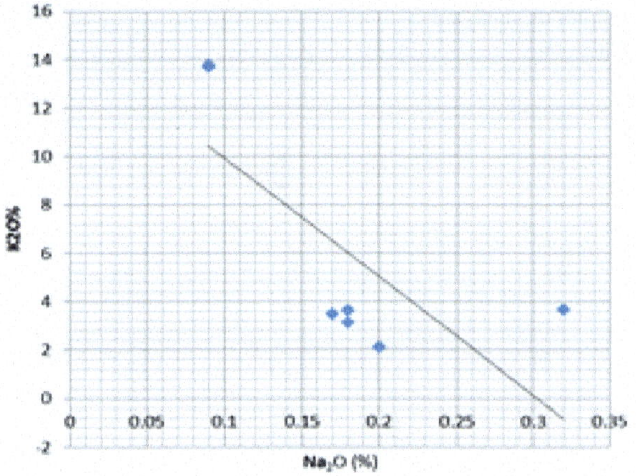

Fig. 21 Scatter plot, indicating negative correlation of K_2O with Na_2O in bed rock samples of glauconitic sandstone

b. **Ore Beneficiation Studies Results**

Two number of bulk samples were analysed from Mineral Processing Division, Indian Bureau of Mines, Nagpur (IBM) for ore beneficiation studies. Sample no. OB-1was collected from Chutia area. The thin section of glauconitic sandstone from Chutia area contains dominant quartz, minor feldspar (microcline), glauconite and biotite. The percentage of glauconite is approximately 3–4.5% and the mineral grains are medium grained with size variation of about 1–2 mm. The glauconitic grains are greenish to yellowish in colour with dull earthy lustre. Biotite occurs in trace amount. Sample no. OB-2 was collected from Jadunathpur area. As per the results of Ore beneficiation samples, the composite magnetic concentrate of Sample OB-1 assayed 4.86% K_2O, 10.1% Fe_2O_3, 18.1% Al_2O_3, 56.9% SiO_2, and 5.2% LOI with 75–80% of Glauconite content.

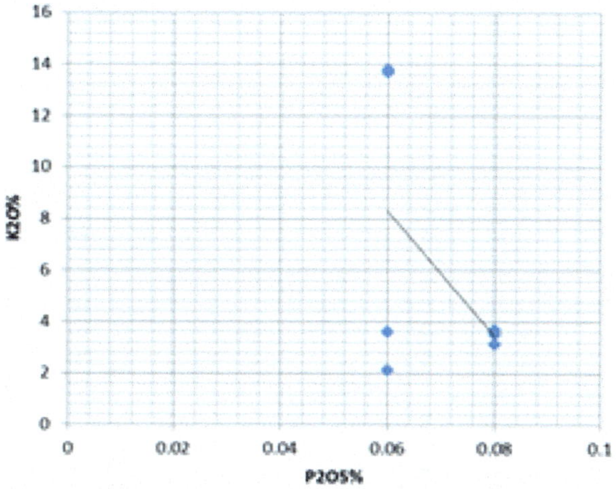

Fig. 22 Negative correlation of K_2O with P_2O_5 in bed rock samples of glauconitic sandstone

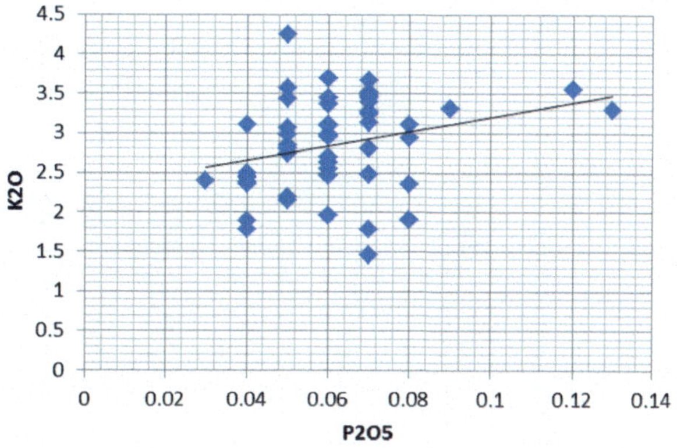

Fig. 23 Positive correlation of K_2O with P_2O_5 in 50 bed rock samples of glauconitic sandstone

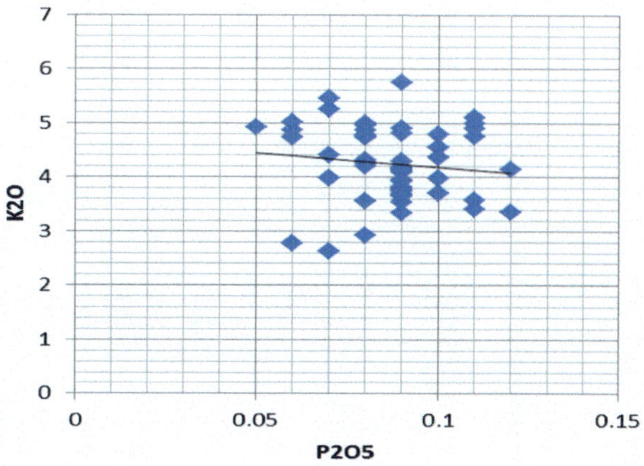

Fig. 24 Marginal negative correlation of K_2O with P_2O_5 in 54 bed rock samples of welded tuff & acid volcanoclastic sandstone

Table 6 Correlation Matrix (Values of Pearson's coefficient of correlation) of major oxides in seven bed rock samples of glauconitic sandstone from Bandu-Chutia area

	SiO_2 (%)	Al_2O_3 (%)	Fe_2O_3 (%)	MnO (%)	TiO_2 (%)	CaO (%)	MgO (%)	Na_2O (%)	K_2O (%)	P_2O_5 (%)
SiO_2 (%)	1									
Al_2O_3 (%)	−0.932	1								
Fe_2O_3 (%)	0.320	−0.539	1							
MnO (%)	−0.019	−0.239	0.913	1						
TiO_2 (%)	−0.052	0.321	−0.914	−0.885	1					
CaO (%)	0.276	−0.521	0.817	0.858	−0.754	1				
MgO (%)	−0.832	0.849	−0.642	−0.366	0.496	−0.593	1			
Na_2O (%)	0.734	−0.574	0.320	0.076	0.057	0.086	−0.544	1		
K_2O%	−0.922	0.944	−0.655	−0.337	0.409	−0.508	0.918	−0.734	1	
P_2O_5%	0.772	−0.633	−0.226	−0.430	0.515	−0.072	−0.337	0.574	−0.504	1
Correlation type	Value range									
Strong	>0.75									
Moderately strong	0.50–0.75									
Moderate	0.25–0.50									
Low	0–0.25									
Very low	<0									

Table 7 Descriptive statistics of glauconitic sandstone

	SiO_2 (%)	Al_2O_3 (%)	Fe_2O_3 (%)	Mn (%)	TiO_2 (%)	CaO (%)	MgO (%)	Na_2O (%)	K_2O (%)	P_2O_5 (%)
Mean	70.828	12.155	3.981	0.064	0.324	0.184	1.24	0.176	6.223	0.069
Standard error	2.764	1.555	1.088	0.036	0.043	0.021	0.108	0.029	1.954	0.004
Median	72.87	11.62	3.34	0.03	0.36	0.16	1.24	0.18	3.63	0.06
Mode	–	–	–	0.01	–	0.14	–	0.18	–	0.06
Standard deviation	7.314	4.116	2.880	0.096	0.113	0.056	0.287	0.078	5.169	0.011
Variance	53.497	16.946	8.300	0.009	0.013	0.003	0.082	0.006	26.723	0.0001
Kurtosis	−1.229	−1.190	3.320	6.491	1.466	1.307	−1.394	1.556	−0.865	−2.8
Skewness	−0.815	0.748	1.710	2.518	−1.281	1.452	0.542	0.877	1.183	0.374
Range	17.4	10.06	8.4	0.27	0.34	0.15	0.72	0.23	11.68	0.02
Minimum	60.67	7.84	1.48	0.01	0.11	0.14	0.92	0.09	2.11	0.06
Maximum	78.07	17.9	9.88	0.28	0.45	0.29	1.64	0.32	13.79	0.08
Count	7	7	7	7	7	7	7	7	7	7

Glauconite weight % yield is 7.0. The composite magnetic concentrate of Sample OB-2 assayed 4.61% K_2O, 9.74% Fe_2O_3, 18.55% Al_2O_3, 58.11% SiO_2 and 4.38% LOI with 60% of glauconite content. Glauconite weight % yield is 14.3. On the basis of beneficiation result from IBM, Nagpur, glauconite recovered from OB-1 may be suitable for use as potash fertilizer.

c. **Recreating Climatic Past**

Glauconite formed in the Vindhyan Supergroup in a warm and humid climatic regime. A review of global occurrences of Paleogene glauconites broadly corresponds to warm climatic intervals that witnessed multiple hyperthermal events. The widespread occurrence of glauconite across the globe in the late Paleocene and early Eocene relates to a combination of factors including global sea-level rise, hypoxic shelf, and warm and humid climate. A slow rate of sediment accumulation within the transgressive shallow seas facilitated the formation of glauconite on the seafloor.

Although the occurrence of authigenic deposits was largely influenced by global climatic conditions, factors like upwelling current and palaeolatitude led to the formation of phosphorite and lignite regionally. A detailed facies and paleogeographic analysis establish that the glauconite formed in a shallow, sub-tidal environment, unlike its Phanerozoic counterpart which forms in the deep marine condition. Glauconite may have formed on shallow suboxic shelve conditions. Rapid transgression and abundant supply of nutrients may have facilitated glauconitization showing unique geochemistry.

The presence of subtidal facies give an indication of subtidal environments with predominant energy tides and high wave action. This results in fining upward sequence with coarse sediments at the base and progressively finer sediments at the top defining a continuous vertical sequence. This also indicates decrease in energy as environment changes from sub-tidal to intertidal. Hummocky cross stratification present in the sequence are testimony to the subtidal facies.

Enrichment of glauconite does give unmistakable evidence of marine conditions. The shallow water and even subaerially exposed beds give inferences to shallow sea (Mahadevan 2002). Sedimentological attributes indicate that the environment of deposition of the coarser siliciclastic facies fluctuated among alluvial fans, braided plain, fan delta, Aeolian, shallow marine, and lacustrine environment. The carbonates were deposited in a ramp setting varying from intertidal to deep offshore (Chakraborty and Bhattacharya 1996).

The present climatic regime is thus drastically in contrast to the depositional conditions which may have led to the thick deposition of Vindhyan Group of rocks covering an area of ~3700 km^2 (Singh 1998). Thus, it unveils stratigraphic sequence of climate change observed across the stratigraphic time scale.

d. **Conclusion**

In the area of interest, Potash bearing mineralized zones is mainly confined in the rocks of Glauconitic Sandstone Formation of Semri Group of Vindhyan Supergroup.

In Chutia-Bandu area, toposheet no. 63P/14, 4 km strike length of the glauconite mineralized zone is delineated. The Bed rock samples analysed by XRF for K_2O % ranges from 2.14 to 4.55% in glauconitic sandstone, 0.68–5.12% in acid volcanoclastic and 1.38–5.40% in welded tuff (glauconitic). In Tiura area, three samples from the glauconitic siltstone unit from different depth of dug wells yielded K_2O value of 13.79, 13.71, and 13.44%, respectively.

To the south of Tipa (Banahi), one sample of the glauconitic siltstone, near the contact of Fawn limestone yielded 13.02% K_2O and one sample of Fawn limestone intercalated with green shale yielded 6.69% K_2O. In Tenwa area, samples of Fawn limestone intercalated with green shale K_2O% ranges from 2.59 to 6.69%. Kheinjua Shale K_2O% ranges from 4.03 to 12.15%.

In Kantudanr-Jaradag area, mineralized zones were delineated in Paruka-Mujuraha area, Jadunathpur and Belduria–Jaradag area. In Paruka-Mujuraha area, glauconite mineralized zone has a strike length of 1 km and a width of 500 m and it further extended up to 2 km in the west of mapped area. Bed rock samples analysed by XRF for K_2O% ranges from1.79 to 4.25% in glauconitic sandstone, 1.79–4.25% in acid volcanoclastic and 2.62–5.45% in welded tuff (glauconitic). Tuffaceous shale yielded 5.29% K_2O in south of Paruka area. Fawn limestone intercalated with green shale yielded 6.92% K_2O.

In Jadunathpur area, glauconite mineralization is observed with strike length of 1.3 km and width of 80–100 m. Bed rock samples analysed by XRF for K_2O % ranges from 2.50 to 3.14% in glauconitic sandstone, 2.77–5.00% in acid volcanoclastic and 3.42–4.99% in welded tuff (glauconitic). In Belduria–Jaradag area, Glauconite mineralization is observed with strike length of 4.75 km and width of 80–100 m. Bed rock samples analysed by XRF for K_2O% ranges from 1.46 to 3.69%in glauconitic sandstone, 4.73–5.01% in acid volcanoclastic and 3.54–4.40% in welded tuff (glauconitic).

Tentative correlation of lithologs determined from geological map of the study area reveals facies variation, thickness of litho-units, glauconite concentration and flow regime. Glauconite concentration increases towards west and it decreases towards east. Argillaceous facies increase towards east and decreases towards west likewise arenaceous facies increases towards west and decreases towards east.

In the mapped area, glauconite of glauconitic sandstone is nascent to slightly evolved, maturity is low to moderate and colour of glauconite is pale to light green having yielded 4 to <6% K_2O whereas glauconite in glauconitic siltstone yielded <13% K_2O.

As per the results of Ore beneficiation samples send to IBM, Nagpur the composite magnetic concentrate of Sample OB-1 assayed 4.86% K_2O, 10.1% Fe_2O_3, 18.1% Al_2O_3, 56.9% SiO_2, and 5.2% LOI with 75–80% of Glauconite content. Glauconite weight % yield is 7.0. The composite magnetic concentrate of Sample OB-2 assayed 4.61% K_2O, 9.74% Fe_2O_3, 18.55% Al_2O_3, 58.11% SiO_2, and 4.38% LOI with 60% of glauconite content. Glauconite weight % yield is 14.3.

On the basis of beneficiation result from IBM, Nagpur, glauconite recovered from OB-1 was found suitable for use as potash fertilizer. Glauconite as a fertilizer can either be directly applied in the field or as a source of refined potash by processing to KCl (potassium chloride) or even purer forms of K (McRae 1972; McConchie 1978; Odin and Fullagar 1988). Glauconite as a fertilizer can easily be separated from the host rock or host sediment by de-sliming and magnetic separation to produce a 99% pure glauconite product (Bell and Goodell 1967; Compton 1989; Potash West NL 2011). Apart from the recognized source of potassium, glauconite also provides a beneficial source of magnesium and iron, and small amounts of phosphate, as a soil fertilizer (McRae 1972; Coles et al., 2002; Payne 2008). The magnetic separation process may lead to some carbonate and quartz impurities being present within the glauconitic fertilizer, but small amounts of these may actually be beneficial to the plants within the soil (Coles et al. 2002).

Acknowledgements Authors are indebted to the Director General, Geological Survey of India for his support. They are also thankful to the Deputy Director General, Geological Survey of India, SU: Bihar for support, constant reassurance, supervision and valuable suggestions during the work. The authors also acknowledge the support received from Chemical Division, Eastern Region, Kolkata, and XRD laboratory for facilitating the chemical analyses of submitted samples. The authors are also thankful to the ore beneficiation studies from IBM, Nagpur. The authors are thankful to all their colleagues and individuals, who helped them directly or indirectly during the investigation.

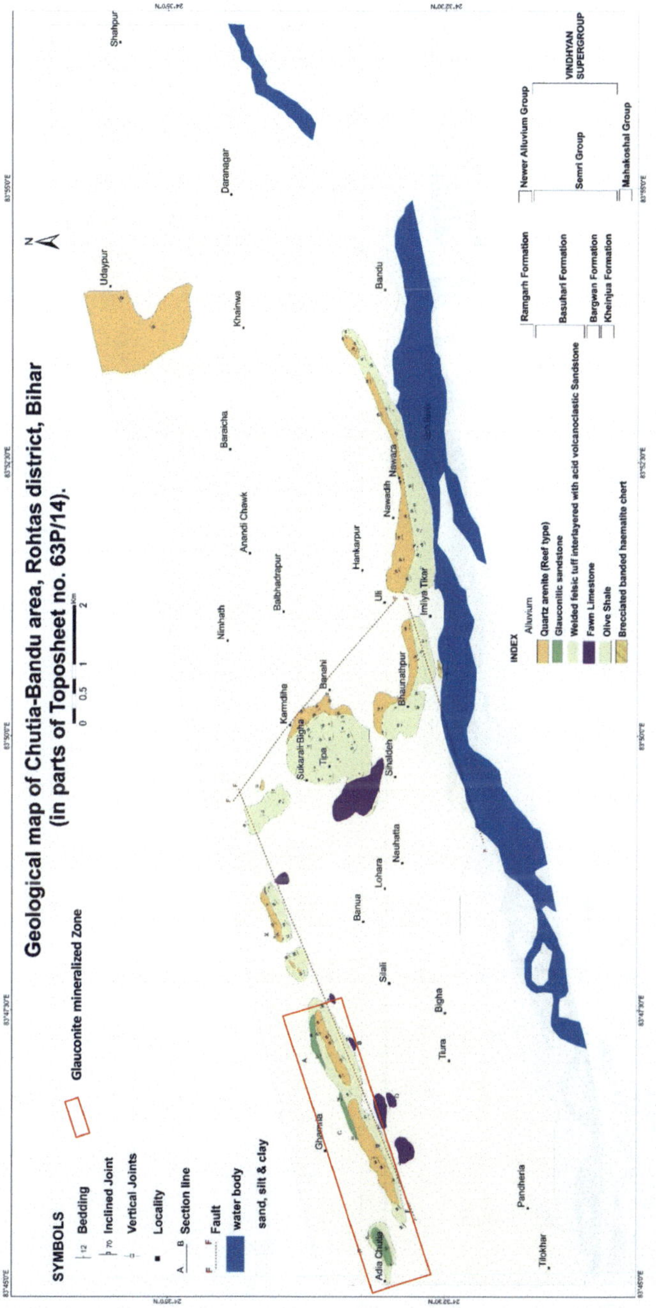
Geological map of Chutia-Bandu area, Rohtas district, Bihar
(in parts of Toposheet no. 63P/14).
SYMBOLS
Bedding
Inclined Joint
Vertical Joints
Locality
Section line
Fault
water body
sand, silt & clay
Glauconite mineralized Zone
INDEX
Alluvium
Quartz arenite (Reef type)
Glauconitic sandstone
Welded felsic tuff interlayered with acid volcanoclastic Sandstone
Fawn Limestone
Olive Shale
Brecciated banded haematite chert
Newer Alluvium Group
Ramgarh Formation
Basuhari Formation
Bargwan Formation
Kheinjua Formation
Semri Group
Mahakoshal Group
VINDHYAN SUPERGROUP
Shahpur
Daranagar
Udaypur
Khainwa
Baraicha
Anandi Chawk
Nimhath
Balbhadrapur
Karmdihe
Banahi
Sukarail-Bigha
Tipa
Sihaldeh
Nauhatta
Lohara
Banua
Silali
Bigha
Tiura
Ghamha
Adla Chutia
Pandheria
Tilokhar
Hankarpur
Uli
Nawadih
Nawara
Bandu
Imiliya Tikar
Bhaunathpur

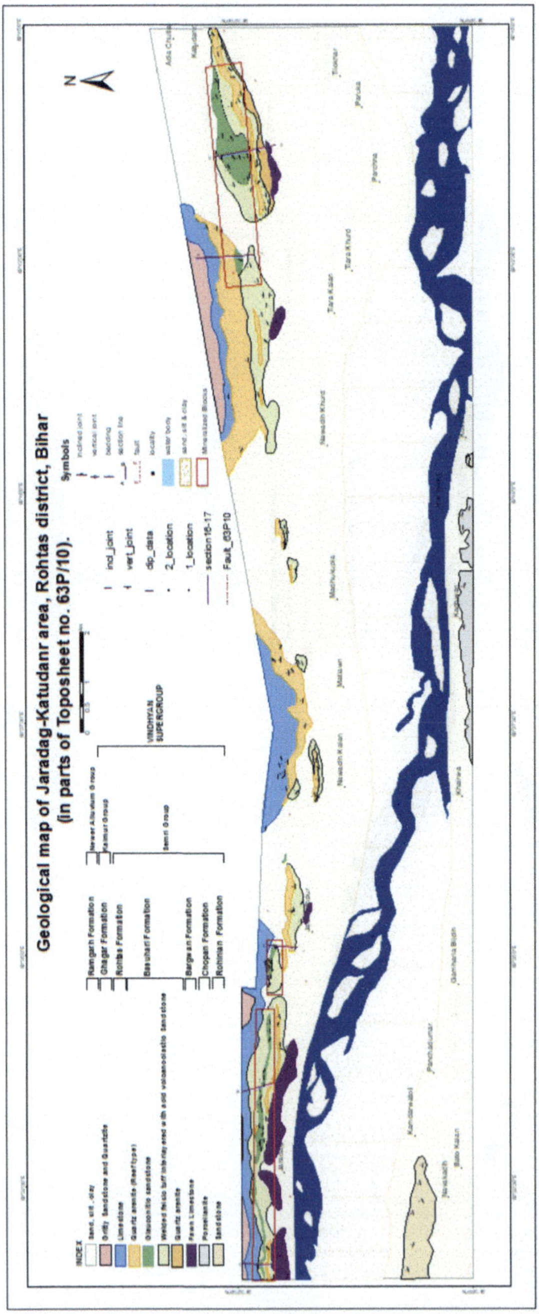
Geological map of Jaradag-Katudanr area, Rohtas district, Bihar
(in parts of Toposheet no. 63P/10).
INDEX
Symbols
Rampurh Formation
Ghagar Formation
Rohtas Formation
Basuhari Formation
Bargwan Formation
Chopan Formation
Rohinian Formation
VINDHYAN SUPERGROUP
ind_joint
vert_joint
dip_data
2_location
1_location
section16-17
Fault_63P10
N

References

Amorosi, A., 1997, Detecting compositional, spatial, and temporal attributes of glaucony: a tool for provenance research: Sedimentary Geology, v. 109, p. 135–153.

A. Amorosi, The occurrence of glaucony in the stratigraphic record: distribution patterns and sequence-stratigraphic significance, Rendiconti Online della Società Geologica Italiana (2012)

Anderson, A., Jonas, E.C., and Odum, H.T., 1958, Alteration of clay minerals by digestive processes of marine organisms: Science, v. 127, p. 190–191.

Auden J.B., 1933 Vindhyan Sedimentation in the son Valley, Mirzapur District. Mem. Geol. Surv. of India, Vol. 62, Pt. Pp. 141–250.

Auden, J. B. 1940. Paleoclimates during the deposition of the Vindhyan and related systems. Report of the Session International Geological Congress, 6: 137–152.

Bell KG, Goodell HG (1967) Glauconite and celadonite. In: Brown G (ed) The X-ray identification and crystal structures of clay minerals. Mineralogical Society, London, pp 393–450.

Bengtson S et al. (2017) The controversial "Cambrian" fossils of the Vindhyan are real but more than a billion years older. Proc Natl Acad Sci USA 114(41):10715–10720. https://doi.org/10.1073/pnas.1711292114

Bose P K, Sarkar S, Das N G, Banerjee S, Mandal A and Chakraborty N, 2015 Proterozoic Vindhyan Basin: Configuration and evolution; In: Precambrian basins of India: Stratigraphic and tectonic context (eds) Mazumder R and Eriksson P G, Geol. Soc. London Memoir 43(1): 85–102.

Chakraborty PP, Bhattacharya A (1996) Tidal sedimentation in the Proterozoic Kaimur Group, Vindhyan Supergroup, central India. Sediment Geol 105:211–227.

Coles DW, Sanders JV, Payne TE (2002) Glauconite as a nutrient source and soil conditioner for plant growth. Aust J Soil Res 40:1061–1076.

Compton JS (1989) Degree of glauconitization, a possible measure of the oceanographic conditions on continental shelves. Clay Miner 24:287–301.

M. Delamette, Trace fossil assemblages from the Albian phosphate-rich sandstones of the Helvetic Shelf (Western Alps), Cretaceous Research (1989).

Garzanti, E., Haas, R., Jadoul, F., 1989. Ironstones in the Mesozoic passive margin sequence of the Tethys Himalaya (Zanskar, Northern India): Sedimentology and metamorphism. In: Young, T.P., Taylor, W.E.G. (Eds.), Phanerozoic Ironstones, Geological Society Special Publication, 46, pp. 229–244.

C. V. Jeans, R. J. Merriman, J. G. Mitchellt and D. J. Bland., 1982, Volcanic Clays in the Cretaceous of Southern England and Northern Ireland, Clay Minerals (1982) 17, 105–156.

Jimenez-Millan, J., Molina, J.M., Nieto, F., Nieto, L., Ortiz, P.A.R., 1998. Glauconite and phosphate peloids in Mesozoiccarbonate sediments (Eastern Subbetic Zone, BeticCordilleras, SE Spain). Clay Minerals 33, 547e559.

Kelly, J.C., and Webb, J.A., 1999, The genesis of glaucony in the Oligo-Miocene Torquay Group, southeastern Australia: petrographic and geochemical evidence: Sedimentary Geology, v. 125, p. 99–114.

Kumar S. and Sharma Mukund, PSI-Field Guide, Vindhyan Basin, Son Valley Area, Central India, First Edition 2012, The Palaeontological Society of India, 2012.

Mahadevan TM (2002) Geology of Bihar and Jharkhand. Geological Society of India, Bangalore.

Mallet, 1869 On the Vindhyan Series exhibited in northwestern and Central Provinces of India. Mem. Geol. Surv. Ind., Vol. 7, Pt-1.

Mandal S, Choudhuri A, Mondal I, Sarkar S, Chakraborty P P and Banerjee S (2019) Revisiting the boundary between the Lower and Upper Vindhyan, Son Valley, India Journal of Earth System Science 128 222 https://doi.org/10.1007/s12040-019-1250-2.

Mathur, S.M., 1981: A revision of the stratigraphy of the Vindhyan Supergroup in the Son valley, Mirzapur district, U.P. Misc. Pub. Geol. Surv. India, No.50, pp. 7–20.

McConchie DM (1978) Glauconite as a possible potassium fertilizer. Miner Sci Eng 10:98–106.

McRay, S.G., 1972, Glauconite: Earth Science Review, v. 8, p. 397–440.

Odin, G.S., and Matter, A., 1981, De glauconiarum origin: Sedimentology, v. 28, p. 611–641.

Odin GS, Fullagar PD (1988) Geological significance of the glaucony facies. In: Odin GS (ed) Green Marine Clays. Elsevier, Amsterdam, pp 295–332.
Oldham, R.D., Vredenberg, E. and Datta, P.N. (1901): Geology of the Sone valley in the Rewa State and part of adjoining districts of Jabalpur and Mirzapur. Mem. Geol. Surv. India, Vol. 31(1), pp. 1–178.
Potash West NL (2011) Potential for significant potassium deposits. Breakaway Investment Group Research Report, Perth, Western Australia.
Pryor, 1975, Biogenic sedimentation and alternation of argillaceous sediments in shallow marine environments: Geological Society of America Bulletin, v. 86, p. 1244–1254. 95.
Singh, S. P., 1998: Precambrian Stratigraphy of Bihar-An overview in B.S. Paliwal (Ed). The Indian Precambrian (Scientific Publisher, Jodhpur Indian pp. 376–408).
Unified Legend India, CHQ, 2017, (for 1: 50,000 Database; upto Litho-unit level), Geological Survey of India.
Vinogradov, A.P., Tugarinov, A.I., Zikhov, C.I., Stanikova, N.I., Bibikova, E.V. and Khorre, K., 1964: Geochronology of Indian Precambrian; Report 22nd IGC New Delhi 10 553–567.

Chapter 16
Paleoclimate and Mass Extinction: Insights from the Gondwana Basin in Jharkhand

Shreya Shrey and Akhouri Bishwapriya

1 Introduction

Climate change is a natural process. Climatic conditions prevailing on the Earth at present have dynamically evolved over the evolutionary history of the planet. Such changes are inevitable. The earliest life to emerge on Earth was simple and a reflection of the "paleo," i.e., past, climatic conditions. The primordial life adapted and evolved in consequence to these ever-evolving conditions. That which exists on Earth today is the best reflection of the climatic conditions of the present. However, during this evolutionary process, landmass and ocean assumed various dispositions (Fig. 1) in the course of which, innumerable forms of life appeared on the planet. Their adaptability to change in response to the ambient conditions governed the duration of their residence on the Earth. Closely linked to changes in paleoclimatic conditions and the emergence and flourishing of newer, more suitable life forms, the Earth has witnessed five distinct occasions where a considerable percentage of species that existed throughout the planet were wiped off owing to their inability to adapt and survive. These events of massive and widespread extinctions are known as "mass extinction." The five major mass extinction events are known as the "*Big Five.*" Following the extinction, the species that emerged were resilient and adapted to the changed conditions. These events of mass extinctions provide the realization that climate change and loss of species are natural evolutionary processes; however, the rate of such changes and the impact of anthropogenic activities altering the "rate" are a cause of concern.

S. Shrey · A. Bishwapriya (✉)
Geological Survey of India, SU: Jharkhand, Ranchi, India
e-mail: akhouri.bishwapriya@gsi.gov.in

N. Khare (ed.), *Changing Climate over India*,
https://doi.org/10.1007/978-981-92-1098-5_16

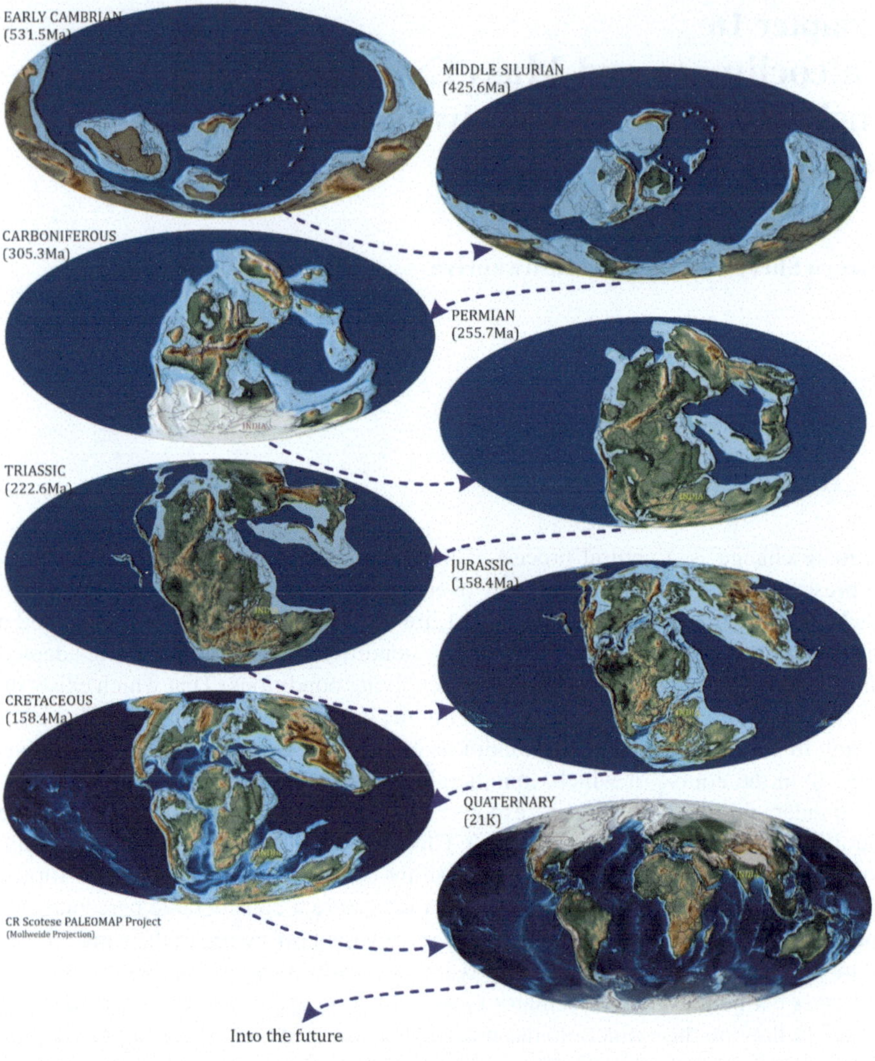

Fig. 1 Paleogeographic reconstruction of the Phanerozoic (Cambrian to Present). Change in the distribution of continents and oceans, presence and absence of ice sheets, etc., depicted over the geological past (Modified after CR Scotese, PALEOMAP Project 2014 and Scotese 2021)

1.1 Proxies to Paleoclimatic Studies

Understanding the past climate, i.e., paleoclimate and its processes, gains significance in order to foresee the future, especially in response to anthropogenic impact on climate. There is a growing consensus that the past is the key to the present and the future. A variety of proxies for making an estimation of the paleoclimatic conditions are being increasingly studied for this purpose. Interestingly, such proxies, biological or physico-chemical, are housed within the geological material, such as fossils or geochemical markers. They, in turn, indicate periods of climatic stress, or, otherwise, which recent studies suggest are generally linked with specific events like magmatism involving a Large Igneous Province, continental-scale plate movements, bolide impact, etc. Such episodes gain greater significance when they result in a sudden increase in the rate of loss of life forms, i.e., over and above the background, on a global scale.

The most widely accepted proxies for estimation of paleoclimatic conditions, often coupled with correlation and age connotations, are the isotopes. For example, insights into the global carbon cycle can be obtained through the determination of changes in $^{13}C_{carb}$ obtained from marine carbonates. It is a representation of change in $^{13}C/^{12}C$ value of the Dissolved Inorganic Carbon (Carmer and Jarvis 2020, and references therein). Similarly, measurements of changes in $^{18}O/^{16}O$ obtained from a wide variety of materials, including Precambrian limestones and cherts, Phanerozoic foraminifers, mollusks, and brachiopods, have generated a reliable database that records changes in the $^{18}O/^{16}O$ ratio of seawater and thereby associated changes in temperature. They provide fine resolution of several thousand years and are particularly useful to identify events, including global climate change, during the Phanerozoic (Grossman and Joachimski 2020, and references therein). Isotopic variation of sulfur records the isotopic composition of sulfate dissolved in seawater (Paytan et al. 2020, and references therein). Osmium isotope (^{187}Os) is a product of the radioactive decay of the long-lived ^{187}Re. It is measured as $^{187}Os/^{188}Os$ and is considered suitable to differentiate among osmium with radiogenic origin (Earth's crust influenced by rivers, groundwater, etc.), non-radiogenic mantle (submarine hydrothermal vents), and even undifferentiated extraterrestrial matter (Peucker-Ehrenbrink and Ravizza 2020, and references therein). Biogenic material like calcite (of belemnites or foraminifers) are deposited from seawater while capturing the prevailing $^{87}Sr/^{86}Sr$ ratio of the sea. The ratio is known to have varied across geological time and is actively incorporated in dating and correlation of suitable marine sediments (McArthur et al. 2020, and references therein). Ice cores also serve as a suitable medium to access physico-chemical conditions, for example, the atmospheric concentration of constituents like carbon dioxide, dust particles, etc. As an allied subject, Tephrochronology, i.e., geochemical signatures of volcanic ash (tephra) frozen within ice cores, are particularly helpful if the actual age of the volcanic event is known (Lowe 2011, and references therein). Similarly, Dendroclimatology is the study of annular rings (i.e., the growth rings) of a tree, which is a reflection of the plant's growth. Plant growth is, in turn, a direct manifestation of complex biochemical processes governed by numerous environmental and ecosystem factors like the climate, availability of nutrients, etc. (Coulthard and Smith 2013, and references therein).

For paleoclimatic studies, sedimentary rock records are second to the precision offered by isotopes. However, they are a direct reflection of a combination of factors which include the depositional setting, paleoenvironment, paleoclimate, and paleogeography. Clay minerals are particularly sensitive to changes in ambient environmental and climatic conditions and are considered second only to isotopes (Mason and Moore 1966). The preservation of fossil remains within terrestrial sediments, especially of index nature, is highly accidental. Nonetheless, volcanism involving a Large Igneous Province, mostly continental flood basalts (Ernst et al. 2020), along with the largest extraterrestrial impacts, shows significant correlations with mass-extinction events in the Phanerozoic geologic record (Rampino et al. 2024). They usher within a short span of geological time anoxic/euxinic conditions in the oceans, increase their acidity, drastically raise the atmospheric pCO_2, increase the UV-B radiation upon ozone-layer destruction, among others, and provide potential immediate causes for widespread environmental and thereby ecological stress. Beloid impacts are known to have triggered impact winters, which in turn contributed to mass extinction events. (Esmeray-Senlet 2020; Rampino et al. 2024) (Fig. 2).

The five most devastating episodes of mass extinction during the Phanerozoic times and the probable causative paleoclimatic factors are discussed. While many insights on the "Big Five" are available from marine sources, the evolution of the terrestrial Gondwana Supergroup of mainland India also offers a perception into changes consequential to the same global factors, albeit resulting in the deposition of distinct lithopackages.

1.2 The Big Five

The large-scale environmental perturbations in marine and terrestrial realms resulting in a distinct rise in the rate of extinction of species during specific times in the Phanerozoic are known to have happened on five occasions, wherein nearly 75% or more species globally were lost within a short span of geological time. Such climatic perturbations also had a significant influence on the development of new and evolved species with greater adaptation. An estimate suggests that almost 99% of the 4 billion species that may have evolved on Earth are extinct (*Novacek, 2001*). However, extinction rates reaching a peak within a short span of geological time and over wide geographical areas, having global ecological and evolutionary effects, are known as "mass extinction."

2 Late Ordovician Mass Extinction

Soon after the Cambrian, which marks the onset of Phanerozoic times, the Ordovician Period, 486.9443.1 Ma (Gradstein et al. 2020), marks the first loss of life since the Ediacaran time. The Great Ordovician Biodiversification Event (GOBE) is an extraordinary event of marine diversification following the Cambrian Explosion (Servais et al. 2010). However, the Ordovician mass extinction that followed resulted in the

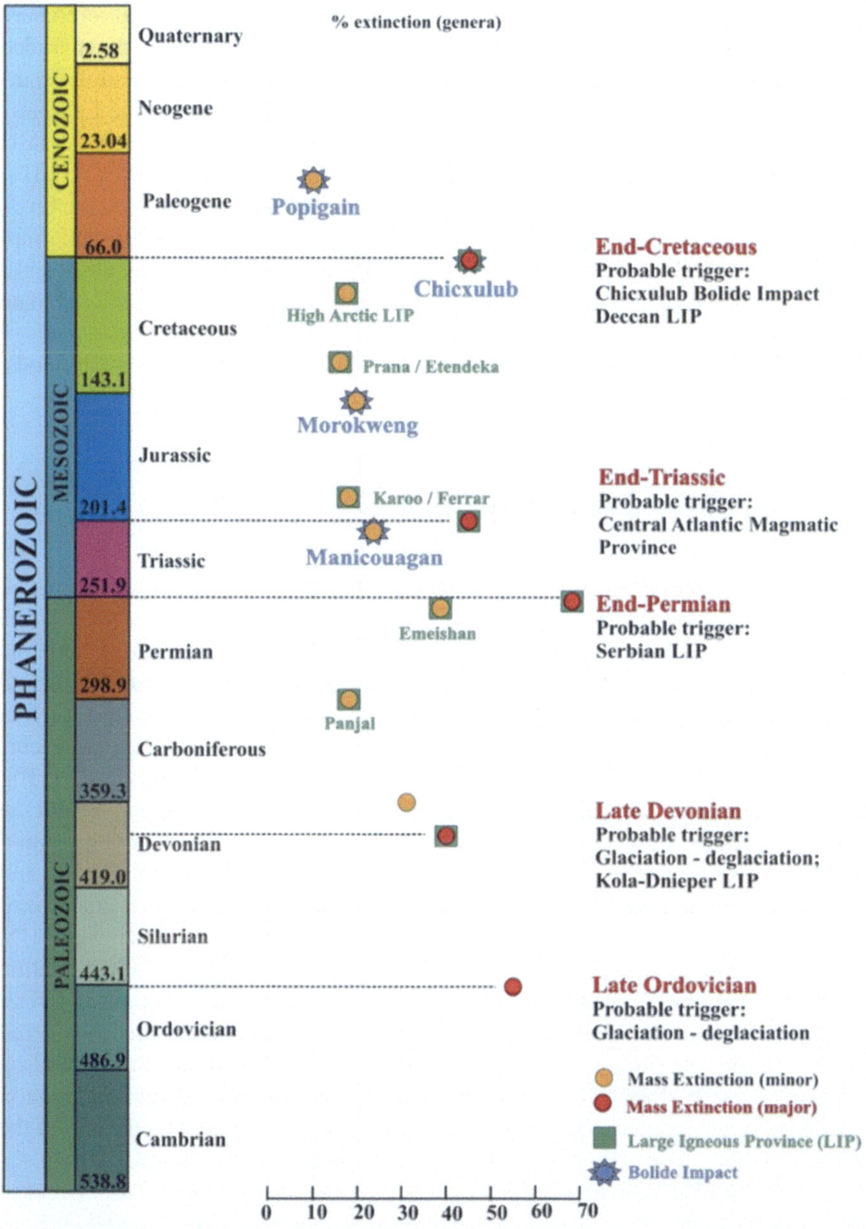

Fig. 2 The "Big Five," and selected minor extinctions, are shown with the plausible correlation with Large Igneous Province and Bolide impact (Modified after Esmeray-Senlet 2020; Ernst et al. 2020 and Rampino et al. 2024)

loss of ~86% species of marine invertebrates (Esmeray-Senlet 2020, and references therein). Benthic and planktonic ecosystems were severely disrupted in a downgraded biocomplexity (Harper et al. 2010) that resulted from glaciation deglaciation occurring in the Gondwanaland that was located around the South Pole, while Laurentia was clustered along the Equator. Significantly altered from a state of greenhouse, the short-lived cooling of ~5 °C during the Hirnantian Stage (Finnegan et al. 2011) brought global climatic change alongside ocean circulation and chemistry, sea level, and euxinia. The inability of life forms, including brachiopods, trilobites, and graptolites, to adapt to glaciation peaking within a short span of half a million years and a similarly quick return to normal was devastating. Diverse siliceous sponges, tabulate corals, and others represent evolutionary radiations developed in the aftermath of the extinction, which aided in ecosystem recovery, allowing forms like brachiopods, corals, and bryozoans to recover rapidly (Botting et al. 2017).

2.1 Late Devonian Mass Extinction

The Devonian represents a geologic time from 419.2 Ma to 359 Ma (Gradstein et al. 2020). It is particularly interesting that most Global Boundary Stratotype Section and Points (GSSPs) markers on the Geologic Time Scale representing Devonian are the First Appearance Datums (FADs) of specific conodont taxa (Becker et al. 2020), and references therein). However, it is also notable for a prominent mass extinction event of the Phanerozoic. While there were multiple globally widespread anoxic events during the Devonian (Becker et al. 2012), that at ~372 Ma is considered one of the Big Five. The Taghanic Event (extinction of goniatites, corals, and brachiopods) and the Hangenberg Event (extinction of phacopid trilobites, several groups of goniatites, and the unusual Late Devonian coiled cephalopods) are major Devonian biotic crises, and the extinction around the Frasnian-Famennian boundary (371.1 Ma) is characterized by two pulses of oceanic anoxia, named the Lower and Upper Kellwasser events (Da Silva et al. 2020). It marks the extinction of beloceratid and manticoceratid goniatite, conodont species, most colonial corals, along with several groups of trilobites. Pulses of global warming and cooling, accompanied by oceanic anoxia, are considered plausible for driving out ~35–40% genera and ~75% species. Against the backdrop of peaking growth of reefs, carbonate production, and diversity of marine fauna in the Paleozoic, contemporaneous with the growth of forests on land by the end of the Middle Devonian, the extinction is often corroborated with exceptionally high sea level and quick glaciation and deglaciation before the onset of mighty ice covers of Carboniferous times. The extinction also draws reference to Viluy (Yakutsk-Vilyui) Continental Flood Basalt (Serbia) and the smaller Kola-Dnieper CFB in Baltica. Global warming, estimated to the tune of ~9 °C (Bond and Sun 2021), might have decreased oceanic oxygen circulation. Negative δ13C anomaly in carbonates and organic sediments (Rampino et al. 2024, and references therein) further ascertains a decrease in primary productivity.

2.2 End-Permian Mass Extinction

Beginning nearly 298.89 Ma, the Permian Period represents a geological history of nearly 47 million years (Gradstein et al. 2020). Rock records of this period are a testament to the climatic transition from icehouse to greenhouse conditions. The end-Permian registers the undisputed greatest of the Big Five mass extinctions, causing a loss of ~85% marine and ~75% terrestrial genera (Esmeray-Senlet 2020, and references therein). However, it is also responsible for influencing evolution, structuring life forms as we see today. Eruption and emplacement of the Serbian Large Igneous Province occurred for at least 500kyr around the time of extinction (Ernst et al. 2020, and references therein; Rampino et al. 2024, and references therein). It is estimated to have injected enormous quantities of greenhouse gases into the atmosphere, significantly increasing the sea-surface temperature by 5–9 °C and atmospheric CO_2 shooting up to 8000 ppm (Cui and Kump 2015). Oceanic anoxia is also linked with enhanced activity of sulfate-reducing bacteria leading to euxinia. Release of H_2S from oceans probably led to ozone depletion coupled with an increase in UV radiation; however, it is weighted against photochemical model calculations. Nonetheless, ocean acidification remains a valid concern due to the rapid release of CO_2, disrupting the source–sink balance.

The twin factors of collapse in primary productivity and injection of ^{13}C-depleted carbon to the ocean and atmosphere contributed to perturbation of the carbon cycle, leading to the unprecedented loss across marine and terrestrial realms (Esmeray-Senlet 2020). These significantly disrupted ecosystems, and the End-Permian mass extinction took millions of years to stabilize. While the main pulse of the extinction lasted not more than a few hundred thousand years (Burgess and Bowring 2015; Esmeray-Senlet 2020), the surviving and new lineages of different marine and terrestrial forms took time until the Middle Triassic to recover, with their adaptations being significantly different from the pre-extinction forms.

2.3 End-Triassic Mass Extinction

The Triassic represents a paleogeography wherein the Pangea occupied the equatorial disposition, spanning well into the tropics of both Northern and Southern Hemispheres and no glacial deposits of this time are known (Ogg and Chen 2020, and references therein). The mass extinction around ~201 Ma, i.e., toward the end-Triassic, is linked with the emergence of the Central Atlantic Magmatic Province (Ernst et al. 2020, and references therein; Rampino et al. 2024, and references therein). It resulted in the loss of ~40% genera and 80% of species. Global warming and ocean acidification are considered to have driven the environmental stress (Esmeray-Senlet 2020, and references therein).

It is postulated that the continental flood basalt may have caused a significant biocalcification crisis resulting from drastic ocean acidification caused by an increase in pCO_2. It is suggested that ocean anoxia was coupled with ocean-surface euxinia. Climatic warming is estimated at an average rise of 6 °C (Sun et al. 2012; Rampino et al. 2024). The disruption of the carbon cycle, along with other causative factors, had resulted in abrupt marine and terrestrial extinction, which took the highly disrupted ecosystem millions of years to recover (Chen and Benton 2012; Esmeray-Senlet 2020, and references therein).

2.4 End-Cretaceous Mass Extinction

The end-Cretaceous mass extinction is unquestionably the most popular extinction among the Big Five. It was catastrophic to terrestrial non-avian dinosaurs and pterosaurs, while it eliminated ~75% species and ~40% genera from the marine realm, including ammonites and belemnites. Bolide impact and volcanism are possible triggers leading to widespread impact on winter, wildfire, soot and sulfur injection, global warming, acid rains, and toxic metal poisoning, bringing about the downfall of giant creatures (Esmeray-Senlet 2020, and references therein). The Cretaceous–Paleogene boundary (K–Pg) is placed around 66 Ma (Gradstein et al. 2020). The Cretaceous Period builds from the opening of the Atlantic Ocean and the disruption of Pangea during the Jurassic. It represents warm seas with calcareous nanoplanktons and foraminifers and proliferated growth of angiosperms on land inhabited by dinosaurs (Gale et al. 2020). Hypothesis explaining the mass extinction revolves around an asteroid/extraterrestrial impact (Alvarez et al. 1980; Smit et al. 1980) ~10 km in diameter colliding with the Earth. It is suggested that the impact created the Chicxulub Crater in Mexico (Hildebrand et al. 1990) and resulted in continent-wide environmental perturbations in the aftermath of massive earthquakes, tsunamis, and submarine landslides, among others.

The impact hypothesis is widely coupled with the emplacement of the Deccan Continental Flood Basalt in India. The volcanism, active over a period of ~750 kyr spanning both sides of the K–Pg boundary, is postulated to have caused environmental stress consequential to the release of climate-modifying gases like sulfur and carbon dioxide (Esmeray-Senlet 2020, and references therein). Nonetheless, kill mechanisms are also linked with a decrease in photosynthesis (Alvarez et al. 1980) and acid rains (Toon et al. 1982) caused by an increase in atmospheric nitric oxide, sulfuric and carbonic acid. Recovery following the extinction took differential trends for different habitats. While the loss of non-avian dinosaurs was not replenished for millions of years, recovery in the neritic and upper bathyal regions was faster (10^4kyr) than the oceanic regions (10^5–10^6kyr) (Sepulveda et al. 2019). Despite significant loss to global forests leading to the complete loss of tree-dwelling birds, the modern tree-dwelling birds are considered descendants of the surviving ground-dwelling lineages (Esmeray-Senlet 2020, and references therein).

3 The Gondwana Supergroup: A Terrestrial Testament to Changes in Paleoclimatic Conditions

Until the breakup of the Pangea and dismemberment during the Cretaceous, India was a part of the supercontinent assembly. The Gondwana Supercontinent, particularly the East Gondwanaland, had Antarctica, Australia, India, and Africa amalgamated into a continuous landform. It occupied different positions around the South Pole right from the onset of the Paleozoic. Covered beneath the mighty ice sheets that existed up until the Carboniferous Period, the first sediments preserved within mainland India since at least the Cambrian times were those of the Gondwana Supergroup. Such sediments are found along four narrow–constricted basins developed along zones suturing the four Precambrian Cratons, namely Dharwar, Bastar, Chhotanagpur, and Bundelkhand.

The (1) Trans-Indian basin belt comprising the ENE–WSW trending Satpura and Son Valley Basins and E–W to WNW–ESE trending Damodar–Koel Valley Basins, (2) NNW–SSE trending Wardha-Pranhita Godavari Valley Basin belt, (3) NW–SE trending Mahanadi Valley Basin belt, and (4) NNW–SSE trending Purnea–Rajmahal–Galsi basin belt constitute the main basins, while those found along the eastern and central Himalayas and offshore Bay of Bengal are considered an extension of the mainland basins (Mukhopadhyay et al. 2010, and references therein; Acharyya 2019).

The sedimentation that began ~302 Ma in the Carboniferous, after a prolonged hiatus since the Cambrian, much like the counterparts of the East Gondwanaland, starts with records of glacial deposits. The following 180 million years or so, up until the outpour of the Rajmahal Volcanics of 118–115 Ma (Kent et al. 2002) and associated sedimentation lasting up to ~105 Ma, constitute the Gondwana Supergroup in the Indian mainland (Mukhopadhyay et al. 2010). Although not as fossiliferous as the rich marine counterparts, yet, the distinct terrestrial formations of the Gondwana Supergroup are a testament to changes in paleoclimatic conditions. Among other plausible factors, the climatic variations are sequential to the movement of Gondwanaland across the south rotational pole (Acharyya 2019).

The two-fold classification of the Gondwana Supergroup in India is based on faunal turnover; the Lower Gondwana Group (Talchir, Barakar, Barren Measure, and Raniganj formations and their equivalents) is characterized by *Gangamopteris–Glossopteris* flora, while the Upper Gondwana Group (Panchet, Suprapanchet, Rajmahal Traps, and their equivalents) is defined by *Dicroidium–Lepidopteris–Ptilophyllum* flora (Jaiswal et al. 2020, and references therein). This classification coincides with the end-Permian mass extinction.

4 Glacial Imprints of the Cretaceous Period

Evidences of mighty glaciers of the Carboniferous Period dating ~302 Ma (Mukhopadhyay et al. 2010) on the Indian Plate are immaculately preserved at the Dudhi Nala section, a geotourism site in Ramgarh and Hazaribagh districts of Jharkhand State. These deposits uncover nearly ~20 Ma of Earth's transition, from icehouse toward greenhouse conditions (Aretz et al. 2020).

At Dudhi Nala, scars of glacial scouring and scratching on the ancient Chhotanagpur Granite Gneiss rocks reflect the forceful movement of mighty continental glaciers. Sub-horizontal fractures filled with fine sediments reveal the power of thermal contraction and the mighty glaciers that once moved over them. As the glaciers ground against the hard Basement, they created jagged breccias through downslope creep. Their downhill descent toward the supercontinent's periphery, in response to the gradual warming of the climate, causes melting, leaving behind the inversely graded beds of laminated sandstone, large pebbles, and well-sorted sandstone interbeds, marking the flows influenced by gravity and kinetic sieving. Among the most fascinating features at Dudhi Nala are the dropstones. Much like icebergs, the frozen masses laden with fine sediments and rock fragments of huge sizes released their sediment load suddenly, dropping off gigantic blocks of basement rock into the fine sediments below. Multistory sand bodies along this stretch indicate climatic changes and reveal the influence of the Tethyan Sea located to the north. As the glacier advanced and retreated, signs of the globally evident Sakmarian transgression (Mukhopadhyay et al. 2010) appeared in the form of bi-directional cross-stratified sandstone, bifurcating ripples, bundle up-buildings, and mud drapes.

Storms of the Tethyan Sea created hummocky cross-stratified beds as a rare signature of wave surges (Bhattacharya et al. 1989). However, turbulence or seismicity on the landmass created pene-contemporaneous/ soft-sediment deformation features like the Bouma sequence, ball and pillow structures, flame structures, convolute laminations, and sand dykes. Syn-sedimentary faults in sandstone, formed by gravity tectonics or seismite slumps, and convolute bedding in coarse silt and sand layers, suggest the land being in constant upheaval (Bhattacharya and Mukherjee 2020). However, as the glaciers retreated and the climate warmed, channels filled with sandstone-siltstone associations began to appear. It marked the end of the great continental glaciation of the Carboniferous Period on the Indian mainland. These glacial deposits are interwoven with signatures depicting episodic influence of shallow marine, lacustrine, and fluvial geomorphic forces (Mukhopadhyay and Bhattacharya 1996; Mukhopadhyay et al. 2010). Trace fossils like *Phycodes* isp. suggest survival activities. Burrows, tracks, and feeding marks are valuable insights into how ancient organisms interacted with their environment, including their strategies for survival (Pollard 1981; Boggs Jr 2006).

The section represents a complete succession of transition from icehouse and a gradual warming of the climate (Fig. 3). The Geological Survey of India designates this world-class section as a Geotourism Site.

Fig. 3 Sedimentary features created under glacial–marine–lacustrine–fluvial conditions at Dudhi Nala, Jharkhand. **A.** Lodgement tillite. **B.** Graded bedding (normal and reverse). **C.** Varve. **D.** Dropstone. **E.** Convolute laminations. **F.** Bifurcating ripples. **G.** *Phycodes* (burrows). **H.** Channel beds

4.1 Coal Deposits of the Permian

Humongous coal deposits of the Permian times are found in all Gondwana basins throughout the globe. Transitioning from icehouse conditions of the Carboniferous times (Henderson and Shen 2020), the Sakmarian transgression introduced fluvial systems in marginal parts of the basin under a periglacial climate (Mukhopadhyay et al. 2010). Coupled with tectonism at the Cape Fold Belt (Halbich et al. 1983), the opening of the Natal Trough and East African Rift Valley, the combined factors contributed to the development of major Coal deposits over the entire Gondwanaland under warmer environmental conditions (Mukhopadhyay et al. 2010). Despite uniformity between basins, the Damodar–Koel Valley Basin registers fluvial fronts oscillating probably in response to a change in the rate of sedimentation and subsidence. Nonetheless, the presence of swamps with an abundant supply of luxuriant vegetal matter and tranquility allowed for the development of humongous Coal deposits. The deposits are also an indicator of climatic conditions conducive to both the growth of plants and their transformation into Coal. The Barakar Formation represents the acme of the development of low-sulfur Coal, deposited in a mildly reducing environment (Acharyya 2019), which is in contrast with the significantly altered environmental conditions that followed and led to the deposition of Barren Measures.

4.2 Guadalupian Transgression

The Barren Measure Formation from the Damodar–Koel Valley Basin consists of gray to black colored shale, bioturbated heterolithic deposits, and laterally impersistent sideritic ironstone shales. In a marked change from the Coal forming conditions prevailing during the Barakar times, the Barren Measure Formation represents the prevalence of humid climatic conditions (Mukhopadhyay et al. 2010) and deposition in a pro-deltaic environment (Bhattacharya and Banerjee 2015). The ichnofossil assemblage (*Phycodes curvipalmatum, Palaeophycus isp., Phycodes templus, Thalassinoides isp.*) at the Ara–Dumerbera locality of the West Bokaro Coalfield in Damodar–Koel Valley Basin provides credence to the ideas of Barren Measures representing transgressive facies. The transgression, a rarity for interior basins of mainland India, probably resulted from rifting and opening of the Meso–Tethys Ocean along the northern edges of the Greater India and brought about the separation of Sibumasu and Qiangtang from Gondwanaland, impacting the deposition of Barren Measures (Metcalfe 1996; Mukhopadhyay et al. 2010). However, the end-Guadalupian is a notable mass extinction event; although not as severe as the Big Five, it marks a biodiversity crisis, altering Permian diversity trends (Jin et al. 1994). It is often related to the Emeishan LIP (Bond et al. 2010), bringing about environmental changes, including rapid fluctuation in seawater temperature (Chen and Shen 2019). The end-Guadalupian or Guadalupian–Lopingian boundary ~259.5 Ma is characterized by regression, depicting the largest Paleozoic sea-level fall (Haq and Schutter

2008). Following it, preserved as the Raniganj Formation (and its equivalents), these changes signify the return of coal-forming conditions in the Indian mainland and counterparts of the East Gondwanaland.

4.3 Triassic Aridity

The advent of the Triassic Period is manifested by the development of a new set of microfloral assemblages, including *Densisporites, Lunatisporites, Klausipollenties, Allisporites, Falcisporites, Lundbladispora,* and *Playfordiaspora* that evolved in arid climatic conditions. Coarse clastic-dominated quartzo-feldspathic sandstone with conglomerate and thin, persistent (red) clay-rich sediments are represented as the Panchet Formation in the Damodar–Koel Valley Basin. Clastic nature of sediments restricts preservation of body fossils, yet it is presumed that sedimentation had ceased in most basins throughout India by Early Norian (Mukhopadhyay et al. 2010).

Globally, the Triassic resumes with progressive transgression and the spread of the arid climate of the western Gondwanaland into the eastern counterpart. Glaciers were nowhere on the Pangea (Ogg and Chen 2020). The end-Triassic–Jurassic is characterized by major global deformations (Veevers 2004), however are not well recognizable in the Indian mainland (Mukhopadhyay et al. 2010). However, the angular unconformity between the Panchet and overlying Supra-Panchet formations (Lower–Middle Jurassic) at Lugu Hill in West Bokaro Coalfield is among the counted few sites with signatures of the deformation and erosion (Raja Rao 1987). The onset of Triassic is synchronous with the Serbian Traps and the deadliest among the Big Five, while the end-Triassic is associated with unimaginable quantum of volcanism along the Central Atlantic Magmatic Province (Esmeray-Senlet 2020), global regression accompanied with deep erosion within the India basins (including Damodar–Koel Valley Basin) and neighbors like the western Australia and also, one of the Big Five.

4.4 Cretaceous Traps

Between the dismemberment of the supercontinent Pangea and the emplacement of the Deccan Traps at the end of the Cretaceous, India was located around a temperate latitude in the Southern Hemisphere (paleo-latitude 47°S; max. 50.4°S–min. 43.8°S) (Klootwijk 1971; Gradstein et al. 2020). By this time, sedimentation had ceased in most Gondwana basins of mainland India; however, the Dubrajpur Formation and the Infra-Trappean units of the Rajmahal Basin are unique with thin Coal beds and a rich floral assemblage depicting a return to coal-forming conditions since the end-Permian times. Cycadales (*Ptilophyllum, Pterophyllum, Williamsonia*), Coniferales (*Elatocladus, Pagiophyllum, Araucaria*), Pteridosperm (*Dicroidium, Lepidopteris*), etc., and the advent of angiosperms denote temperate conditions with warm summers and cold winters (Jaiswal et al. 2020, and references therein) and the evolution of a floral assemblage different from the preceding Glossopteris and *Dicrodium* flora (Fig. 4).

However, extensive eruption of basic magma under the influence of the Kerguelen Hotspot brought an end to the tranquility. Manifested as ~600 m thick basaltic rocks at the Rajmahal Basin, the Traps also contain thin and laterally impersistent inter-Trappean beds of sedimentary origin. These remnants of the Rajmahal Large Igneous Province are co-relatable with those of the Sylhet Traps (Meghalaya) and are known to occur beneath the cover of recent Gangetic alluvium in the north and east (Mukhopadhyay et al. 2010). Globally, the trail of the Hotspot can be traced down to the Kerguelen Island in the south Indian Ocean and with signatures across Australia and Antarctica (Kent et al. 2002).

Interestingly, the inter-trappean beds of the Rajmahal Traps are not exclusively sedimentary in nature. Pumice and differentially welded tuff constituting ignimbrite are known to host floral impressions and petrified wood (Shrey et al. 2023, and references therein) at numerous localities throughout the Rajmahal Basin. One such locality, the Geological Heritage Site at Mandro in Sahibganj District is a remarkable place to observe the phenomena. Considering the girth of petrified trees measuring nearly 1.25 m and the variety of floral impressions found in the Rajmahal Traps, extreme environmental conditions causing distress to plant growth are not obvious (Fig. 5). However, their association with ignimbrite suggests a sudden downfall. Nonetheless, the Traps denote the end of the Gondwana sedimentation in mainland India. It also symbolizes the detachment of the Indian plate from the rest of Gondwanaland (Mukhopadhyay et al. 2010) and independent movement toward the north.

5 Conclusions

Climate change is a longstanding fact, witnessing the changing moods of evolving and disappearing faces from the face of the Earth. In present times, human beings can only decipher such time-warranted shifts and can vouch for a few short-term climatic changes, comparing them with times from the recent past.

Influenced by a variety of local, continental, global, and even extraterrestrial factors, there are multiple proxies to access paleoclimatic conditions. Changes consequential to climate modification, including adaptation and evolution, in life forms remain a reliable indicator; however, they are not exclusive or independently conclusive.

Marine sediments hold a better prospect to record and preserve such changes compared to the terrestrial formations; however, the latter should not be considered as unaffected. Mass extinctions are usually the culmination of the combined effect

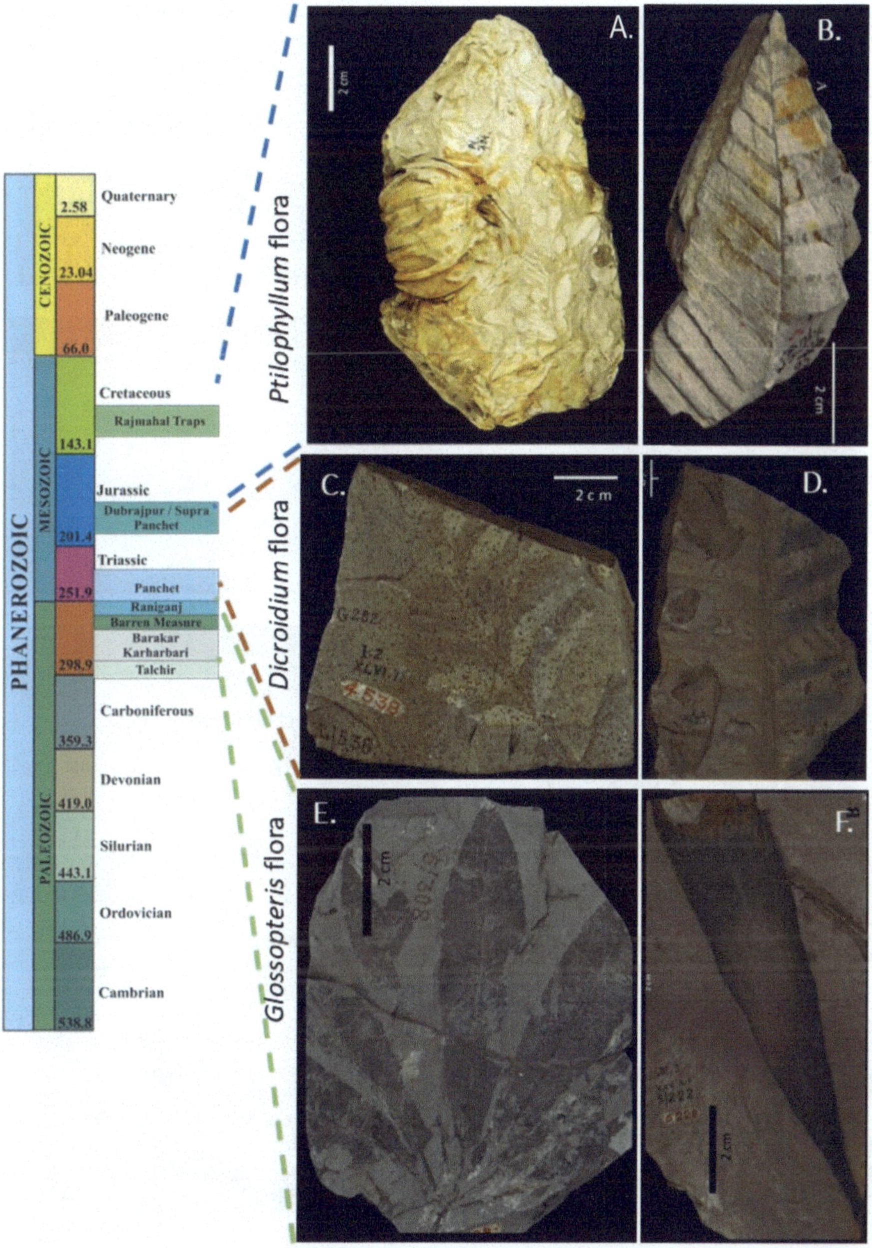

Fig. 4 **a** *Williamsonia* (Fossil flower) from Rajmahal Traps (GSI Type No. 4424). **b** *Pterophyllum morrisianum* (Oldham) from Rajmahal Traps (GSI Type no. 20148). **c** *Dicroidium indica* (Feist.) from Rajmahal Formation (Infra-trappean?) (GSI Type No. 4538). **d** *Taeniopteris musaefolia* (Oldham) from Rajmahal Formation (Infra-trappean?) (GSI Type No. 4362). **e** *Glossopteris retifera* (Feist.) from Raniganj Formation (GSI Type No. 5157). **f** *Glossopteris communis* (Feist.) from Raniganj Formation (GSI Type No. 5222) (Photographs from the Geological Survey of India, Special Publication compiled by Jaiswal et al. 2020)

Fig. 5 Remains of ignimbrite and associated fossils at the Geological Heritage Site, Mandro, Sahibganj, Jharkhand. **a** Pumaceous mudflow. **b** Petrified wood embedded within welded tuff. **c** Leaf impression—*Elatocladus tenerrima (Feistmantle) Sahni* on tuff. **d** Welded tuff (fine to medium ash) with leaf impression. **e** Leaf impression—*Ptilophyllum cutchense*. **f** Close-up view of petrified (silicified) tree trunk showing well-preserved annular rings. **g** Petrified (silicified) tree stump preserved immaculately from bark to pith, capturing detailed anatomy of the plant (trunk)

of a variety of factors bringing about rapid change, particularly degradation, in environmental conditions, causing ecosystem stress, resulting in a sharp decline in the number and diversity of life forms over a very short geological time but over a large geographic area.

The Gondwana Supergroup is effectively a realization of climate-induced correlation (Acharyya et al. 2018). It is characterized by Coal-bearing Lower Group, registering the presence of *Glossopteris* flora, unconformably overlapped by the Upper Group containing *Ptilophyllum* flora. In between the two, Dicrodium flora with reptilian fossils and red beds depict climatic transitions and the loss and the advent of life forms. Witnessing the acme of very large reptiles, dinosaurs, and their extinction, the Upper Gondwana time and the end-Cretaceous mass extinction created conditions for the transition, rather abrupt, of primitive life into forms that resonate with the present biodiversity of the Earth.

References

Acharyya SK, Bhattacharya S, Dutt AB, Ghosh SC, Mukhopadhyay A, Mukhopadhyay S, Parui PK, and Roychowdhury M: Tectonic setting and Gondwana Basin architecture in Indian Shield. 2018. Elsevier Inc.

Alvarez L, Alvarez W, Asaro F, Michel H: Extraterrestrial cause for the Cretaceous-Tertiary extinction. Science, 208: 1095 1108 (1980).

Aretz M, Herbig HG and Wang XD. Chapter 23: The Carboniferous Period. Geologic Time Scale 2020. Elsevier BV (2020).

Becker RT, Gradstein FM, and Hammer O: The Devonian Period. The Geologic Time Scale 2012. Elsevier Publisher, 2: pp.559 601 (2012).

Bhattacharya B, and Banerjee PP. Record of Permian Tethyan transgression in eastern India: A reappraisal of the Barren Measures Formation, West Bokaro Coalfield. Marine and Petroleum Geology 67, 170 – 179 (2015).

Bhattacharya HN, Mukhopadhyay G, Bose PK: Hummocky cross-stratification and its hydraulic and climatic implications in Talchir formation, Dudhi Nala, Hazaribagh, Bihar., J. Geol. Soc. India, 34, 398 – 404 (1989)

Bhattacharya HN and Mukherjee A: Soft-sediment deformation structures in a Permo-carboniferous glacio-marine setting, Talchir Formation, Dudhi Nala, India; J. Earth Syst. Sci. 129:87 (2020).

Boggs Jr. S. Principles of sedimentology and stratigraphy. Fourth edition. Pearson (2006).

Botting JP, Muir LA, Zhang Y, Ma X, Ma J, Wang L, Zhang J, Song Y and Fang X: Flourishing Sponge-Based Ecosystems after the End-Ordovician Mass Extinction. Current Biology, Volume 27, Issue 5, 6 March 2017, Pages 772. https://doi.org/10.1016/j.cub.2016.12.061

Ernst RE. Bond DPG and Zhang SH: Chapter – 12: Influence of Large Igneous Provinces. Geologic Time Scale 2020. Elsevier BV (2020). 345–356

Becker RT, J.E.A. Marshall JEA, and Da Silva AC. Chapter 22: The Devonian Period. Geologic Time Scale 2020. Elsevier BV (2020).

Bond DPG and Sun Y: Global warming and mass extinctions associated with large Igneous Province volcanism. In: Ernst, R.E., Dickson, A.J., Bekker, A. (Eds.), Large Igneous Provinces: A Driver of Global Environmental and Biotic Changes, Geophysical Monograph, vol. 255, pp. 87–102 (2021).

Burgess SD, and Bowring SA: High-precision geochronology confirms voluminous magmatism before, during, and after Earth's most severe extinction. Science Advances, 1: e1500470 (2015). https://doi.org/10.1016/j.epsl.2009.02.012

Chen ZQ and Benton MJ: The timing and pattern of biotic recovery following the end-Permian mass extinction. Nature Geoscience, 5: 375 383 (2012). https://doi.org/10.1038/ngeo1475
Chen, J., and Shen, S. Mid-Permian (End-Guadalupian) extinctions. Reference Module in Earth Systems and Environmental Sciences (2019). Elsevier Inc, http://doi.org/10.1016/B978-0-12-409548-9.11976-1
Coulthard BL and D.J. Smith: Dendrochronology. Encyclopaedia of Quaternary Science (Second Edition). Pages 453–458. Elsevier B.V. (2013)
Cramer BD and I. Jarvis I: Chapter – 11: Carbon Isotope Stratigraphy. Geologic Time Scale 2020. Elsevier BV (2020).
Cui Y and Kump LR: Global warming and the end-Permian extinction event: proxy and modelling perspectives. Earth Science Reviews, 149:5 22 (2015). https://doi.org/10.1016/j.earscirev.2014.04.007
Da Silva, AC., Sinnesael, M., Claeys, P. et al. Anchoring the Late Devonian mass extinction in absolute time by integrating climatic controls and radio-isotopic dating. Sci Rep 10, 12940 (2020). https://doi.org/10.1038/s41598-020-69097-6
Esmeray-Senlet S: Chapter 3(L): Three major mass extinctions and evolutionary radiations in their aftermath. Geologic Time Scale 2020. Elsevier BV (2020). 125–137
Finnegan S, Bergmann K, Eiler JM, Jones DS, Fike DA, Eisenman I, et al.: The magnitude and duration of Late Ordovician-Early Silurian glaciation. Science, 331: 903 906 (2011). https://doi.org/10.1126/science.1200803
Gale AS, Mutterlose J and Batenburg S. Chapter 27: The Cretaceous Period. Geologic Time Scale 2020. Elsevier BV (2020).
Gradstein FM, Ogg JG, Schmitz MD, Ogg GM: Geologic Time Scale 2020. Elsevier BV (2020).
Grossman EL and Joachimski MM: Chapter – 10: Oxygen Isotope Stratigraphy. Geologic Time Scale 2020. Elsevier BV (2020).
Henderson CM and Shen SZ. Chapter 24: The Permian Period. Geologic Time Scale 2020. Elsevier BV (2020).
Halbich IW, Fitch FJ and Miller JA. Dating the Cape Orogeny. In: A.P.G. Sohnge and I.W. Halbich, (Eds.), Geodynamics of the Cape Fold Belt. Spec. Publ. Geol. Soc. S. Africa (1983), No.12, pp.149–164.
Haq B, and Schutter S. A chronology of Paleozoic sea-level changes. Science (2008), 322:64 68.
Harper DAT, Hammarlund EU, and Rasmussen CMO: End Ordovician extinctions: a coincidence of causes. Gondwana Research, 25: 1294 1307 (2014), https://doi.org/10.1016/j.gr.2012.12.021
Hildebrand A, Penfield G, Kring D, Pilkington M, and Camargo Z: Chicxulub crater: a possible Cretaceous/Tertiary boundary impact crater on the Yucatan Peninsula, Mexico. Geology, 19: 867 871 (1990).
Jaiswal S, Roy P, Kumari S, Chakraborty A, Kumar J. Gondwana life across age in India. Geological Survey of India (2020). (Unpublished)
Jin YG, Zhang J, and Shang QH. Two phases of the end-Permian mass extinction. In Embry, A.F., Beauchamp, B., and Glass, D.J. (eds), Pangea: Global Environments and Resources. Canadian Society of Petroleum Geologists, Memoir, (1994) 17:813 822.
Kent, RW, Pringle, MS, Muller, R.D., Saunders, AD, Ghose, N.C. 40Ar/39Ar Geochronology of the Rajmahal Basalts, India, and their Relationship to the Kerguelen Plateau. Journal of Petrology 43(7): 1141–1153 (2002).
Klootwijk, C.T. Palaeomagnetism of the Upper Gondwana Rajmahal Traps, Northeast India, Tectonophysics, 12, 449-467 (1971).
Lowe DJ: Tephrochronology and its application: A review. Quaternary Geochronology, 6 (2), pp. 107-153 (2011).
Mason B and Moore CB: Principles of Geochemistry. Fourth Edition. John Wiley and Sons Inc. (1966)
McArthur JM, Howarth RJ, Shields GA, Zhou Y: Chapter – 7: Strontium Isotope Stratigraphy. Geologic Time Scale 2020. Elsevier BV (2020).

Metcalfe I. Gondwanaland dispersion, Asian accretion and evolution of eastern Tethys. Australian Jour. Earth Sci. (1996), v.43, pp.605–623.

Mukhopadhyay G and Bhattacharyya HN. Facies analysis of Talchir sediments (Permo-Carboniferous), Dudhi Nala, Bihar, India- A glaciomarine model. Gondwana Nine, Oxford & IBH Publishing Co. Pvt. Ltd., V-2 (1996).

Mukhopadhyay G., Mukhopadhyay S. K., Roychowdhury Manas and Parui P. K. (2010). Stratigraphic correlation between different Gondwana Basins of India. Journal Geological Society of India, Vol.76, September, pp.251–266(2010).

Ogg JG and Chen ZQ. Chapter 25: The Triassic Period. Geologic Time Scale 2020. Elsevier BV (2020).

Paytan A, Yao W, Faul KL and Gray ET: Chapter – 9: Sulfur Isotope Stratigraphy. Geologic Time Scale 2020. Elsevier BV (2020).

Peucker-Ehrenbrink B and Ravizza GE: Chapter – 8: Osmium Isotope Stratigraphy. Geologic Time Scale 2020. Elsevier BV (2020).

Pollard JE. A comparison between the Triassic trace fossils of Cheshire and South Germany. Palaeontology 24, 555–588 (1981).

Raja Rao CS. Coalfields of India: Coal resources of Bihar (excluding Dhanbad District). Bull. Geological Survey of India, V-4 (1987).

Rampino MR, Caldeira K, Rodriguez S: Sixteen mass extinctions of the past 541 My correlated with 15 pulses of Large Igneous Province (LIP) volcanism and the 4 largest extraterrestrial impacts. Global and Planetary Change 234 (2024) 104369. https://doi.org/10.1016/j.gloplacha.2024.104369

Scotese, C.R., 2021. An Atlas of Phanerozoic Paleogeographic Maps: The Seas Come In and the Seas Go Out. Annual Review of Earth and Planetary Sciences, 49(1), 679-728. https://doi.org/10.1146/annurev-earth-081320-064052

Scotese, C.R., 2014. Atlas of Cambrian and Early Ordovician Paleogeographic Maps (Mollweide Projection), Maps 81–88, Volumes 5, The Early Paleozoic, PALEOMAP Atlas for ArcGIS, PALEOMAP Project, Evanston, IL.

Scotese, C.R., 2014. Atlas of Silurian and Middle-Late Ordovician Paleogeographic Maps (Mollweide Projection), Maps 73 – 80, Volumes 5, The Early Paleozoic, PALEOMAP Atlas for ArcGIS, PALEOMAP Project, Evanston, IL.

Scotese, C.R., 2014. Atlas of Permo-Carboniferous Paleogeographic Maps (Mollweide Projection), Maps 53 – 64, Volumes 4, The Late Paleozoic, PALEOMAP Atlas for ArcGIS, PALEOMAP Project, Evanston, IL.

Scotese, C.R., 2014. Atlas of Middle & Late Permian and Triassic Paleogeographic Maps, maps 43 - 48 from Volume 3 of the PALEOMAP Atlas for ArcGIS (Jurassic and Triassic) and maps 49 – 52 from Volume 4 of the PALEOMAP PaleoAtlas for ArcGIS (Late Paleozoic), Mollweide Projection, PALEOMAP Project, Evanston, IL.

Scotese, C.R., 2014. Atlas of Jurassic Paleogeographic Maps, PALEOMAP Atlas for ArcGIS, volume 4, The Jurassic and Triassic, Maps 32–42, Mollweide Projection, PALEOMAP Project, Evanston, IL.

Scotese, C.R., 2014. Atlas of Late Cretaceous Maps, PALEOMAP Atlas for ArcGIS, volume 2, The Cretaceous, Maps 16 – 22, Mollweide Projection, PALEOMAP Project, Evanston, IL.

Scotese, C.R., 2014. Atlas of Neogene Paleogeographic Maps (Mollweide Projection), Maps 1–7, Volumes 1, The Cenozoic, PALEOMAP Atlas for ArcGIS, PALEOMAP Project, Evanston, IL.

Sepulveda J, Alegret L, Thomas E, Haddad E, Cao C and Summons RE: Stable isotope constraints on marine productivity across the Cretaceous-Paleogene mass extinction. Paleoceanography and Paleoclimatology, 34: 1195 1217 (2019). https://doi.org/10.1029/2018PA003442

Servais, T., Owen, A.W., Harper, D.A.T., Kroger, B., and Munnecke, A. The Great Ordovician Biodiversification Event (GOBE): The palaeoecological dimension. Palaeogeography, Palaeoclimatology, Palaeoecology, 294:99 119. (2010)

Shrey S, Chaubey RS, Jaiswal P. Geological Heritage Site for the Rajmahal Flora at Mandro, Sahibganj, India: An Opportunity for an Integrated–Sustainable Tourism

Circuit.GeoconservationResearch.6(20):450–464(2023). https://doi.org/10.57647/j.gcr.2023.0602.27

Smit, J., and Hertogen, J., 1980, An extraterrestrial event at the Cretaceous Tertiary boundary. Nature, 285: 198 200 (1980).

Sun Y, Joachimski MM, Wignall PB, Yan C, Chen Jiang H, Wang L: Lethally hot temperatures during the early Triassic greenhouse. Science 338, 366–370 (2012). https://doi.org/10.1126/science.1224126

Toon OB, Pollack JB, Ackerman TP, Turco RP, McKay CP, and Liu MS. Evolution of an impact-generated dust cloud and its effects on the atmosphere. Geological Society of America Special Papers (1982), 190: 187 200.

Veevers JJ. Gondwanaland from 650-500 Ma assembly through 320 Ma merger in Pangea to 185-100 Ma breakup: supercontinental tectonics via stratigraphy and radiometric dating. Earth Science Rev. (2004), v.68, pp.1–132.

Glossary

Acceptable limit It is the maximum or minimum value of something that is considered safe, suitable, or within agreed standards.

Acid rains Acid rain is rainwater that has become unusually acidic due to the presence of certain pollutants in the air, mainly sulfur dioxide (SO_2) and nitrogen oxides (NO_x).

Acidic rocks Acidic rocks are igneous rocks that contain a high amount of silica (more than about 63%). Examples include granite and rhyolite.

Aeolian It refers to processes, landforms, or sediments that are caused by the action of the wind.

Aerosol It refers to tiny solid particles or liquid droplets suspended in the atmosphere. These can be natural (e.g., volcanic ash, sea salt, dust) or human-made (e.g., industrial emissions).

Albedo Albedo is the proportion of incoming solar radiation that a surface reflects back into space.

Alkaline rocks Alkaline rocks are igneous rocks rich in sodium and potassium, compared to silica. They typically contain minerals such as nepheline, leucite, or alkali feldspar.

Alluvial fans An alluvial fan is a fan- or cone-shaped deposit of sediments formed where a high-gradient stream flows out from a narrow mountain valley onto a flatter plain.

Alluvial Alluvial refers to loose, unconsolidated sediments such as sand, silt, clay, and gravel deposited by running water, usually in riverbeds, floodplains, or deltas.

Amygdaloidal A volcanic rock texture where gas bubbles (vesicles) are later filled with minerals, forming almond-shaped cavities.

Angiosperms Angiosperms are flowering plants whose seeds are enclosed within a fruit.

Angular unconformity An angular unconformity occurs when younger rock layers are deposited on top of older layers that have been tilted or folded, creating an angle between the two sets of beds.

N. Khare (ed.), *Changing Climate over India*,
https://doi.org/10.1007/978-981-92-1098-5

Anorthosite Anorthosite is a coarse-grained intrusive igneous rock composed predominantly of plagioclase feldspar, with minimal amounts of mafic minerals.

Anthropogenic activities Anthropogenic activities are human-induced actions that alter the Earth's surface and processes.

Aquifer An aquifer is a body of rock and sediment that holds groundwater.

Arenaceous rock Arenaceous rock is a clastic sedimentary rock composed mainly of sand-sized particles (0.0625–2 mm in diameter).

Argillaceous rock Argillaceous rock is a type of sedimentary rock rich in clay minerals, usually containing more than 50% clay-sized particles.

Arsenic Arsenic is a naturally occurring element (As) often found in trace amounts within minerals such as arsenopyrite, realgar, and orpiment.

Arsenicosis Arsenicosis is a chronic health condition caused by prolonged ingestion of arsenic-rich groundwater.

Asteroid An asteroid is a small rocky body orbiting the Sun, mostly found in the asteroid belt between Mars and Jupiter.

Atlantic meridional overturning circulation (AMOC) It is a large system of ocean currents in the Atlantic that moves warm surface water northward and returns cold, deep water southward, playing a key role in regulating Earth's climate.

Banded Iron Formation Layered sedimentary rock made up of alternating bands of iron-rich minerals and silica (chert), formed mainly in the Precambrian era.

Benthic It describes organisms or processes associated with the bottom of seas, lakes, or other water bodies, typically living on or within the sediment.

Biodiversity Biodiversity is the variety of living organisms, including plants, animals, and microorganisms, found in a particular area or across the planet.

Biotope A biotope is an area with uniform environmental conditions that supports a specific community of plants and animals.

Bioturbated heterolithic deposits It is a sediment layer made of alternating sand and mud that has been mixed or disturbed by the activity of organisms living within it.

Braid plain A braid plain is a flat area on a riverbed or floodplain where multiple shallow, intertwining channels separate and rejoin around small islands or bars.

Carbon cycle The carbon cycle is the natural process through which carbon moves between the atmosphere, living organisms, oceans, and the Earth's crust.

Catchment A catchment is the area of land where all rainfall and surface water naturally flow into a common outlet, such as a river, lake, or reservoir.

Cement Cement is the natural mineral material that binds sediment grains together, filling the pore spaces and forming a solid rock.

Central Himalaya The Central Himalaya is the middle segment of the Himalayan mountain range, extending mainly through Nepal and adjoining parts of India.

Chemical weathering Chemical weathering is the breakdown or alteration of minerals in rocks through chemical reactions, often involving water, oxygen, acids, or other substances.

Chemogenic sediments They are formed by the direct precipitation of minerals from solution, usually in water, through chemical processes rather than from pre-existing rock fragments.

Chronological It refers to arranging events, formations, or processes in the order in which they occurred through time.

Clastic It describes sediments or rocks made up of broken pieces of other rocks, bound together through compaction or cementation.

Climate change Climate change is the long-term alteration of Earth's temperature and weather patterns.

Coastal plains A coastal plain is a low, flat landform that lies adjacent to a coastline, formed by the deposition of sediments from rivers, waves, or marine processes.

Coastline The boundary where land meets the sea or ocean, marking the edge of the shore.

Conchoidal fracture A conchoidal fracture is a curved, shell-like break that occurs when brittle materials like quartz or glass are broken.

Continental flood basalts Continental flood basalts are vast, thick sequences of basaltic lava flows erupted over short geological periods, covering large continental regions.

Convolute laminations They are irregular, folded, or contorted layers within sedimentary rocks formed by soft-sediment deformation.

Coprolites Coprolites are fossilized animal droppings that provide evidence of ancient diets and ecosystems.

Cosmic Rays High-energy particles from outer space that interact with Earth's surface and rocks.

Creep Creep is the very slow, continuous downslope movement of soil or rock under the influence of gravity.

Cretaceous extinction A major geological event around 66 million years ago that caused the mass extinction of many species, including dinosaurs.

Cretaceous A geologic period that lasted from about 145 to 66 million years ago, known for dinosaurs and the formation of chalk deposits.

Cumulonimbus clouds Towering, dense clouds associated with heavy rain, thunderstorms, and atmospheric instability.

Cut-off grade It is the minimum concentration of a mineral in ore that can be economically mined and processed.

Deccan Trap A vast sequence of layered volcanic basalt flows in western and central India formed during the Late Cretaceous.

Deccan Traps The Deccan Traps are a large volcanic region in western India formed by massive lava flows about 66 million years ago.

Devonian It is a geological period of the Paleozoic Era, about 419 to 359 million years ago, often called the "Age of Fishes" due to the great diversity of fish that evolved then.

Diagenesis Diagenesis is the physical, chemical, and biological changes that transform sediment into sedimentary rock after deposition.

Diatoms Diatoms are microscopic, single-celled algae with silica-based cell walls, commonly preserved in sedimentary rocks.

Diffusivity It is the measure of how quickly a substance spreads or moves through another medium.

Digital Elevation Models A Digital Elevation Model (DEM) is a digital representation of the Earth's surface showing elevation values at specific points.

Dissolution Dissolution is the process in which a solid substance breaks down and disperses uniformly into a liquid to form a solution.

Drainage basin A drainage basin is an area of land where all precipitation collects and drains into a common river or outlet.

Drought A drought is a prolonged period of below-average precipitation resulting in water scarcity.

Earthquake An earthquake is the sudden shaking of the Earth's crust caused by the release of accumulated stress along faults.

Eastern Himalaya The Eastern Himalaya is the segment of the Himalayan Mountain range extending from eastern Nepal through Bhutan and northeastern India to northern Myanmar.

Ecology Ecology is the study of how living organisms interact with each other and their environment.

El Niño El Niño is a Spanish term that means "the Christ Child". It refers to the ocean surface warming in the central and eastern tropical Pacific Ocean. El Niño is part of the El Niño-Southern Oscillation (ENSO) cycle, a cycle of warming and cooling events in the Pacific Ocean.

Eocene A geological epoch of the Paleogene Period, about 56 to 34 million years ago, when modern mammals and early forms of many present-day plants appeared.

Erosion Erosion is the process by which natural forces like water, wind, or ice wear away and transport soil and rock from the Earth's surface.

Estuary A coastal area where a river meets the sea, mixing fresh and saltwater.

Ethnomedicinal plants These plants are species traditionally used by local communities for healing purposes.

Eutrophication It is the excessive enrichment of water bodies with nutrients, leading to increased algae growth and sediment deposition.

Extra peninsular Parts of India situated outside the Peninsular plateau, like the Indo-Gangetic plains and Himalayan region.

Fan delta A fan delta is a sedimentary deposit formed where an alluvial fan enters a standing body of water (like a lake or sea), spreading out into a delta-shaped feature.

Fault breccia Fault breccia is a type of rock made up of broken, angular rock fragments formed by crushing and grinding along a fault zone during movement.

Faults A fault is a fracture or zone of fractures in the Earth's crust along which rocks have slipped or moved relative to each other due to tectonic forces.

Fauna Fauna refers to all the animal life found in a particular region, habitat, or geological period.

Flame structures Flame structures are soft-sediment deformation features formed when less dense, water-saturated sediment is pushed upward into overlying denser layers, creating flame-like shapes.

Flora Flora refers to the collection of all plant life (trees, shrubs, grasses, flowers, etc.) found in a particular region, habitat, or during a specific period of time.

Fold A fold is a bend or curve in rock layers that forms when they are compressed by tectonic forces.

Foliation Foliation is the planar arrangement of mineral grains or structural features in a rock, formed mainly by pressure, resulting in layered or banded textures.

Foothill A foothill is a low hill found at the base of a mountain or mountain range, forming the transition between flat land and higher mountains.

Foraminifera Foraminifera are tiny single-celled marine organisms with shells (tests), usually made of calcium carbonate, that live on or within the seafloor sediments or float in the ocean.

Foreland sedimentary basin It is a low-lying region that forms in front of a growing mountain range, where sediments eroded from the mountains accumulate due to the bending (flexure) of the Earth's crust under the mountain load.

Gabbro Coarse-grained, dark-colored igneous rock formed from the slow cooling of magma beneath the Earth's surface.

Gangetic plain It is a large, fertile flatland in northern India, formed by the deposits of the Ganga River and its tributaries.

Geogeny It is the branch of geology that deals with the origin, history, and development of the Earth as a whole.

Geographic datum It is a reference framework consisting of a coordinate system and a reference surface (like the Earth's ellipsoid) used to define locations on Earth.

Geomorphology It is the scientific study of landforms, their origin, evolution, and the processes that shape the Earth's surface.

Geothermal The natural heat stored within the Earth, originating mainly from the planet's interior and used as a source of energy.

Global warming It is the long-term rise in Earth's average surface temperature mainly due to increased greenhouse gas emissions from human activities like burning fossil fuels and deforestation.

Gondwanaland It was a supercontinent in the Southern Hemisphere that existed from the Late Precambrian to the Jurassic, made up of present-day South America, Africa, Antarctica, Australia, the Indian subcontinent, and Madagascar.

Grain size profile A grain size profile is the vertical variation of sediment particle sizes within a soil or rock layer, showing how grain size changes with depth.

Greenhouse gas A greenhouse gas is a gas in Earth's atmosphere, like carbon dioxide, methane, or water vapor, that traps heat by absorbing and re-emitting infrared radiation, thereby warming the planet.

Groundwater Groundwater is the water that fills the pores, cracks, and spaces within soil, sediments, and rocks beneath the Earth's surface.

High mountain Asia (HMA) region It refers to the interconnected mountain ranges of the Himalaya, Karakoram, Hindu Kush, Pamir, and Tien Shan, which together form the highest and most glacier-rich region outside the polar areas.

Higher Himalaya The Higher Himalaya is the highest part of the Himalayan Mountain range, composed mainly of metamorphic rocks like gneisses and schists, forming the core crystalline zone between the Lesser Himalaya and the Tethys Himalaya.

Hillocks Hillocks are small, naturally raised areas of land or low hills, usually isolated and less prominent than mountains, often formed by erosion or deposition.

Himalayan geogeny It refers to the study of the origin, formation, and tectonic evolution of the Himalayas, including processes like mountain building, rock deformation, and plate interactions that shaped the region.

Holocene period The Holocene is the current geological epoch, starting around 11,700 years ago after the last Ice Age, marked by the rise of human civilizations and relatively stable climate.

Hotsprings Hot springs are natural sources of groundwater that emerge at the Earth's surface, heated by geothermal energy from the Earth's interior.

Hummocky cross stratification It is a sedimentary structure characterized by gently undulating, concave-up and convex-up laminae, usually formed by combined flow of storm waves in shallow marine settings. It indicates deposition from storm-influenced suspension flows.

Hydrodynamic changes Variations in the movement, pressure, or flow of groundwater or surface water within rocks and sediments over time, often affecting erosion, sediment transport, or aquifer behavior.

Hydrology Study of water on, under, and above the Earth's surface, including its distribution, movement, and properties.

Hypolimnetic oxygen It is the amount of dissolved oxygen present in the bottom, colder layer of a stratified lake (the hypolimnion), which is isolated from surface oxygen replenishment.

Hypoxic conditions Hypoxic conditions refer to environments where oxygen levels are extremely low or nearly absent, often leading to limited survival of most aerobic organisms.

Ice sheets Ice sheets are massive, thick layers of glacial ice that cover large land areas, typically over 50,000 km^2, and flow slowly under their own weight. They shape landscapes through erosion and deposition.

Icebergs Icebergs are large pieces of freshwater ice that have broken off from glaciers or ice shelves and float freely in the ocean. They are mostly submerged, with only a small portion visible above water.

Icelandic low It is a semi-permanent low-pressure system located near Iceland in the North Atlantic Ocean, influencing the climate and storm tracks of the North Atlantic and surrounding regions.

Ichnofossil assemblage An ichnofossil assemblage is a group of trace fossils—like burrows, tracks, or feeding marks—found together in a rock, reflecting the behavior and activity of ancient organisms rather than their physical remains.

Ignimbrite Ignimbrite is a volcanic rock formed from the rapid cooling and solidification of hot ash, pumice, and gas flows during explosive eruptions.

Indian Ocean Dipole (IOD) It is a climate phenomenon in the Indian Ocean characterized by an abnormal difference in sea surface temperatures between the western and eastern parts, affecting rainfall and weather patterns in surrounding regions.

Infra-Trapped units They are rock sequences that are confined or "trapped" beneath overlying impermeable layers, preventing fluid or sediment movement.

Inter Tropical Convergence Zone It is a low-pressure area near the equator where the trade winds from the Northern and Southern Hemispheres meet, causing rising warm, moist air and frequent rainfall.

Interfluvial zone An interfluvial zone is the elevated land area or ridge that lies between two rivers or streams, separating their drainage basins. It is typically higher than the surrounding floodplains.

Intertidal The intertidal zone is the area of the shoreline that is exposed to air at low tide and submerged under water at high tide. It lies between the high and low tide marks.

Inter-Trappean beds Inter-Trappean beds are sedimentary layers (like sandstone, shale, or limestone) that were deposited between successive lava flows (traps) in a volcanic region.

Intraformational conglomerate An intraformational conglomerate is a sedimentary rock made up of rounded or angular fragments derived from the same or closely related sedimentary formation in which it is found.

Islands Islands are landmasses completely surrounded by water, typically formed by volcanic activity, tectonic uplift, or sediment accumulation.

Lacustrine Lacustrine refers to sediments, deposits, or environments associated with lakes.

Landscape A landscape is the visible physical features of an area of the Earth's surface, shaped by natural processes like erosion, deposition, tectonics, and weathering.

La-Nina La Niña is a climate phenomenon characterized by unusually cool ocean surface temperatures in the central and eastern Pacific, often causing changes in global weather patterns such as increased rainfall in some regions and droughts in others.

Large Igneous Province It is an extensive region of the Earth's crust formed by massive volcanic eruptions that produce huge volumes of basaltic lava over a relatively short geological time. It typically covers >100,000 km^2 and can influence climate and ecosystems.

Laurentia Laurentia is an ancient geological craton that forms the core of North America, consisting of stable Precambrian rocks that have remained relatively unchanged for billions of years.

Levee A levee is a natural or artificial embankment along the edge of a river, formed by the deposition of sediment during floods, which helps contain or control the river's flow.

Lightning Lightning is a sudden electrostatic discharge in the atmosphere, usually during a thunderstorm, that produces a bright flash of light and often a thunderclap.

Little Ice Age The Little Ice Age was a period of temporary global cooling between roughly the 14th and 19th centuries, marked by colder winters, advancing glaciers, and climatic changes affecting agriculture and societies.

Loess Fine-grained, loosely packed, wind-blown deposit mainly made of silt.

Madden-Julian Oscillation (MJO) It is a large-scale, eastward-moving disturbance of clouds, rainfall, winds, and pressure near the equator that cycles roughly every 30–60 days, influencing tropical weather and monsoon patterns.

Mafic rocks Igneous rocks that are rich in magnesium (Mg) and iron (Fe), typically dark-colored, and include rocks like basalt and gabbro. They generally have low silica content.

Mafic/felsic magmas Mafic magmas are rich in magnesium and iron, typically dark-coloured, and form rocks like basalt and gabbro. Felsic magmas are rich in silica and aluminium, usually light-coloured, and form rocks like granite and rhyolite.

Magmatic complex It is a region where multiple types of igneous rocks—such as intrusive (plutonic) and extrusive (volcanic) rocks—form from the same magma system, often showing different stages of crystallization and emplacement.

Mass balance It refers to the accounting of material entering and leaving a system (like a glacier, basin, or volcano) to determine whether it is gaining, losing, or maintaining mass over time.

Mass extinctions Mass extinctions are events in Earth's history when a large number of species across different groups suddenly die out in a relatively short geological time.

Master joints Master joints are major, persistent fractures in rock that control the orientation and formation of smaller joints or fractures in the surrounding rock mass.

Matrix It is the fine-grained material (clay, silt, or sand) that surrounds and binds together larger grains, clasts, or crystals in a rock. It fills the spaces between the coarser components.

Mean Annual Rainfall It is the average amount of rainfall a region receives over a year, usually calculated by averaging the yearly precipitation over several years.

Meghalayan stage The Meghalayan Stage is the most recent subdivision of the Holocene Epoch, starting about 4,200 years ago, marked by a global drought event that affected ancient civilizations.

Merck test kit Used for rapid field detection of specific minerals or ions in rocks and soils, often to identify carbonate, sulfates, or iron compounds.

Metasedimentary Sedimentary rocks that have been altered by heat, pressure, or chemical processes (metamorphism) but still retain some original sedimentary features.

Meteoric water Meteoric water is natural water derived from precipitation, such as rain or snow, that infiltrates the ground and contributes to groundwater.

Meteorite A meteorite is a fragment of rock or metal from space that survives its passage through Earth's atmosphere and lands on the ground.

Mineralization It is the geological process in which valuable minerals are deposited and concentrated in rocks, veins, or sediments, often forming ore bodies that can be mined.

Mineralized zone A mineralized zone is a region in rock where economically valuable minerals are concentrated due to geological processes, often making it a target for mining.

Miocene The Miocene is a geologic epoch of the Neogene Period that lasted from about 23 to 5.3 million years ago.

Morphometric It refers to the measurement and analysis of the shape, size, and dimensions of landforms or terrain features, often using quantitative data like slopes, elevations, and drainage patterns.

Morphostratigraphic It refers to the study and classification of landforms or surface features (morphology) in relation to their geological layers (stratigraphy), helping to understand the sequence and history of landscape development.

Morphotectonic The study of landforms (morphology) shaped by tectonic forces, focusing on how Earth's surface features reflect underlying structural and tectonic activity.

Mudflats Mudflats are coastal or estuarine areas of flat, muddy land that are regularly flooded and exposed by tides, often formed by the deposition of fine sediments like silt and clay.

Nanoplanktons Nanoplanktons are microscopic planktonic organisms, typically less than 20 micrometers in size.

Neotectonic processes They are the recent or ongoing movements and deformations of the Earth's crust, such as faulting, folding, and uplift, that shape the landscape and influence seismic activity.

North Atlantic Oscillation (NAO) Index It measures the difference in atmospheric pressure between the Azores High and the Icelandic Low, reflecting changes in wind patterns, storm tracks, and climate variability over the North Atlantic region.

Northern Hemisphere It is the half of the Earth that lies north of the Equator, containing most of the Earth's land and the majority of the human population.

Ocean acidification It is the ongoing decrease in the pH of the Earth's oceans caused by the absorption of excess carbon dioxide (CO_2) from the atmosphere, making seawater more acidic.

Oligocene The Oligocene is a geologic epoch that occurred about 34 to 23 million years ago, marking a stage of Earth's history when modern mammals and grasses began to flourish.

Ordovician A geologic period of the Paleozoic Era, about 485 to 444 million years ago, when diverse marine life flourished.

Ore beneficiation The process of improving the quality of a mined ore by removing impurities and increasing the concentration of valuable minerals.

Oscillation zone An oscillation zone is a shallow coastal area where water motion is dominated by wave-induced back-and-forth (oscillatory) movement, rather than strong currents or tides, affecting sediment transport and beach formation.

Ox-bow lakes A crescent-shaped water body formed when a meander of a river is cut off from the main channel, creating a standalone lake. It typically forms in the floodplain of a river.

Paleo-biogeography It is the study of the past distribution of organisms and ecosystems across geological timescales.

Paleocene The Paleocene is a geological epoch that lasted from about 66 to 56 million years ago, marking the time just after the extinction of dinosaurs.

Paleoclimate Paleoclimate refers to the study and reconstruction of Earth's past climates using geological, biological, and geochemical records.

Paleoenvironment It refers to the past environmental conditions and settings in which ancient organisms lived and sediments were deposited.

Paleogene The Paleogene is the first part of the Cenozoic Era, when mammals and birds rapidly evolved and became dominant.

Palynological Relating to the study of pollen, spores, and other microscopic organic particles for geological, environmental, and archaeological interpretations.

Pangea Pangea was a supercontinent that existed during the late Paleozoic and early Mesozoic eras, comprising almost all of Earth's landmasses joined together.

Pediplain A pediplain is a broad, gently sloping lowland surface formed by the coalescence of several pediments due to prolonged erosion in arid and semi-arid regions.

Peneplain A peneplain is a nearly level, low-relief land surface formed by long-term erosion approaching base level.

Peninsular plateau The Peninsular Plateau is a large, stable, and ancient landmass of igneous and metamorphic rocks forming the oldest geological region of India.

Periglacial Relating to the cold, non-glacial environments near glaciers, characterized by intense freeze–thaw processes.

Permian The last period of the Paleozoic Era, about 299–252 million years ago, marked by the formation of Pangaea and ending with the largest mass extinction in Earth's history.

Permissible limit It is the maximum level of a substance or factor allowed by standards or regulations without causing harmful effects.

Petrography It is the branch of geology that studies rocks in detail, mainly through microscopic examination, to determine their composition, texture, and origin.

Phanerozoic Phanerozoic is the longest geological eon in Earth's history, during which most major animal and plant groups appeared and evolved.

Phenocrysts Phenocrysts are large, well-formed crystals embedded in a finer-grained groundmass of an igneous rock.

Photogrammetry Photogrammetry is the science of obtaining accurate measurements and 3D information about objects or terrain from photographs.

Phyllite Phyllite is a fine-grained metamorphic rock, intermediate between slate and schist, characterized by a silky sheen due to microscopic mica flakes.

Physical weathering Physical weathering is the breakdown of rocks into smaller fragments without any change in their chemical composition.

Phytoecological zones Geographic regions classified based on distinct vegetation types shaped by climate, soil, and ecological conditions.

Phytoliths Phytoliths are tiny, rigid silica bodies formed inside plant cells that remain preserved in soil after the plant decays.

Phytoplankton Phytoplankton are microscopic, photosynthetic organisms that float in aquatic environments and form the base of the marine food web.

Piedmont A piedmont is a gently sloping area formed at the base of mountains where they meet the adjoining lowlands or plains.

Piezometric surface It is an imaginary surface that represents the level to which groundwater in an aquifer would rise in tightly cased wells.

Pillow lava Pillow lava is a volcanic rock formation created when basaltic lava erupts underwater, forming rounded, pillow-shaped masses.

Planktonic Pillow lava is a volcanic rock formation created when basaltic lava erupts underwater, forming rounded, pillow-shaped masses.

Pleistocene The Pleistocene Epoch is best known as a time during which extensive ice sheets and other glaciers formed repeatedly on the landmasses and has been informally referred to as the "Great Ice Age.

Pliocene The Pliocene is a geologic epoch that lasted from about 5.3 to 2.6 million years ago, when early human ancestors first appeared.

Point bars Point bars are crescent-shaped deposits of sand and gravel that form on the inner bends of meandering river channels due to sediment deposition.

Pollen Pollen is a fine powder made up of tiny grains produced by flowers that help plants reproduce.

Proterozoic age The Proterozoic age is a geological eon spanning from about 2.5 to 0.54 billion years ago, marked by the stabilization of continents and the rise of atmospheric oxygen.

Pyroclasts Fragmented volcanic materials ejected during explosive eruptions and deposited as loose clasts

Quartzite Quartzite is a hard, non-foliated metamorphic rock formed from the recrystallization of quartz-rich sandstone under heat and pressure.

Quaternary Alluvium Unconsolidated sediments such as sand, silt, clay, and gravel deposited by rivers and streams during the Quaternary period.

Quaternary deposit Unconsolidated sediments (like alluvium, glacial, fluvial, or aeolian deposits) formed during the Quaternary Period (the last ~2.6 million years).

Radiocarbon Dating Used to determine the age of once-living materials by measuring the amount of radioactive carbon-14 they contain.

Rare earth elements (REE) Group of 17 chemically similar metallic elements, including the 15 lanthanides plus scandium and yttrium, that are vital for advanced technologies and clean energy applications.

Reducing environment A reducing environment is a condition where there is little or no oxygen, so substances gain electrons instead of losing them.

Reservoir A reservoir is a natural or artificial storage of water, usually formed by damming a river, used for water supply, irrigation, hydropower, or flood control.

Resistivity survey Geophysical method that measures the electrical resistance of subsurface materials to investigate variations in soil, rock, and groundwater conditions.

Rhyolite Rhyolite is a fine-grained volcanic rock rich in silica, often light-colored and similar in composition to granite.

Rhyolite Rhyolite is a fine-grained, silica-rich (felsic) volcanic rock, typically the extrusive equivalent of granite.

Sand dykes Vertical or near-vertical intrusions of sand injected into overlying sediments, typically formed by liquefaction during seismic or rapid loading events.

Sediment maturity A measure of the extent to which sediment has been physically and chemically processed, reflecting its texture, composition, and rounding.

Sedimentation The process by which particles settle out of a fluid (water, air, or ice) and accumulate as a layer of sediment.

Siberian high The Siberian High is a massive, cold, and dry high-pressure system that forms over Siberia during winter, influencing the climate of Asia.

Silicification The process by which original rock or sediment is replaced or impregnated by silica (SiO_2), often forming chert or quartz-rich rocks.

Siwalik groups A sequence of Neogene to Quaternary sedimentary rocks forming the outermost Himalayan foothills, mainly composed of conglomerates, sandstones, and mudstones.

Slickenside striations Polished, grooved surfaces on a fault plane formed by frictional sliding of rocks, indicating the direction of past movement.

Southern Hemisphere The Southern Hemisphere is the half of Earth that lies south of the Equator.

South-west monsoonal season The period, typically from June to September, when moist winds from the Arabian Sea bring heavy rainfall to the Indian subcontinent.

Speleothem A speleothem is a mineral deposit formed inside caves from the dripping or flowing of mineral-rich water.

Spheroidal weathering Spheroidal weathering is a geological process in which angular rock blocks gradually become rounded due to chemical weathering acting along their edges and corners.

Stable isotope Non-radioactive form of an element that does not decay over time and is used in geology to trace processes like paleoclimate, geochemistry, and fluid-rock interactions.

Stone Age A prehistoric period characterized by the use of stone tools before the advent of metalworking.

Strata Layers of sedimentary rock or soil that were originally deposited horizontally and distinguishable by composition, texture, or color.

Stratification The layering of sedimentary rocks or soils into distinct strata based on differences in composition, grain size, or depositional conditions.

Stratigraphy The branch of geology that studies the layers (strata) of rocks, their age, distribution, and formation sequence.

Sub tidal The zone of the seabed that remains permanently underwater, even at the lowest tide.

Subduction zone Region where one tectonic plate sinks beneath another into the mantle, often causing earthquakes and volcanic activity.

Suture Zone A suture zone is a linear or narrow region where two distinct tectonic plates or terranes have been joined together by past collisional tectonics.

Symbiotic relationship A symbiotic relationship refers to two minerals or rock-forming processes that coexist and influence each other's formation or growth.

Syn-sedimentary Formed or occurring at the same time as sediment deposition, often influencing or being influenced by the depositional processes.

Tambora eruptions The eruption of Mount Tambora in Indonesia in April 1815 was one of the most potent volcanic events in recorded history.

Taxonomic groups Hierarchical classification of organisms in the fossil record based on shared morphological and evolutionary traits.

Tectonic Relating to the structure, deformation, and movement of the Earth's crust and lithospheric plates.

Temporal resolution The ability to distinguish and resolve events or changes in the geological record over specific time intervals.

Tension gashes Tension gashes are fractures or cracks in rocks that form due to extensional stress.

Terrestrial plants Plants that grow on land rather than in water, often preserved in the geological record as fossils or organic matter.

The Himalayan orogenic belt The Himalayan orogenic belt is a mountain range formed by the collision and ongoing convergence of the Indian and Eurasian tectonic plates.

Thermal conductivity The ability of a rock or soil to conduct heat through its material.

Thermal diffusion The process by which heat energy spreads through rocks or sediments due to temperature gradients.

Thermal gradient The rate at which temperature increases with depth below the Earth's surface.

Thrust zone A thrust zone is a region where rocks are displaced over one another along low-angle reverse faults due to compressional tectonic forces.

Topographic Relating to the arrangement, shape, and features of the Earth's surface, including its elevations and landforms.

Toposheet A detailed map showing the natural and man-made features of the Earth's surface, including elevation contours, rivers, roads, and vegetation.

Trace elements Chemical elements present in rocks or minerals in very small amounts, usually less than 0.1 % by weight, but useful for geochemical and petrogenetic studies.

Trace fossils Geological records of the activity or behavior of organisms, such as burrows, tracks, or feeding marks, rather than the organisms themselves.

Transgression A geological event in which sea level rises relative to the land, causing marine waters to advance over previously exposed terrestrial areas.

Tree-ring chronology A dated sequence of tree-ring patterns used to reconstruct past environmental and climatic conditions.

Triassic The Triassic is a geologic period that lasted from about 252 to 201 million years ago, marking the beginning of the Mesozoic Era.

Tsunami A series of large ocean waves generated by sudden underwater disturbances, such as earthquakes, landslides, or volcanic eruptions.

Upper Continental Crust The shallow, less dense, and felsic-rich portion of the continental crust that typically extends from the surface to ~20–35 km depth.

Vertical Electrical Sounding A geophysical method that measures subsurface electrical resistivity variations with depth to infer geological layering and composition.

Vesicular A rock texture that contains small cavities or holes formed by trapped gas bubbles during solidification.

Vitrophyric A volcanic rock texture characterized by large crystals (phenocrysts) embedded in a glassy matrix.

Volcanic hotspot A volcanic hotspot is a localized, long-lived area of intense volcanic activity in the mantle, producing magma that rises to form volcanoes independent of plate boundaries.

Volcano-sedimentary sequences Layers of rock formed from alternating volcanic deposits (like ash or lava) and sedimentary materials, reflecting both volcanic activity and sedimentation.

Water table The subsurface boundary below which the ground is fully saturated with groundwater.

Watershed A watershed is a land area that channels all precipitation and surface runoff into a common outlet, such as a river, stream, or lake.

Welded tuff A pyroclastic rock formed when hot volcanic ash and fragments settle and fuse together due to their own heat.

Western Disturbance A Western Disturbance is a weather system originating in the Mediterranean region that brings sudden rainfall or snowfall to the northwestern parts of the Indian subcontinent.

Western Ghat Discontinuous Mountain range along the western coast of India, formed mainly of ancient Precambrian rocks and known for their high biodiversity and geomorphological significance.

Western Himalaya The segment of the Himalaya extending from the Indus River in the west to the Kali River in the east, characterized by distinct tectonic zones, high-grade metamorphic rocks, and active thrust faulting.

Xenoliths Fragments of rock enclosed within another rock, typically foreign to the host rock in origin.

Index

N. Khare (ed.), *Changing Climate over India*,
https://doi.org/10.1007/978-981-92-1098-5

The manufacturer's authorised representative in the EU is Springer
Nature Customer Service Centre GmbH, Europaplatz 3, 69115 Heidelberg,
Germany. If you have any concerns regarding our products, please
contact ProductSafety@springernature.com

Printed and bound by CPI Group (UK) Ltd, Croydon, CR0 4YY
07/10/2026
02229349-0013